NETWORK NEUROSCIENCE

FLAVIO FRÖHLICH

School of Medicine, University of North Carolina at Chapel Hill, Chapel Hill, NC, United States

AMSTERDAM • BOSTON • HEIDELBERG • LONDON
NEW YORK • OXFORD • PARIS • SAN DIEGO
SAN FRANCISCO • SINGAPORE • SYDNEY • TOKYO

Academic Press is an imprint of Elsevier

Academic Press is an imprint of Elsevier
125 London Wall, London EC2Y 5AS, United Kingdom
525 B Street, Suite 1800, San Diego, CA 92101-4495, United States
50 Hampshire Street, 5th Floor, Cambridge, MA 02139, United States
The Boulevard, Langford Lane, Kidlington, Oxford OX5 1GB, United Kingdom

Notices
Knowledge and best practice in this field are constantly changing. As new research and experience broaden our understanding, changes in research methods, professional practices, or medical treatment may become necessary.

Practitioners and researchers must always rely on their own experience and knowledge in evaluating and using any information, methods, compounds, or experiments described herein. In using such information or methods they should be mindful of their own safety and the safety of others, including parties for whom they have a professional responsibility.

To the fullest extent of the law, neither the Publisher nor the authors, contributors, or editors, assume any liability for any injury and/or damage to persons or property as a matter of products liability, negligence or otherwise, or from any use or operation of any methods, products, instructions, or ideas contained in the material herein.

Library of Congress Cataloging-in-Publication Data
A catalog record for this book is available from the Library of Congress

British Library Cataloguing-in-Publication Data
A catalogue record for this book is available from the British Library

ISBN: 978-0-12-801560-5

For information on all Academic Press publications
visit our website at https://www.elsevier.com/

  Working together
to grow libraries in
developing countries

www.elsevier.com • www.bookaid.org

Publisher: Mara Conner
Acquisition Editor: Mara Conner
Editorial Project Manager: Kathy Padilla
Production Project Manager: Chris Wortley
Designer: Victoria Pearson

Typeset by TNQ Books and Journals

Dedication

To my wife Anita and my children Sophia, Galileo, Amalia, and Leonardo

Contents

II

MEASURING, PERTURBING, AND ANALYZING BRAIN NETWORKS

9. Unit Activity

10. LFP and EEG

11. Optical Measurements and Perturbations

12. Imaging Structural Networks With MRI

13. Imaging Functional Networks With MRI

14. Deep Brain Stimulation

15. Noninvasive Brain Stimulation

16. Network Interactions

III

CORTICAL OSCILLATIONS

17. Low-Frequency Oscillations

IV

NETWORK DISORDERS

V

TOOLBOXES

Preface

When I first proposed to develop and teach a course entitled *Network Neuroscience* at the University of North Carolina at Chapel Hill, it was unclear if it would resonate with our trainees. After all, I was known for emphasizing the quantitative aspects of neuroscience, not always to the delight of students who did not have such a background. Will I be able to reach out to them and build an interdisciplinary community with the shared interest of understanding brain networks? And the stakes only got higher when I was approached by my senior faculty colleagues to see if I, in my second year as assistant professor, would accept them as students in my course. Using the Socratic method (or some less punishing form thereof), I made the course a dialog in which we explored the topics covered by this book. The salient features were that we did include many mathematical concepts but introduced and developed them in a way that the students started to become fluent in quantitative thinking and mathematical modeling without the usual growing pains. We covered a broad range of topics. The initial version did not include what is now Unit 1 (in essence cellular and synaptic neurophysiology), which forced us to loop back so often that I decided to include that material in the course. We got inspired, had sweeping conversations about network neuroscience, and had a lot of fun! But at the end of the day there was no text to refer the students to. In other words, what we needed was a textbook in the style of the course—a broad and interdisciplinary overview of network neuroscience!

Little did I realize how much work it would be to transform my notes from the course into a textbook. The only reason why this book exists is that so many people have contributed to its making. I am incredibly grateful for all their help and support. First, I would like to thank the students who took the course over the years and provided me with heartwarming encouragement and feedback. In particular, Steve Schmidt, Zhe (Charles) Zhou, Logan Brown, Katie Valeta, Amanda Moawad, Brittany Katz, and Amy Webster who shared their notes with me and edited early drafts of several chapters. As I was writing the book, the Fröhlich Lab felt my absence and I know that it was not always easy to deal with me being absorbed in the book-writing process. Perhaps the most important contributions came from the lab members. In a certain way, the book is the result of intense collaboration with them. Over many months, a large number of lab members edited drafts of the chapters and provided feedback on content, style, and grammar. I would like to thank Kristin Sellers, Caroline Lustenberger, and Steve Schmidt for editing the entire book from cover to cover. Many other lab members have also dedicated significant time and resources to edit chapters: Angela Pikus, Betsy Price (who has a cameo appearance in the "LFP and EEG" chapter), Zhe (Charles) Zhou, Courtney Lugo, Craig Henriquez, Ehsan Negahbani Franz Hamilton, Guoshi

Li, Iain Stitt, Jessica Page, Juliann Mellin, Lauren Lickwar, Michael Boyle, Sankar Alagapan, and Yuhui Li. All of them have worked tirelessly to help improve both content and presentation. In addition, they provided the material for a large number of illustrations in this book, in particular Kristin Sellers, Michael Boyle, Steve Schmidt, Caroline Lustenberger, Sankar Alagapan, Zhe (Charles) Zhou, and Chunxiu Yu. Besides the Fröhlich Lab, other colleagues generously helped to make this book a success. In particular, Garret Stuber, Michael Murias, and Serena Dudek have edited chapters in detail. Hae Won Shin and Josh Trachtenberg kindly donated materials for several illustrations. Angel Peterchev is not only a highly valued collaborator but he also volunteered to have a cameo appearance in the book.

I would like to emphasize that despite all the generous help I have received, all shortcomings of the book are exclusively mine. I consider the book a living document and am looking forward to receiving your feedback on my website (www.networkneuroscientist. org). I promise to reply to every comment and include the feedback in the next edition. I urge the reader not to misunderstand the book as a series of in-depth reviews of the topics covered. Such an endeavor would be impossible and require tens of thousands of citations. Rather it aims to highlight the fundamental concepts and provide some examples in an easy-to-digest manner. As such it will never be complete and I present my apologies to all the world-class scientists whose work I have not had the space to discuss and cite. Sorry!

I had a fantastic experience collaborating with Elsevier. I am grateful to Mica Haley who shared my enthusiasm for the idea of writing this book, Kathy Pedilla and Mara Connor who made sure that the idea turned into a book, and all the other staff that worked behind the scenes on this book.

I would like to thank my mentors and friends who have always encouraged me to pursue my dreams, as crazy as they sometimes seemed. David Rubinow has not only recruited me to the University of North Carolina but has mentored and supported me (in addition to playing a vital role in our clinical trials work). I am incredibly indebted to him for supporting my unusual quest to write a single-author textbook in my early years as an assistant professor. It is very clear that his support has been the foundation of not only my science but also my teaching and writing. I would like to thank John Gilmore, who is a collaborator and mentor but more importantly also a close friend. That I was able to share my dreams, my concerns, and my happiness with him has given me the energy to complete this book. I would also like to thank my mentors in my early years who have believed in me and taught me how to creatively think about big problems in neuroscience, in particular Igor Aleksander and Terry Sejnowski. Along the way, I have had the privilege to benefit from training by the outstanding scientists David McCormick, Massimo Scanziani, and Maxim Bazhenov.

I would like to thank the different funding agencies that have supported the Fröhlich Lab over its last few years. The research performed in the lab has inspired this book. I thank the National Institute of Mental Health, the Foundation of Hope, the Brain and Behavior Research Foundation, the Human Frontier Science Program, the University of North Carolina at Chapel Hill (who also supported the writing of this book through a Junior Faculty Award), and the generous donors who share my vision of changing the world through fusing neurobiology, engineering, and medicine.

The people who paid the greatest price and deserve the biggest credit for the successful completion of the book are the members of my family. I thank my sister Carla Fröhlich, who I hope will one day write a textbook on astrophysics, as she excels as a researcher and teacher in that field. I thank my parents who have dedicated their lives to raising my sister and me. Their love and support surpasses what is imaginable. They have supported us both in the most selfless way in our journeys, even when it meant that we both moved to a different continent to pursue our scientific passions. I can trace who I am today back to the days of my childhood, filled with fondest memories. Finally, I can say with absolute certainty that the reason why not only this book and the Fröhlich Lab exists, but also why I am the happiest person on this planet is my wife Anita. Her strength and support surpasses the imaginable. Her love has made all this possible. She is in her own right an accomplished legal scholar and teacher who has dedicated her life and energy to our family. Everything I draw strength from is her making. I have been an absent husband and father in many ways over the last 2 years. Our children Sophia, Galileo, Amalia, and Leonardo have accompanied me on my journey of writing this book, all in their own ways ("Are you done with the book?"). I can only hope that the many times I was unavailable to them will be somehow compensated by what they have learned in terms of that there are no limits to what one can achieve when one lives by the most powerful combination of love, respect, and hard work.

Yours,
Flavio Fröhlich
Chapel Hill, February 2016

Introduction

Understanding the human brain is one of the most formidable and urgent scientific challenges of our time. The complexity of the brain has made it very difficult for us to understand how the tiny electric impulses elegantly orchestrated by billions of neurons make us *who we are, how we perceive the world*, and *how we interact with our environment*. In addition to appreciating the challenge of being at the cutting edge of studying the perhaps most-complex biological system, it is important to recognize that we are frustratingly limited in available treatments for disorders of the brain. A growing fraction of society is severely affected by neurological and psychiatric illnesses that we poorly understand. According to the National Institute of Mental Health, an estimated 44 million adults in the United States suffer from a mental illness [1]. The National Institute of Neurological Disorders and Stroke estimates that there are 50 million people with a neurological disorder in the United States [2]. All generations are touched by this problem, ranging from the rising rate of autism diagnoses in young children to the growing number of elderly people with neurodegenerative disorders such as dementia. Only an in-depth understanding of both physiological and pathological brain function will enable significant progress in these fields.

We are stumped by the complexity of the nervous system and despite decades of efforts many seemingly basic questions remain unanswered. The few established "facts" have a significant chance of being wrong or at least oversimplified. This fundamental lack of understanding of how the brain works has broad and severe consequences for our lives and our society. In particular, mental illness has a long and dark history of stigma and lack of adequate medical care for most patients, issues that sadly have persisted until today. For too long, society has failed to understand and accept that mental illness arises from biological pathology in the brain (in complex interaction with the environment) as much as heart disease arises from biological pathology in the cardiovascular system. A lack of understanding of the disease process and the underlying physiology and pathology of the brain severely hampers the development of not only effective treatments but also appropriate policies to help patients and families affected by neurological and psychiatric disorders.

Today, whether your motivation is to contribute to solving the intellectual puzzle of how the brain works or to perform translational research to develop novel treatments for neurological and psychiatric illnesses, we are at a point where new and major breakthroughs can and will happen in neuroscience. The recent advent of revolutionary tools to probe the nervous system has energized the scientific community and policy makers. For example, imaging of brain function, noninvasive human brain stimulation, and optogenetics have provided unprecedented abilities to record and perturb neuronal activity. Not surprisingly, neuroscience has become a large field of study with many different subfields that focus on the very small scale (molecular and cellular neuroscience) all the way to the final outcome of brain activity (behavioral neuroscience). These subfields have developed their own scientific communities of graduate student training, scientific meetings, and peer-reviewed journals. At the same time, scientists with quantitative backgrounds such as physicists and mathematicians have started to work on neuroscience problems and have given rise to their own communities of computational and theoretical neuroscience. However, only convergence of these subfields and methodological approaches will enable significant progress in our understanding of how the brain works and our ability to treat disorders of the central nervous system.

This point of convergence is the interdisciplinary study of networks in the brain—neuronal networks. Networks are fundamental to most technologies invented over the last few decades. Networks are also omnipresent in nature. The brain is comprised of large networks of neurons connected to each other in sophisticated patterns. Electric activity patterns in these networks mediate behavior and are impaired in neurological and psychiatric illnesses such as epilepsy and schizophrenia. Despite this knowledge, understanding how the structure and the dynamics of brain networks mediate behavior has remained an elusive challenge.

The premise of this book is that the intermediate, *mesoscopic* scale of networks provides an ideal level of granularity for studying the brain (Fig. 1). This focus is motivated by the fact that there is a gap in research and understanding between the *microscopic* scales of molecules and cells and the *macroscopic* scale of brain areas and their relationship to behavior and disease. At the microscopic scale, molecular mechanisms of central nervous system disorders have been

extensively studied. But leveraging the resulting insights into effective treatments for symptoms that arise through large-scale dysfunction in neuronal electrical activity continues to be a daunting challenge. At the macroscopic scale, the study of patients with focal brain lesions has helped to localize specific brain functions such as executive control in the prefrontal cortex. But learning *how* such brain function arises in specific areas remains unaddressed.

Welcome to *Network Neuroscience*, the emerging scientific discipline that focuses on understanding brain networks. In my opinion, the most promising approach to study neuronal networks is to combine tools from different disciplines to develop a multifaceted array of approaches that span medicine, engineering, biology, mathematics, and physics. Acquiring such an interdisciplinary toolkit is a difficult challenge for students and both junior and senior scientists. This is because academic training is historically designed to provide depth in one specific discipline rather than to convey an interdisciplinary, broader perspective. Clearly, no one can be an expert in a large number of methods, but a solid understanding of the fundamentals of network neuroscience will empower scientists to effectively collaborate across disciplines since they share a common foundation and are able to use a joint scientific language.

I recognized the challenge of getting started in network neuroscience; as a result, I started teaching a course with the same title as this book. The broad interest in the course motivated me to write a book that can reach a larger audience. In essence, the aim of this book is to provide you with an interdisciplinary toolset to study the structure and function of brain networks. The book will prepare you for working on fundamental questions of *how electric activity is organized in the brain, how complex activity patterns mediate*

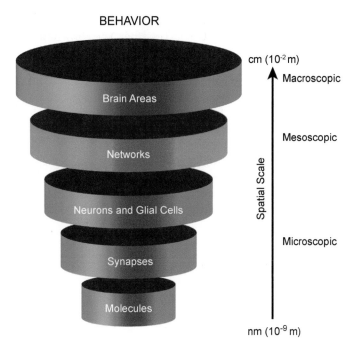

BEHAVIOR

FIGURE 1 The brain can be studied at many different spatial scales. This book focuses on the scale of networks, also referred to as the *mesoscopic* scale, which is bracketed by the *microscopic* scales of molecules, synapses, and cells and by the *macroscopic* scale of brain areas. Conceptualizing the brain as a hierarchical, multiscale system was inspired by Churchland and Sejnowski [3].

behavior, and *how networks can be targeted for the treatment of brain disorders*. This book makes no assumption about your field of expertise and aims to equally serve neurologists, psychiatrists, neuroscientists, cognitive scientists, cell biologists, psychologists, engineers, mathematicians, physicists, and anyone else looking for a comprehensive, multidisciplinary introduction to one of the most fascinating and rapidly developing areas of scientific inquiry in the 21st century. There are a few sections within individual chapters that require more mathematical background and may thus be less approachable despite my best efforts at simplifying complex matter. These sections are marked with an asterisk (*) and can be skipped.

The book covers a broad range of topics and is organized in functional units that correspond to larger themes within network neuroscience:

- The unit *Neurons, Synapses, and Circuits* introduces neurons and synapses as the basic building blocks of networks.
- The unit *Measuring, Perturbing, and Analyzing Brain Networks* provides an overview of the main experimental and analytical techniques used to study networks of neurons.
- The unit *Cortical Oscillations* presents the functional roles and mechanisms of rhythmic activity patterns in the neocortex and the hippocampus.
- The unit *Network Disorders* covers several examples of brain disorders for which there

is an emerging understanding of the network-level pathology.

At the end of the book, *toolboxes* are included that provide background information on relevant terminology and conceptual frameworks that aim to bridge gaps in the academic backgrounds of the readers. By design, there is no need to read all chapters in order, and not all chapters will be equally relevant for every reader. Every chapter in this book aims to motivate, explain, and apply key concepts and tools in network neuroscience.

It is my hope that this book will provide you with the fundamentals you will need to be ready for the truly interdisciplinary and very exciting field of network neuroscience. Together, we can change the world!

References

[1] National Institute of Mental Health: Any mental illness (AMI) among adults. 11/27/2015. Available from: http://www.nimh.nih.gov/health/statistics/prevalence/any-mental-illness-ami-among-adults.shtml.

[2] National Institute of Neurological Disorders and Stroke: Overview. Available from: http://www.ninds.nih.gov/about_ninds/ninds_overview.htm.

[3] Churchland PS, Sejnowski TJ. The computational brain. Computational neuroscience. Cambridge, Mass: MIT Press; 1992. xi, 544 p.

NEURONS, SYNAPSES, AND CIRCUITS

In this first unit, we will learn about the elements that form neuronal networks: neurons, synapses, and circuits. Together, these elements form large-scale brain networks that generate sophisticated physiological and pathological activity patterns. By knowing these building blocks, we will be able not only to *describe* structure and function of large-scale brain networks but also to *mechanistically understand* how network function arises from cellular and synaptic interactions. By discussing cellular and synaptic neurophysiology, the chapters focus on setting the stage to discuss methods and applications of network neuroscience in the ensuing units of the book.

The unit begins with an introduction to how electric signaling is measured in individual neurons (chapter: Membrane Voltage). We then learn about the mechanisms that give rise to the main electrical signal in neurons, the action potential (chapter: Dynamics of the Action Potential). With this understanding of intrinsic electrical signaling, we expand our discussion to how neurons interact with each other. We discuss point-to-point communication by synapses (chapter: Synaptic Transmission). Then we look at the mechanisms by which synapses and neurons change their behavior as a function of neuronal activity (chapter: Synaptic Plasticity). We also dedicate a chapter to endogenous neuromodulators that define the overall state of neuronal networks and represent the basis for sophisticated, state-dependent information processing in the brain (chapter: Neuromodulators) and to neuronal interactions that do not require synapses (chapter: Neuronal Communication Beyond Synapses). We then review the basic circuitry in both the evolutionarily new cortex (chapter: Microcircuits of the Neocortex) and the old (chapter: Microcircuits of the Hippocampus).

Together, the chapters in this unit provide the required preparation for the discussion of methods to study networks of neurons in the next unit: Measuring, Perturbing, and Analyzing Brain Networks.

1

Membrane Voltage

Before we can understand neuronal networks, we need to understand the electrical activity of individual neurons. In this chapter, we will focus on the main experimental approaches used to measure the electrical activity of individual neurons. The *membrane voltage* of a neuron describes its electrical state; brief spikes in the membrane voltage, called *action potentials* or simply *spikes*, represent the main electrical signal in the nervous system. These transient deflections in the membrane voltage are caused by ionic currents across the cell membrane. First, we will look at the two fundamental and complementary ways of measuring neuronal activity within a single cell. The first method is the *current clamp* that measures the membrane voltage. The second is the *voltage clamp* that measures ionic currents. We will also discuss the dynamic clamp, a derivative of these techniques. Then, we will discuss the specific experimental techniques to perform current and voltage clamp measurements and how these measurements differ from the theoretical case. The toolboxes "Neurons," "Electrical Circuits," and "Differential Equations" are particularly helpful in preparation for this chapter.

MEMBRANE VOLTAGE AND IONIC CURRENTS

The *membrane voltage* V_m (or, more formally, the voltage across the cell membrane) is the most fundamental electrophysiological property of a neuron (Fig. 1.1). V_m is defined as the difference in electric potential between inside the cell, φ_{intra}, and outside the cell, φ_{extra}:

$$V_m = \varphi_{intra} - \varphi_{extra} \tag{1.1}$$

Eq. (1.1) reminds us of the fact that any voltage—by definition—is the difference in electric potential between two points. Most of the time, we will assume that φ_{extra} remains constant and that any change to V_m is the result of changes to φ_{intra}. Since both the fluid in the extracellular space and the cytosol are good conductors of electric currents because of their ionic content, we assume that there is no change in electric potential between two points in the extracellular space or between two points within the cell. We will revisit this assumption later when we include the complex morphology of neurons in our considerations (see toolbox: Neurons).

If no input impinges on the neuron and no fluctuations of its membrane voltage are observed, the value of V_m is called the *resting potential*. This is a confusing term for several

3

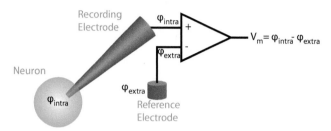

FIGURE 1.1   The membrane voltage is defined as the difference in electric potential between the inside and the outside of the neuron (φ_{intra} and φ_{extra}, respectively). The membrane voltage is measured by positioning a recording electrode in contact with the intracellular space and positioning a reference electrode in the extracellular space. A differential amplifier (shown as a triangle with a "+" and a "−" input) measures (and often amplifies) the difference between these two potentials.

reasons. First, technically spoken, V_m is a voltage (an electric potential difference) and not strictly an electric "potential." Second, a neuron is virtually never "at rest," but rather is constantly bombarded with synaptic input. Nevertheless, the resting potential, which describes the state of the cell in the absence of input, is a useful concept.

Ionic currents that flow through pores in the cell membrane, *ion channels*, modulate the membrane voltage because they change the electric potential φ_{intra} inside the cell (Fig. 1.2). For now, we assume that the extracellular space is so large that the change in charge caused by ion currents across the cell membrane does not alter the extracellular electric potential φ_{extra}. We will revisit this assumption in the chapter "Neuronal Communication Beyond Synapses." Ion channels are typically selectively permeable to one or more types of ions (eg, sodium, potassium, chloride, calcium). Channels are selectively permeable to either anions (negatively charged) or cations (positively charged) ions. The concentrations of these ions differ between the inside and the outside of the cell.

This concentration gradient and the membrane voltage together determine the direction of the ion flow (diffusion). In absence of an electric force, ions flow from points from high concentration to low concentration. However, the membrane voltage creates an electric force that

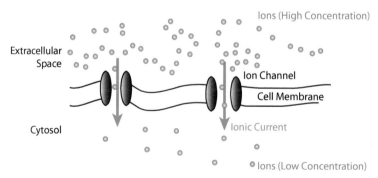

FIGURE 1.2   Ionic currents flow through *ion channels*, which can be thought of pores in the cell membrane. Diffusion across the concentration gradient between the extracellular space and the cytosol moves the ions through the ion channels.

moves the ions, which are electrically charged. The membrane voltage for which there is no net flow of ions since diffusion and drift currents cancel each other is called the *equilibrium potential* V_{eq}. The equilibrium potential is also referred to as the *reversal potential* since the current flow actually changes its direction when comparing values of V_m above and below V_{eq}. The equilibrium potential is determined by the charge and concentration gradients of the ions to which a given ion channel is permeable. The Nernst equation defines V_{eq}:

$$V_{eq} = \frac{RT}{zF} \ln \frac{[\text{Ion}]_o}{[\text{Ion}]_i}, \tag{1.2}$$

where $[\text{Ion}]_i$ and $[\text{Ion}]_o$ are the concentrations in the cytosol and in the extracellular space, respectively. In addition, the equation contains three constants: the ideal gas constant $R = 8.31 \text{ J/(mol K)}$, the Faraday constant $F = 96{,}485.33 \text{ C/mol}$, and the absolute temperature T in kelvin (37°C corresponds to 310.15 K). Finally, z is the valence of the ion (+1 for potassium, sodium, +2 for calcium and magnesium, and −1 for chloride). Typical values for ion concentrations and the resulting values for V_{eq} are provided in Table 1.1. Each time an ion channel opens and ions flow across the cell membrane, the intra- and extracellular ion concentrations change accordingly. *Ion pumps* are proteins in the cell membrane that move ions across the cell membrane, in the opposite direction to the ion current. These pumps help to maintain the ion concentration gradients across the cell membrane and require significant amounts of energy to do so. In the later chapter "Neuronal Communication Beyond Synapses," we will consider a scenario where these pumps fail to maintain the ion concentration gradients.

Note that the values for $[\text{Ion}]_i$ and $[\text{Ion}]_o$ vary between studies and that typical examples are shown here. When using Eq. (1.2) it is important to remember to convert the result from volts [V] to millivolts [mV] by multiplying with the factor 1000.

These concentration gradients enable the flow of ions through the channel similar to how a battery or voltage source enables the flow of electrons through an electric resistor (see toolbox: Electrical Circuits). The electric circuit model of an ion channel is therefore a resistor with resistance R and a voltage source in series (Fig. 1.3). The strength of the voltage source is the equilibrium potential. Additionally, the conductance G is defined as the inverse of the resistance R:

$$G = \frac{1}{R} \tag{1.3}$$

TABLE 1.1 Concentrations and Equilibrium Potentials V_{eq} for the Ion Species That Mediate Electric Signaling in the Brain

	$[\text{Ion}]_i$	$[\text{Ion}]_o$	V_{eq}
Sodium (Na$^+$)	10 mM	145 mM	71 mV
Potassium (K$^+$)	140 mM	3 mM	−103 mV
Calcium (Ca^{2+})	0.1 μM	1 mM	123 mV
Chloride (Cl$^-$)	4 mM	110 mM	−89 mV

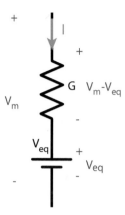

FIGURE 1.3   Electric circuit equivalent of an ion channel. The ion channel is modeled as a conductance G in series with a voltage source that represents the equilibrium potential V_{eq}. Using Eq. (1.4), the ionic current caused by the concentration gradient is cancelled such that there is no net ionic current if the membrane voltage V_m equals the equilibrium potential V_{eq}.

Using Ohm's law and replacing R with G, a given ion current I is then determined by:

$$I = G(V_m - V_{eq}) \qquad (1.4)$$

The term $V_m - V_{eq}$ is called the *driving force*, and represents the net voltage across the ionic conductance. The driving force (and thus the net ion current) is zero if the membrane voltage creates an electric force that cancels out diffusion caused by the ion concentration gradient. As we will see in the next chapter, conductance G often depends on the membrane voltage in quite sophisticated ways.

In summary, the membrane voltage and ion currents represent a dynamical system where voltage and current are linked together. Any change in the membrane voltage will lead to a change in the state of the ion channels and the ion currents. These changes in the ion current change the membrane voltage. Disentangling these feedback dynamics is challenging. Next, we will introduce the main methods of doing so—current clamp and voltage clamp—with a focus on the theoretical and practical aspects. The following chapter "Dynamics of the Action Potential" explains how these feedback interactions lead to the generation of action potentials.

MEASURING THE MEMBRANE VOLTAGE: CURRENT CLAMP

Measuring the membrane voltage V_m with a recording electrode inserted into the neuron is called *current clamp*. To probe the electric behavior of a neuron, electric current is injected through the recording electrode while the membrane voltage is measured. Typically, the amount of current injected into a neuron in the current clamp is less than 1 nA. Abstracting for now from the (sometimes severe) deviations from an ideal measurement, the injected current passes into the cell and changes the measured membrane voltage. The name of this technique comes from the fact that the experimentalist specifies the amount (or to be more precise, the time course) of the injected current. The measured V_m in the current clamp represents the response of the neuron to the injected current and can include *depolarization* (moving the membrane voltage in the direction of 0 mV, triggering action potentials if sufficiently strong) or *hyperpolarization* (moving the membrane voltage to more negative values, away from the threshold for action potential initiation). The response to a current injection enables the measurement of the basic electric properties of a neuron. For example, the waveform and amplitude of the current that is required to trigger an action potential give us important clues about how a neuron will behave when receiving synaptic input from other cells in the network. Furthermore, the number and frequency of action potentials that occur in response to a depolarizing current step reveal the intrinsic firing patterns of the neuron. Typically, current clamp recordings include measurement of the resting potential (in the presence of zero injected current), response to pulses of negative current that hyperpolarize the cell, and responses to (longer) pulses of positive current to depolarize the cell. We will now discuss the interpretation of the resulting changes in membrane voltage to hyperpolarizing and depolarizing current injections.

Negative current injections are used since the resulting hyperpolarization closes (technically "deactivates," see chapter: Dynamics of the Action Potential) voltage-gated ion channels. Thus the response of V_m to hyperpolarizing current reveals the *passive* properties of a neuron. The behavior of the passive cell membrane is mediated by the properties of the cell membrane, cell morphology, and the *leak ion channels* that are open independent of V_m. In response to a brief negative current pulse, V_m changes over time to reach a new, more hyperpolarized value (Fig. 1.4, see also toolboxes: Electrical Circuits and Differential Equations).

Two key properties can be extracted from such a hyperpolarizing pulse: how different the new, hyperpolarized value of V_m is from the resting potential, and how long it takes the cell to reach this new value. Together, these two numbers provide fundamental information about how much (amplitude) and how fast (time course) cells can respond to input. Once V_m has settled after the onset of the injected current pulse I_{inj}, the change ΔV_m caused by the current injection is given by Ohm's law (see toolbox: Electrical Circuits):

$$\Delta V_m = I_{inj} R_m \qquad (1.5)$$

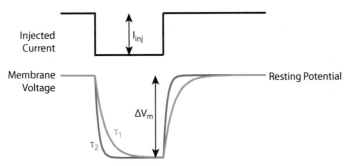

FIGURE 1.4 Injection of hyperpolarizing current pulse I_{inj} into passive cell membrane model. The membrane voltage changes by ΔV_m. The time constant τ quantifies how fast the membrane voltage reaches the steady-state value ΔV_m. Larger time constants indicate that the cell takes longer to reach the steady-state membrane voltage (τ_1 in *red*). Smaller time constants indicate fast dynamics of the membrane voltage (τ_2 in *blue*) in response to input.

FIGURE 1.5 An exponential decay function with time constant τ has undergone 63% of its change after time τ has elapsed.

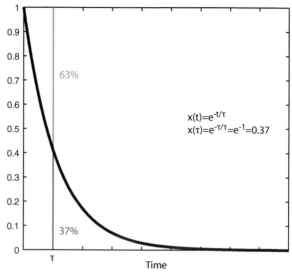

where R_m is the *input* or *membrane resistance* of the cell. Thus we can compute R_m by simply dividing the induced change in voltage by the amount of current injected. The input resistance of the cell provides information about how V_m changes in response to both artificially injected and physiological currents. The transient in V_m before it reaches the new steady-state value is dictated by the resistance and capacitance of the cell; the time constant denotes the period of time it takes for V_m to reach 63% of its final value (Fig. 1.5).

The time constant, denoted as τ, has units of milliseconds and is given by the product of the input resistance R_m and the *membrane capacitance C_m* of the neuron:

$$\tau = R_m C_m \tag{1.6}$$

The intuition behind Eq. (1.6) is that the larger the capacitance C that needs to get charged, the longer it takes for the membrane voltage to change in response to a current injection. Also using Ohm's law, the larger the resistance, the larger the change induced in the membrane voltage by a given I_{inj} and therefore the longer it takes for V_m to reach the final value. By measuring ΔV_m and τ, we can use Eqs. (1.5) and (1.6) to determine the numeric values for R_m and C_m. The capacitance C_m reflects the electric properties of the bilipid cell membrane and is proportional to the total surface area (and therefore size) of the neuron.

Injecting depolarizing current pulses enables the determination of the threshold for action potential generation. Typically, a series of brief current pulses (as short as a few tens of milliseconds) with increasing amplitudes are injected into the neuron until the value of V_m for which an action potential occurs on a defined fraction of trials is found. Longer current pulses (eg, 1 s long, often referred to as *current steps*) are used to determine the action potential response pattern to a sustained depolarization. The characterization of the firing pattern in response to current steps is motivated by the fact that in vivo, synaptic current often arrives as barrages that can last up to hundreds of milliseconds. Typically, the rate of occurrence of action potentials slows down during prolonged current injections. This phenomenon is called *spike frequency adaptation* (often simply referred to as *adaptation*) and can be quantified by determining the initial firing rate at the onset of the depolarizing current step (f_{init}) and the steady-state firing rate at the end of the current step (f_{ss}), and computing the degree of adaptation F_{adapt} as:

$$F_{adapt} = \frac{f_{init} - f_{ss}}{f_{init}} \tag{1.7}$$

In reality, any current clamp measurement is affected by nonideal properties of the measurement system. First, the measurement electrode would ideally exhibit zero electric resistance, but in reality it assumes values up to tens of megaohms because of its small tip size. Second, real world amplifiers affect V_m while measuring it. Both deviations from the theoretical current clamp measurement can be included in the electric circuit diagram of the passive cell membrane (see toolbox: Electrical Circuits). To understand the effect of a nonideal recording electrode, we add an extra electric resistor R_{el} in our equivalent circuit diagram such that there is now a resistor in series with the model of the passive cell membrane (membrane resistance R_m and membrane capacitance C_m). To simplify the analysis, we will first ignore C_m to determine the effect of the injected current I_{inj} on V_m (Fig. 1.6).

In this case, the voltage measured at the amplifier, V_{amp}, is now the sum of the voltage drop across the electrode V_{el} and the cell membrane (membrane voltage V_m). This circuit is referred to as a *voltage divider,* and the relative fraction of the voltage across the two resistors is directly proportional to their relative resistances. This can be determined by lumping the two resistors together by adding them since they are in series and then applying Ohm's law:

$$V_{amp} = I_{inj}(R_m + R_{el}) \tag{1.8}$$

Therefore the current is (by dividing by the sum of the resistances):

$$I_{inj} = \frac{V_{amp}}{(R_m + R_{el})} \tag{1.9}$$

And therefore again by Ohm's law applied to the membrane resistance R_m:

$$V_m = I_{inj}R_m = V_{amp}\frac{R_m}{(R_m + R_{el})} \tag{1.10}$$

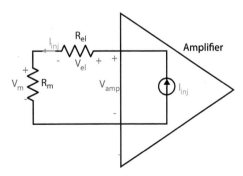

FIGURE 1.6 Electric circuit equivalent of the current clamp with electrode resistance R_{el}.

or:

$$V_{\text{amp}} = V_{\text{m}}\frac{(R_{\text{m}} + R_{\text{el}})}{R_{\text{m}}} \tag{1.11}$$

Therefore the measured voltage V_{amp} is an overestimation of the true membrane voltage and is only fully accurate in the absence of any electrode resistance ($R_{\text{el}} = 0$). In this ideal case, V_{amp} equals V_{m}. Amplifiers provide a means to compensate for the electrode resistance, a method referred to as *balancing the bridge*. This term stems from an early implementation of this functionality using a specific circuit referred to as the *Wheatstone bridge*. Today, balancing the bridge is achieved by analog electronics that subtract a voltage from the measured voltage that is proportional to the injected current.

Real amplifiers do not provide an exact measurement of the voltage of interest since a small amount of current flows into the amplifier. This is in contrast to an ideal amplifier that exhibits zero current flow and thus infinite *amplifier input resistance*, R_{amp}, which defines how hard it is for current to flow into the amplifier. As we will see, the measured signal is not affected by noninfinite input resistance as long as no current is injected. Once a current is injected, a part of the current is not passed through the electrode into the cell but rather flows through the amplifier. We model this by adding a resistor in parallel to the current source that provides the injected current (Fig. 1.7).

We find that the injected current splits into two currents, the current going into the cell and the current going through the amplifier. The voltages across both resistors equal the membrane voltage (by Kirchhoff's voltage law), therefore the two currents are:

$$I_{\text{m}} = \frac{V_{\text{m}}}{R_{\text{m}}}$$

$$I_{\text{amp}} = \frac{V_{\text{m}}}{R_{\text{amp}}} \tag{1.12}$$

Together, the two currents add up to the injected current I_{inj} (by Kirchhoff's current law):

$$I_{\text{inj}} = I_{\text{m}} + I_{\text{amp}} = \frac{V_{\text{m}}}{R_{\text{m}}} + \frac{V_{\text{m}}}{R_{\text{amp}}} \tag{1.13}$$

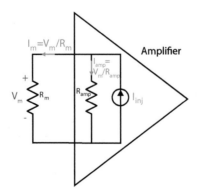

FIGURE 1.7   Electric circuit equivalent of the current clamp configuration with amplifier input resistance R_{amp}. The injected current I_{inj} is split between R_{m} and R_{amp}.

We then solve Eq. (1.13) for V_m and find:

$$V_m = I_{inj} \frac{R_m R_{amp}}{R_m + R_{amp}} \tag{1.14}$$

For large values of R_{amp}, adding R_m to R_{amp} has little effect ($R_m + R_{amp} \approx R_{amp}$), and Eq. (1.14) reduces to:

$$V_m \approx I_{inj} \frac{R_m R_{amp}}{R_{amp}} = I_{inj} R_m \tag{1.15}$$

Eq. (1.15) is in essence the same as Eq. (1.5), which describes the behavior of the cell membrane in the theoretical case of ideal measurement equipment.

MEASURING IONIC CURRENTS: VOLTAGE CLAMP

Voltage clamp is fundamentally different from the current clamp since it enables control of the membrane voltage of the cell. The value (ie, time course) of the membrane voltage is specified by the experimentalist (called command voltage V_{cmd}) and the circuit in the amplifier injects the required current to counteract any change to the membrane voltage that would occur without the voltage clamp. The amount (and the time course) of injected current is a direct measurement of the sum of all ionic currents in case of a theoretical, "ideal" voltage clamp. Keeping the membrane voltage constant by voltage clamp is key to understanding the behavior of ion channels. Many ion channels enable the flow of ionic currents as a function of the membrane voltage, that is, they are voltage-gated. With the voltage clamp, we can determine the time course of the ionic current as a function of the membrane voltage [1]. This would not be possible without the use of the voltage clamp since ionic current flow would cause a change in membrane voltage and thus alter channel activation.

The challenge in the case of the voltage clamp is the previously discussed inaccuracy in measuring membrane voltage. In particular, the electric resistance of the recording electrode presents an issue (*series resistance* or *access resistance*, named after the fact that the electrode resistance is in series with the cell membrane). Several solutions have been developed to deal with this issue. In the case of high impedance electrodes, two separate electrodes can be used for measuring voltage and injecting current (*two-electrode voltage clamp technique*). This is beneficial since no current needs to be passed through the electrode used to measure the membrane voltage; the series resistance only becomes a problem once current is passed through it. The drawback of this approach is that it is not always possible to position two recording electrodes on the same cell. Alternatively, a single electrode can be used, whereby measuring voltage and applying current is rapidly interleaved (*discontinuous single-electrode voltage clamp*). In the case of relatively low impedance electrodes, the amplifier can compensate for the series resistance (*series-resistance compensation*).

The voltage clamp can be modeled as a voltage source that is connected to the cell membrane (and the series resistance, Fig. 1.8). The series resistance introduces a voltage error and a temporal error.

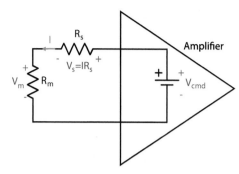

FIGURE 1.8   Electric circuit equivalent of the voltage clamp configuration with membrane resistance and series resistance.

Using the same approach as just described for the current clamp (ignoring the membrane capacitance), we find that the actual membrane voltage V_m and the command voltage V_{cmd} prescribed by the experimenter relate to each other as:

$$V_m = \frac{R_m}{R_m + R_s} V_{cmd} \tag{1.16}$$

The temporal error is caused by the current limitation imposed by the series resistance. The amplifier cannot provide sufficient current to instantaneously change the membrane voltage. For determining the temporal error, we now need to take into account C_m, the capacitance of the cell membrane (Fig. 1.9).

We first use Kirchhoff's voltage law to relate the command voltage V_{cmd} to the actual membrane voltage V_m:

$$V_{cmd} = IR_s + V_m \tag{1.17}$$

Eq. (1.17) states that V_{cmd} is the sum of the voltage across the series resistance (by Ohm's law the current I multiplied with the resistance R_s) and the membrane voltage. By Kirchhoff's current law, the current I equals the current that flows through the passive cell membrane,

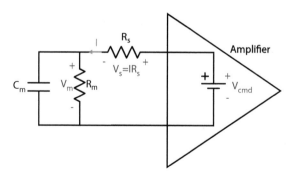

FIGURE 1.9   Electric circuit equivalent of the voltage clamp of the cell membrane with series resistance R_s introduced by electrode resistance.

which is the sum of the current flowing through the cell resistance and the current flowing through the cell capacitance:

$$I = \frac{V_m}{R_m} + C_m \frac{dV_m}{dt} \qquad (1.18)$$

By combining Eq. (1.17) with Eq. (1.18), we find that:

$$\frac{R_s R_m}{R_s + R_m} C_m \frac{dV_m}{dt} = -V_m + V_{cmd} \frac{R_m}{R_m + R_s} \qquad (1.19)$$

By comparing Eq. (1.19) to the first-order linear differential equation (see toolbox: Differential Equations), we see that the time constant of the voltage clamp is:

$$\tau = \frac{R_s R_m}{R_s + R_m} C_m. \qquad (1.20)$$

Usually, the input resistance of the cell, R_m, is significantly larger than R_s. In this case, Eq. (1.20) reduces to:

$$\tau = R_s C_m \qquad (1.21)$$

since $R_s + R_m$ in the denominator of Eq. (1.20) can be approximated by R_m. Therefore the time constant τ is proportional to the series resistance of the recording electrode.

Also, aside from the errors introduced by the series resistance, other aspects need to be considered in the voltage clamp. Most prominently, each time the command voltage is stepped to a new value, the membrane capacitance needs to be charged. This happens very rapidly with a strong transient current pulse. This artifact can be compensated by the amplifier with the injection of a matched opposite current to suppress the artifact.

In addition to these technical issues that amplifiers can (at least partially) compensate utilizing electronic circuits, there is one major concern for which there is no technical solution. In conflict with the assumptions stated at the beginning of the chapter that the electric potential inside a neuron is the same everywhere, the electric potential can differ within a cell with complex shape. Such differences in electric potential within the cell are caused by nonzero electric resistance between distant points within the same cell. As a result, controlling the membrane voltage at the soma with the voltage clamp only partially controls the membrane voltage in locations distant from the soma. This issue is referred to as poor *space clamp*, and needs to be taken into account when interpreting voltage clamp results.

DYNAMIC CLAMP

Dynamic clamp is a technique that enables the introduction of artificial ion currents into a cell membrane at the location of the electrode [2,3]. Therefore the dynamic clamp is used to create biological–computational hybrid systems. A dynamic clamp system measures the membrane voltage V_m, computes the current required to mimic an ion channel ("inject a conductance G_{cmd}") and injects the computed current into the cell. The dynamic clamp computes the desired driving force and multiplies it with conductance G_{cmd} to determine the injected current I_{inj}:

$$I_{inj}(t) = G_{cmd}(t)(V_m(t) - V_{eq}) \tag{1.22}$$

Both synaptic and intrinsic conductances can be inserted using this technique. Note that for voltage-dependent conductances more computation is required at each time step since the new conductance value depends on the measured voltage. The resulting current is then injected into the cell. Dynamic clamp requires a hardware/software solution that is fast enough to compute the injected current in real time. The dynamic clamp is a powerful tool to dissect the role of individual conductances in shaping the overall electric excitability of neurons. For example, an elegant experimental approach to identify the role of a specific ion channel is first to pharmacologically block that channel and then recover the original behavior of the cell by reintroducing that ion channel via the dynamic clamp. The dynamic clamp can also be used to inject synaptic conductances to mimic in vivo-like conditions in a single cell by reproducing the bombardment of excitatory and inhibitory inputs through the dynamic clamp in vitro.

INSTRUMENTATION

Sharp Micropipette Intracellular Recordings

Intracellular recordings with sharp micropipettes [4] have been mostly replaced by whole-cell patch-clamp recordings (see later). However, a pioneering body of membrane voltage recordings in the intact animal (in vivo) was accomplished with sharp micropipettes, including some of the most seminal work on cortical activity during sleep, anesthesia, and waking [5]. Sharp glass microelectrodes are glass pipettes filled with an electrolyte solution (called *internal solution*); they are prepared in the laboratory using a *pipette puller*. This device heats the middle of a fine glass capillary (outer diameter: ~1 mm) to a specified temperature and then pulls at both ends of the capillary until it breaks into two equal pieces. Both pieces have one sharp end from the original capillary broken into two pieces. The resulting tip is very small (order of magnitude: tens of nanometers) and is therefore ideally suited for penetrating the cell wall without causing too much damage. The electrode is backfilled with the internal solution that interfaces with the cytosol at the sharp end of the electrode and with a silver wire with a silver-chloride coating that connects to the amplifier at the other end of the electrode. The sharp tip is so fine that it cannot be readily visualized in tissue under light microscopy, and therefore these recordings are performed "blind." In essence, the electrode is advanced into the tissue under continuous observation of the voltage at the tip of the electrode (measured relative to a reference electrode, typically positioned somewhere else in or near the brain). By passing current pulses through the tip of the electrode and measuring the resulting voltage, the resistance of the electrode is measured (typically 60−150 MΩ). The resistance is mostly defined by the geometry of the tip and is used to assess the quality of an electrode. Once the tip of the electrode approaches a cell membrane (slowly advanced using a micromanipulator), the impedance increases because of the hindered flow of ions. At this point, brief but strong electric pulses are applied to help the pipette enter the cell. Typically, it takes several minutes until the cell has recovered from the penetration, likely because the cell membrane fuses around the micropipette. Under ideal circumstances, it is not atypical to record for an hour from a single neuron in the intact brain using this technique. However, intracellular recordings have several drawbacks that have spurred the widespread adoption of whole-cell patch-clamp recordings as an alternative. First, the high impedance of the electrode represents a challenge since the impedance is similar or even higher than the input resistance of neurons. We have seen in our discussion of current clamp and voltage clamp that high values for the electrode/series resistance represent a problem. In addition, the seal of the cell membrane around the micropipette is not perfect and introduces additional ionic current that flows into the cell and changes the membrane voltage.

Whole-Cell Patch-Clamp Recordings

In contrast to recordings with sharp micropipettes, whole-cell recordings are performed with patch pipettes [6], which are glass micropipettes with a much larger opening at the tip (1–2 μm, typically). As a result, the electrode resistance is lower. In saline, the resistance is typically a few megaohms (at least one order of magnitude lower than the resistance of sharp electrodes). The patch electrode is also filled with an internal solution, similar to the case of sharp intracellular electrodes. In the case of brain slices or cultured neurons (in vitro studies), patch recordings are performed under visual guidance using infrared video microscopy. The sample is illuminated with an infrared light source and visualized with an infrared camera. The complete setup for patch-clamp recordings (*patch rig*, Fig. 1.10) also includes micromanipulators to position the electrodes and an amplifier for current and voltage

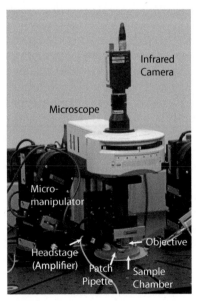

FIGURE 1.10 Typical setup for whole-cell patch-clamp recordings in vitro ("patch rig"). The microscope is equipped with an infrared camera to visualize the sample and the patch pipette. Micromanipulators are used to move the patch pipette to the targeted cell. Amplification of the signal occurs in two stages; the first stage (called *headstage*) is directly connected to the patch pipette.

clamp recordings. To record ("patch") a cell, the patch pipette is moved to the target cell using a micromanipulator. Visual guidance is possible because of the larger tip size of the patch pipette.

Patch pipettes are not inserted into the cell but rather sealed to the intact cell membrane in a first step (the so-called *gigaohm seal*). In a second step, the membrane covering the tip of the electrode is ruptured, typically by application of brief negative pressure through the tip of the electrode. At that point, the *whole-cell configuration* is achieved and the solution in the patch pipette and the cytosol are in direct contact. The content of the cell and the pipette are rapidly mixed (referred to as *dialysis*), which enables filling the cell with dyes both for imaging and post-hoc histological procedures. Patch-clamp recordings overcome the limitations of sharp electrode intracellular recordings since the low electrode impedance allows for good control of the membrane voltage with minimal electrical distortions. Also no additional leak current is introduced. However, the dialysis of the cytosol can also represent a disadvantage, since it degrades the actual ion concentrations in the cell and therefore likely alters electrical signaling. Also disruption of the cytosol affects the health of the neuron and reduces stability (especially for recording times longer than a few minutes). Typically, several extra compounds are added to the internal solution in the patch pipette in an attempt to improve cell viability. Other workarounds to avoid dialysis are chemical perforation instead of rupture of the cell membrane at the tip of the electrode (*perforated patch* configuration) and recording of the extracellular spiking signal with the patch pipette sealed to the intact membrane (*cell-attached* configuration).

SUMMARY AND OUTLOOK

In this chapter, we discussed how to measure the most fundamental electric property of neurons, the membrane voltage. The recording strategies presented in this chapter have been originally most applied in in vitro preparation but have been successfully adapted for use in vivo. Current clamp is used to measure the membrane voltage in response to current injections. For example, current steps are injected into neurons to determine how they respond to input. Different cell types exhibit different firing patterns. Voltage clamp allows us to prescribe a membrane voltage for the cell and the amplifier injects the amount of current needed to maintain the membrane voltage at that command voltage. The voltage clamp is crucial for neuroscience. Most ion channels depend on the membrane voltage, that is, they are voltage gated. Thus as the membrane voltage changes, the state of the ion channel changes, which then causes further change in the membrane voltage because of the change in ion currents through these channels. Being able to clamp the voltage and thereby dissect this feedback interaction was fundamental to understanding how ion channels give rise to action potentials, as we will see in the next chapter.

References

[1] Cole KS. Squid axon membrane — impedance decrease to voltage clamp. Annu Rev Neurosci 1982;5:305—23.
[2] Economo MN, Fernandez FR, White JA. Dynamic clamp: alteration of response properties and creation of virtual realities in neurophysiology. J Neurosci 2010;30(7):2407—13.

[3] Prinz AA, Abbott LF, Marder E. The dynamic clamp comes of age. Trends Neurosci 2004;27(4):218—24.

[4] Brown KT, Flaming DG. Advanced micropipette techniques for cell physiology. In: IBRO handbook series. Chichester West Sussex, New York: Wiley; 1986. x, p. 296.

[5] Steriade M, Timofeev I, Grenier F. Natural waking and sleep states: a view from inside neocortical neurons. J Neurophysiol 2001;85(5):1969—85.

[6] Neher E, Sakmann B. Single-channel currents recorded from membrane of denervated frog muscle fibres. Nature 1976;260(5554):799—802.

Dynamics of the Action Potential

In this chapter, we will learn how ion currents mediate *action potentials*, brief spikes in the membrane voltage (Fig. 2.1). The Hodgkin–Huxley equations [1] explain how the dynamics of voltage-gated sodium and potassium channels together enable the neuronal cell membrane to generate action potentials. The toolboxes on "Electrical Circuits" and "Differential Equations" offer the background needed to deal with the mathematics of the Hodgkin–Huxley equations.

A single ion channel is a pore in the cell membrane that can assume an "open" state in which ions (often only of a single given type, such as sodium or potassium) can pass through. Ion channels are transmembrane proteins; there is an entire field of biophysics that studies the details of the channel structure. Here, we focus on the ability of ion channels to enable ion flow (ie, transmembrane current). We will use a simplified model in which a single channel can be in either an open or closed state and the transition between these two states is probabilistic. For voltage-gated ion channels embedded in the cell membrane, the probability of a channel being in the open state is a function of the membrane voltage V_m. Because of the high density of ion channels in the membrane, we can lump all the individual channels together

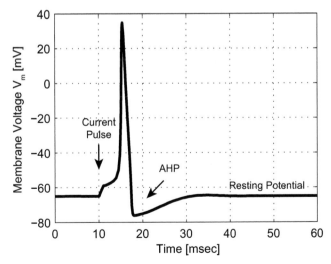

FIGURE 2.1   Action potential triggered by a brief depolarizing current injection. *AHP*, After-hyperpolarization.

and focus on a macroscopic description of the collective behavior of all ion channels of a given type. We will encounter two types of ion channels in this chapter: channels that open and stay open when the cell is depolarized (*persistent conductances*) and channels that open but then close while the cell is still depolarized (*transient conductances*).

POTASSIUM CURRENT

Persistent conductances conceptualize ion channels that have a single mechanism that opens (*activation*, when ions can flow through) and closes (*deactivation*, when ions cannot flow through) the ion channel. Note that this is a simplified model. In reality, the proteins that form the ion channel may need to undergo a number of conformational changes. We here consider the persistent potassium conductance, which, as we will see, mediates the repolarization of the membrane voltage during an action potential.

If we call the probability that a given channel is activated n (and correspondingly $1 - n$ the probability that a channel is deactivated), first-order kinetics model the transition between the two states of the channel:

$$\frac{dn}{dt} = \alpha(1 - n) - \beta n \tag{2.1}$$

We can reinterpret n as a population metric across the entire ensemble of channels that reflects the fraction of channels activated (n is often referred to as the *gating variable*). This implies then that $(1 - n)$ represents the fraction of channels that are deactivated. Eq. (2.1) states that deactivated channels activate with a rate α and activated channels deactivate with rate β. The two terms on the right-hand side of Eq. (2.1) can be conceptualized such that only deactivated channels can activate (thus the product of the fraction of deactivated channels $1 - n$ and rate α) and only the fraction n (activated channels) can deactivate and thus reduce n with rate β. We can rearrange the terms in Eq. (2.1) to get:

$$\frac{1}{\alpha + \beta} \frac{dn}{dt} = -n + \frac{\alpha}{\alpha + \beta} \tag{2.2}$$

We can rewrite this equation in the form of the linear differential equation (introduced in toolbox: Differential Equations) with time constant τ_n and steady-state n_∞. Remember that the time constant defines how fast n changes as a function of time and that the steady-state value is the value that n will assume if we wait long enough (see toolbox: Differential Equations):

$$\tau_n \frac{dn}{dt} = -n + n_\infty \tag{2.3}$$

where:

$$\tau_n = \frac{1}{\alpha + \beta}$$

$$n_\infty = \frac{\alpha}{\alpha + \beta} \tag{2.4}$$

Based on our previous discussion of such differential equations (toolbox: Differential Equations), we know that n will converge to its steady-state value n_∞. Since the channel is voltage-gated, the opening and closing rates α and β are voltage dependent, an important fact that we will denote as $\alpha(V_m)$ and $\beta(V_m)$, where V_m is the membrane voltage. Therefore, both τ_n and n_∞ are voltage dependent. Hodgkin and Huxley experimentally measured this voltage dependence with voltage clamp (Fig. 2.2, green traces the voltage dependence of n_∞, left, and τ_n, right). In essence, the steady-state value n_∞ assumes a sigmoid function of V_m such that for hyperpolarized voltages all channels are deactivated ($n_\infty = 0$), while for depolarized voltages the channels are all activated ($n_\infty = 1$). We will ignore the voltage dependence of the time constant since it is less relevant for understanding the overall dynamics. Now that we have modeled the dynamics of the voltage-gated activation of the potassium channel, we will relate activation variable n to the potassium current I_K.

The total potassium conductance is determined by the product of the maximal potassium conductance g_K (maximal conductance when all ion channels are open) and the probability p of any given channel being open. Hodgkin and Huxley experimentally found that the best fit can be achieved by modeling four identical gates each with an open probability n:

$$p = n * n * n * n = n^4 \tag{2.5}$$

It was later discovered that four independent and identical subunits (ie, proteins that make up the ion channel) need to undergo a voltage dependent structural change for the channel to open. The full equation for the potassium current is:

$$I_K = g_K n^4 \left(V_m - V_{eq}^K\right) \tag{2.6}$$

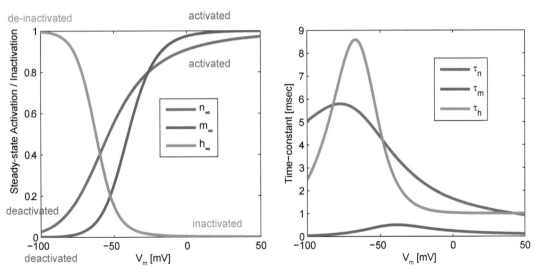

FIGURE 2.2 Steady-state values (*left*) and time constants (*right*) for potassium channel activation n (*green*), sodium channel activation m (*blue*), and sodium channel inactivation h (*red*) as a function of membrane voltage V_m. These curves were determined from the equations provided in the text. Note that deinactivated is *not* the same as activated and will be discussed with sodium channels later.

For potassium, the equilibrium potential is $V_{eq}^K = -77$ mV. The maximal potassium conductance is $g_K = 36$ mS/cm^2. The voltage dependencies of the rates were experimentally measured and modeled as (we omit the units for presentation purposes):

$$\alpha_n = 0.01 \frac{(V_m + 55)}{1 + e^{-\frac{V_m + 55}{10}}}$$

$$\beta_n = 0.125 e^{-\frac{V_m + 65}{80}} \tag{2.7}$$

Importantly, the activation variable n depends on the membrane voltage V_m and responds to a change in membrane voltage by converging to its new steady-state value n_∞ with time constant τ_n as discussed earlier. We now characterize the behavior of the potassium current by studying its time course $I_K(t)$ when we switch V_m from a hyperpolarized value to different levels of depolarization (ie, voltage clamp, see chapter: Membrane Voltage). Before numerically solving the equations, we note that I_K should be zero before we switch the membrane voltage since the channels are all deactivated for hyperpolarized V_m. Furthermore, we know that when we hold V_m constant, Eq. (2.3) is similar to the other linear differential equations we have encountered and thus n will converge to its steady-state value n_∞. As expected, when we hold V_m constant, I_K rises to its steady-state value that depends on the value of V_m, with more depolarized values enabling a stronger I_K (Fig. 2.3).

We now systematically vary V_m, numerically solve the equation for each value [or directly plug the steady-state values of n_∞ into Eq. (2.6)] and determine I_K at steady-state. The plot of an ionic current as a function of the membrane voltage is called a *current–voltage* or *I–V* plot (Fig. 2.4). Note that this representation abstracts from the dynamics since it only shows the steady-state relationship between voltage and current.

Nevertheless, this representation is very helpful in understanding how this ion channel behaves. Let us consider this *I–V* plot in more detail. The potassium current is an *outward current* since the vast majority of the curve is positive (values above the x-axis). Furthermore, the current–voltage relationship does not resemble a straight line, which would be indicative of a linear relationship. Rather, the current assumes values very close to zero unless the cell

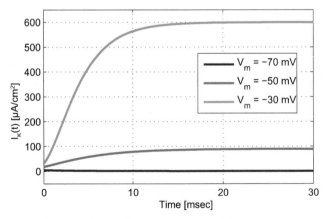

FIGURE 2.3   Voltage-dependent time course of potassium current $I_K(t)$.

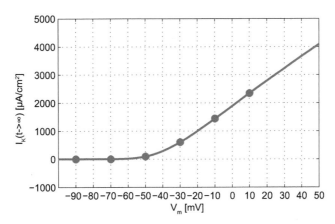

FIGURE 2.4 *Current–voltage* or *I–V* plot of steady-state potassium current I_K.

is depolarized. At that point, for more depolarized voltages the current assumes positive, nonzero values. This is called an outward current, which hyperpolarizes the cell (positive potassium ions leaving the cytosol). The reason for this outward rectifying property is the voltage gating. Remember that n_∞ is zero for hyperpolarized values, which means that the channel is deactivated. Also the $I–V$ relationship looks linear for values of V_m above about -30 mV. At those voltages, all channels are activated and n_∞ is close to its maximum value of one. Since the current is proportional to n^4 times the driving force $V_m - V_{eq}^K$, the voltage dependence of the current is at that point almost exclusively determined by the driving force, which corresponds to a straight line that intersects the x-axis for $V_m - V_{eq}^K$. How does this $I–V$ plot help us understand the role of this outward rectifying potassium channel in the generation of action potentials? Let us assume that V_m is very depolarized (eg, around the peak of an action potential). According to the $I–V$ plot, there will be a strong outward potassium current that will pull the membrane voltage down to more hyperpolarized values. At that point, the potassium current is still an outward current (albeit weaker) that continues to reduce the membrane voltage until the equilibrium voltage V_{eq}^K is reached, at which point the current is zero. Therefore this potassium channel counteracts depolarization and in fact represents one of the two key mechanisms that mediate the downslope of the action potential waveform.

SODIUM CURRENT

We now turn to the voltage-gated sodium (Na) channel that is modeled as a *transient conductance* since it only opens for a short time (permitting Na ion flux) upon depolarization. As in the case of the persistent potassium conductance, for hyperpolarized values of V_m, the channel is deactivated and the channel becomes activated for more depolarized membrane voltages. However, even when the channel is activated, sodium ions are not necessarily able to pass through these sodium channels. Rather, there is a second gating mechanism that is (1) voltage dependent, (2) modeled as independent of activation, and (3) works in the opposite direction of activation such that depolarization hinders ion flux. This mechanism is called *inactivation*. For hyperpolarized voltages, the channels are *deinactivated* (inactivation gate open) and upon depolarization the channels *inactivate*. Together, we now have two gating variables instead of one, describing the opening and closing of these channels. For the transient sodium current, which we consider here, *activation m* (same concept as activation *n* for the potassium current) and *inactivation h* describe the two processes that work in opposite directions to each other. According to the Hodgkin–Huxley model, for a voltage-gated sodium channel to be permeable to sodium ions, three independent activation gates must be activated and the inactivation gate must be open as well (deinactivated). Therefore the equation for the sodium current includes the term m^3h, which corresponds to the probability of all gates being in the correct state for current flow:

$$I_{Na} = g_{Na}m^3h\left(V_m - V_{eq}^{Na}\right) \tag{2.8}$$

The constant g_{Na} denotes the maximal conductance if all channels are both activated and deinactivated. For sodium, the equilibrium potential is $V_{eq}^{Na} = 50$ mV in the Hodgkin–Huxley model. The maximal conductance is $g_{Na} = 120$ mS/cm^2. Both m and h in the sodium current equation are state variables described by the same type of differential equation as for activation n of the potassium channels:

$$\tau_m \frac{dm}{dt} = -m + m_\infty \tag{2.9}$$

$$\tau_h \frac{dh}{dt} = -h + h_\infty \tag{2.10}$$

The time constants τ_m and τ_h and the steady-state values m_∞ and h_∞ are determined from the rates α_m, α_m, β_m, and β_m as for the potassium activation variable n [Eq. (2.4)]. The complete equations for the rates of the sodium current are:

$$\alpha_m = 0.1\frac{(V_m + 40)}{1 - e^{-\frac{V_m+40}{10}}}$$

$$\beta_m = 4e^{-\frac{V_m+65}{18}}$$

$$\alpha_h = 0.07e^{-\frac{V_m+65}{20}}$$

$$\beta_h = \frac{1}{1 + e^{-\frac{V_m+35}{10}}} \tag{2.11}$$

Before we proceed to investigate the time course of the sodium current, we should consider two important facts. First, steady-state activation and inactivation again both exhibit the typical sigmoid shape (Fig. 2.2). However, inactivation converges to a value of one for hyperpolarized values of V_m and to a value of zero for depolarized values of V_m. If we wait long enough, both for hyperpolarized and depolarized cells, one of the two gates is closed and the current is zero since the product of m^3 and h is zero. There is one exception for intermediate values of V_m where both m_∞ and h_∞ are small but nonzero. So how can there ever be significant sodium current if activation and inactivation essentially neutralize each other? Answering this question takes us to a discussion of the activation time scales of different ion channel types. The time scale of the differential equations for state variables m and h are voltage dependent (Fig. 2.2, *blue* and *red traces*); although we will not discuss this voltage dependence, note that for all values of V_m the activation time constant τ_m is much smaller than the inactivation time constant τ_h. Therefore, as we will see next, activation outraces inactivation for a brief period (~ 1 ms) during which there is a substantial net sodium current flowing across the membrane in case of a depolarization.

We can now numerically solve these equations and determine the time course of I_{Na} for different (constant) values of V_m (Fig. 2.5).

Inspecting the time course, we find that the sodium current is indeed transient and very rapidly converges back to values close to zero. Also the sign of the current is negative, indicating that the positively charged sodium ions are flowing into the cell (inward current). Because of the transient nature of the current, determining the $I-V$ plot for the steady-state current makes little sense. Instead, we prepare the $I-V$ plot for the peak current (note that most peaks will be negative peaks) (Fig. 2.6).

Inspection of the $I-V$ plot reveals several important features. Indeed, as expected, the sodium current is zero for hyperpolarized values of V_m. If V_m is depolarized, the channels activate and mediate a depolarizing inward current. Inactivation plays a comparably small role since we measured peak current before inactivation has fully kicked in. As V_m gets closer to the equilibrium potential for sodium (50 mV in the Hodgkin–Huxley model, for which the current is zero), the $I-V$ plot resembles a straight line, since all sodium channels are maximally activated and therefore the voltage dependence of the current is dominated by the driving force (which is a linear equation). With this plot in hand, we can now understand how the sodium current mediates the rising phase of the action potential. As the cell depolarizes, sodium starts to flow into the cell. As a result, the cell depolarizes further, which in turn leads to an increased sodium current. This positive feedback cycle mediates the rapid rise V_m at the onset of the action potential. Several processes limit the sodium current and eventually help to repolarize the cell to its resting value of V_m. As the cell is depolarized further, the current decreases, since V_m approaches the equilibrium potential (and therefore the driving force is reduced). Also, inactivation will start to increase and therefore decrease the sodium current.

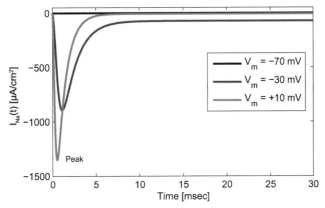

FIGURE 2.5   Voltage-dependent time course of sodium current $I_{Na}(t)$. The sodium current exhibits a rapid peak and then quickly returns back to (almost) zero.

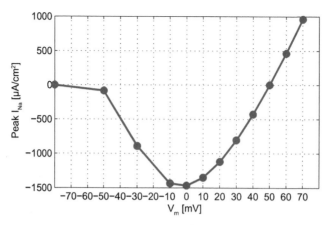

FIGURE 2.6 $I-V$ plot of peak sodium current I_{Na}, the maximum value is reached at $V_m = 0$ mV.

ACTION POTENTIALS

Let us consider the action potential in more detail (Fig. 2.7). Action potentials are transient spikes in the membrane voltage mediated by the interaction of the two currents studied earlier. So far, we have held V_m constant to map the voltage dependence of the sodium and potassium currents. We used the $I-V$ plots to explain how these currents contribute to the genesis of the action potential when we do not hold V_m constant. We will now explore these dynamics further. Suppose that V_m is depolarized by a brief current pulse, which activates the sodium channels.

If a significant number of sodium channels are opened, sodium ions flow into the cell, depolarize it and thereby open even more sodium channels. The sodium current increases rapidly by this positive feedback mechanism (Fig. 2.7, bottom, *purple*). However, we have seen that this current is transient and decreases rapidly (because of inactivation h). Furthermore, with some delay the persistent potassium current is activated (Fig. 2.7, bottom, *green*), which drives the cell back to its negative resting potential (since potassium ions flow out of the cell). This is a negative feedback cycle, since any decrease in the membrane voltage caused by outflowing potassium ions leads to a further decrease in the potassium current. As we have seen in Fig. 2.2, the potassium current is slower than the sodium current. Thus the potassium current will still be active when the sodium current is already shut off. This explains the so-called *after-hyperpolarization* (transient hyperpolarization at the offset of the action potential). We conclude this chapter by looking at the electric circuit diagram of the complete Hodgkin–Huxley model (Fig. 2.8). Note that when we are summing the currents (Kirchhoff's current law), there are three more currents we need to include: a capacitive current (the cell membrane has a capacitance C_m), a leak current (with conductance g_L and equilibrium potential V_{eq}^L), and an externally injected current I_{inj}:

$$C_m \frac{dV_m}{dt} = -g_K n^4 \left(V_m - V_{eq}^K\right) - g_{Na} m^3 h \left(V_m - V_{eq}^{Na}\right) - g_L \left(V_m - V_{eq}^L\right) + I_{inj} \qquad (2.12)$$

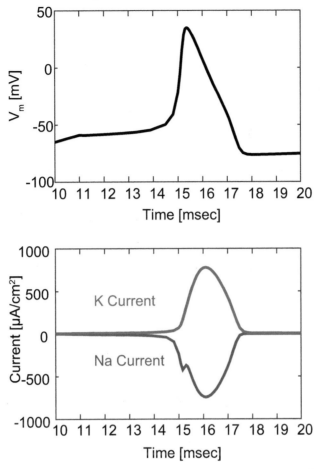

FIGURE 2.7 *Top*: Action potential in response to current pulse that starts at $t = 10$ ms. *Bottom*: Sodium (*purple*) and potassium (*green*) current time course.

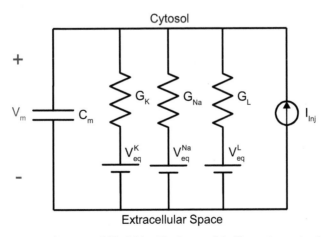

FIGURE 2.8 Electric circuit diagram of Hodgkin–Huxley model. The action potential is generated by the interaction of the sodium and potassium currents. These two currents are mediated by the corresponding voltage-gated conductances G_{Na} and G_K.

SUMMARY AND OUTLOOK

In this chapter, we learned how to use electric circuit modeling to describe the voltage-gated ion channels that give rise to the action potential. We first discussed the potassium channel, which repolarizes neurons and thereby terminates individual action potentials. Then, we discussed the sodium channel, which mediates the rising phase of the action potential. To understand the dynamics of these voltage-gated channels, we differentiated between transient and persistent conductances and familiarized ourselves with different representations of the voltage dependence of ion channels such as the $I-V$ plot. We concluded the chapter by establishing how positive and negative feedback are at work when action potentials are generated.

Reference

[1] Hodgkin AL, Huxley AF. A quantitative description of membrane current and its application to conduction and excitation in nerve. J Physiol London 1952;117(4):500−44.

3

Synaptic Transmission

In the previous two chapters, we have discussed how the membrane voltage is measured and how individual neurons generate action potentials. Along the way, we introduced many fundamental quantitative techniques for studying the brain, including electric circuit modeling and differential equations. In this chapter, we use these strategies to understand communication between neurons. The most prominent model of communication between neurons is through *synapses*, which are sophisticated communication interfaces that combine chemical and electrical signaling. Synapses are the biological substrate of the connections that form neuronal networks. Understanding how electrical signals spread from one neuron to another at synapses will prepare us to study networks. We will first discuss chemical synapses, which use chemical signaling to convey information between the pre- and postsynaptic neurons, and then we will look at electrical synapses, which are a more basic but important second type of synapse that relies on direct electrical signaling. The previous chapters "Membrane Voltage" and "Dynamics of the Action Potential" provide important background for this chapter.

THE CHEMICAL SYNAPSE

A *chemical synapse* is a directional communication link; *synaptic transmission* at a chemical synapse occurs when a *presynaptic* neuron generates an action potential and a *postsynaptic* neuron receives a small synaptic input current caused by the presynaptic action potential. Chemical synapses use a sophisticated chemical signaling scheme to transform the presynaptic action potential into a chemical signal and then back into an electrical signal in the postsynaptic cell. Action potentials are generated in the axon hillock and propagate along the axon. Once an action potential reaches a presynaptic terminal, the depolarization induced by the action potential activates voltage-gated calcium channels. The resulting influx of calcium ions triggers intercellular signaling that leads to the fusion of membrane *vesicles* filled with *neurotransmitter* molecules (of a given type) with the presynaptic cell membrane. As a result, the neurotransmitter molecules are released into the *synaptic cleft*, a tiny gap between the presynaptic axonal terminal and the postsynaptic cell. Through diffusion, the neurotransmitter molecules reach and bind to their specific *synaptic receptors*, which are embedded in the postsynaptic membrane and modulate the electrical state of the postsynaptic neuron (Fig. 3.1).

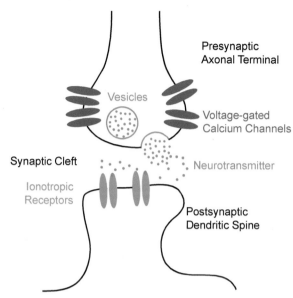

FIGURE 3.1   Basic elements of a chemical synapse with ionotropic receptors.

Two main types of synaptic receptors exist: the first are *ionotropic receptors*, which are receptors that bind neurotransmitter molecules and are also ion channels that directly change the electrical state of the postsynaptic neuron by enabling ion flux (ie, current) across the cell membrane. Ionotropic receptors are ideal for fast synaptic communication. The second type, *metabotropic receptors*, also bind neurotransmitter molecules but then use intracellular biochemical signaling cascades to modulate ion channels in response to neurotransmitter binding. In comparison to ionotropic receptors, metabotropic receptors often provide slower, more complex modulation of the electrical state of the postsynaptic neuron. In addition, binding of neurotransmitter to metabotropic receptors activates signaling pathways that regulate other cellular processes such as synaptic plasticity, gene expression, and apoptosis (cell death).

Presynaptic Voltage-Gated Calcium Channels

There exists a large diversity of voltage-gated calcium channels that exhibit different electrical properties [1]. Traditionally, calcium currents received a single letter determination that abbreviated their main role (eg, "N-type calcium current," standing for "neural type"). However, as the molecular structure of calcium channels became better understood, that naming convention was replaced with the name of the gene that encodes the so-called alpha-1 subunit, an essential component of the ion channel that mediates calcium permeability, voltage sensing, and the response to compounds that block the channel (*antagonists*). We will use both notations in parallel. In terms of voltage gating, there are two main classes of calcium channels: high-voltage-activated (HVA) and low-voltage-activated (LVA) channels (Fig. 3.2). HVA channels require substantial depolarization—caused by an action potential—to be

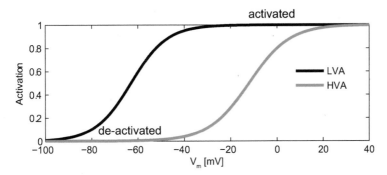

FIGURE 3.2 Activation curves for low-voltage-activated (LVA) and high-voltage-activated (HVA) calcium channels.

activated and mediate the calcium influx in the presynaptic terminals. HVA calcium channels are divided into L-type channels (slow voltage-dependent activation, three different genes in neurons: CaV1.2, CaV1.3, and CaV1.4), P/Q-type channels (CaV2.1), N-type channels (CaV2.2), and R-type channels (CaV2.3). P/Q-, N-, and R-type channels are located at the presynaptic terminal; their rapid kinetics enable precise and brief increase in intracellular calcium to initiate neurotransmitter release with high temporal fidelity. There is only one type of LVA calcium channel, the T-type calcium channel (CaV3.1). This channel not only activates for hyperpolarized membrane voltages but also rapidly inactivates (the letter "T" stands for transient). These properties make this channel ideal for generating intrinsic rhythmic activity—the channel activates and depolarizes the cell, which then inactivates the channel, leading to a subsequent hyperpolarization, from which the next cycle begins [2].

Presynaptic Calcium Dynamics

Calcium influx into the presynaptic terminal regulates the release of synaptic vesicles [3]. Multiple buffering mechanisms tightly control intracellular calcium concentration. Dedicated molecules such as *synaptotagmin* act as *calcium sensors*, detecting changes in calcium concentration and triggering the release of neurotransmitter by fusion of vesicles filled with neurotransmitter molecules with the membrane (*exocytosis*). Synapses typically consist of several release sites, and upon calcium influx in response to a presynaptic action potential, there is a certain probability for a vesicle to be released at each release site. This probability is termed *release probability*, p(release), often abbreviated as p. Every vesicle contains a constant amount of neurotransmitter molecules, denoted as q (abbreviation for quantum of neurotransmitter inside the vesicle). Therefore only a small number of different, discrete quantities of neurotransmitter can be released at a synapse. If a synapse has N release sites, the amount of neurotransmitter that can be released ranges from zero to Nq.

We need to revisit probability theory to understand how we can use measurements of the mean and the variance of synaptic strength to determine p and N. First, let us consider an example. If there are $N = 5$ release sites and we know that $m = 3$ sites actually did release a vesicle for a given presynaptic action potential, there are several different sets of three release sites that may have released neurotransmitter (for example: sites 1, 2, and 3; or sites

2, 3, and 4). The number of possible sets is given by how many different sets of three sites can be drawn from the total set of five. Mathematically, this number is called N *choose* m (*combination*), denoted as:

$$\binom{N}{m} = \frac{N!}{(N-m)!m!} = \frac{N(N-1)...(N-m+1)}{m(m-1)...1} \tag{3.1}$$

where *n factorial* is defined as $N! = N(N-1)(N-2)...1$. In our example, we find that 5 choose $3 = (5*4*3)/(3*2*1) = 10$. In other words, there are 10 different outcomes that all result in three vesicles released. To determine the probability with which any three vesicles are released, we need to determine with what joint probability release of three vesicles occurs and then multiply this probability with 10. If we assume that the release probabilities at individual release sites are independent from each other, we can multiply the individual probabilities of release to get the joint probability. Therefore, the joint probability with what release of three vesicles occurs is given by $p*p*p*(1-p)*(1-p)$, where the first three factors stand for the sites where release occurred and the last two for the sites where no release occurred. Finally, the probability of any three vesicles being released is given by $10p^3(1-p)^2$. In general, the joint probability of m out of N sites releasing neurotransmitter is:

$$\Pr(m) = p^m(1-p)^{N-m} \tag{3.2}$$

Since we do not care which of the N release sites provided neurotransmitter (as long as we know how many did), we can combine these equations to find the *binomial distribution* that describes the probability of any m out of N sites being active:

$$p(m) = \binom{N}{m}p^m(1-p)^{N-m} \tag{3.3}$$

The mean μ and variance σ^2 of this binomial distribution are:

$$\mu = Np \tag{3.4}$$

$$\sigma^2 = Np(1-p) \tag{3.5}$$

Therefore we can use measurements of postsynaptic events and determine their mean and variance to compute N and p (we use "hats" in our notation to indicate that those values are estimates):

$$\widehat{p} = 1 - \frac{\sigma^2}{\mu} \tag{3.6}$$

$$\widehat{N} = \frac{\mu}{\widehat{p}} \tag{3.7}$$

Synapses also release neurotransmitter vesicles (typically one) in the absence of presynaptic action potentials. These so-called *miniature postsynaptic potentials* (often referred to as minis or mPSPs) can be isolated by application of tetrodotoxin, a drug that blocks voltage-gated sodium channels and therefore action potential firing. The functional role of minis is still undetermined. Recording minis provides another way to estimate the value of q since only one vesicle is released at a time.

Postsynaptic Glutamate Receptors

Fast neurotransmission by chemical synapses in the central nervous system is mostly mediated by the neurotransmitters glutamate and gamma-aminobutyric acid (GABA). Under typical circumstances, glutamate mediates *excitatory* synaptic transmission, whereas GABA mediates *inhibitory* synaptic transmission. Most neurons release either glutamate (excitatory neurons) or GABA (inhibitory neurons). We will first discuss excitatory neurotransmission and then consider the inhibitory counterpart.

Glutamate released from presynaptic cells binds to glutamate receptors on the postsynaptic cell. Glutamate receptor ion channels are ionotropic receptors that serve both as a receptor for the ligand (glutamate) and as an ion channel that is activated by binding of the ligand. Ionotropic glutamate receptors are usually classified as *α-amino-3-hydroxy-5-methyl-4-isoxazolepropionic acid* (AMPA), N-*methyl-D-aspartate* (NMDA), *kainate*, or *delta* receptors [4]. These names (with exception of the delta receptors) are derived from the selective agonists that have been found in experiments to activate the corresponding type of ionotropic glutamate receptor (Table 3.1, which also lists the full names of the agonists and antagonists).

Ionotropic glutamate receptors are composed of four subunits that together form an ion channel pore in the membrane (*tetrameric structure*). All four subunits need to be from the same functional class (AMPA, NMDA, kainate, delta) to form a functioning receptor. For AMPA receptors, subunits GluA1, GluA2, GluA3, and GluA4 can form either *homomers* (receptors composed of only one type of subunit) or *heteromers* (mixed composition of different subunit types). For kainate receptors, the five subunit types GluK1−GluK5 form receptors; delta receptors are composed of GluD1 and GluD2 subunits. NMDA receptors are more diverse in terms of their possible subunit compositions (subunits: GluN1, GluN2A−GluN2D, GluN3A, GluN3B). NMDA receptors activate in response to simultaneous binding of both glutamate and glycine. GluN1 and GluN3 provide the glycine binding sites and GluN2 provides the glutamate binding sites.

We will restrict our focus to AMPA and NMDA receptors since they are the most important glutamate receptors for understanding network dynamics. Kainate receptors likely play

TABLE 3.1 Pharmacology of Ionotropic Glutamate Receptors

Compound	AMPA receptor	Kainate receptor	NMDA receptor
Glutamate	Agonist	Agonist	Agonist
AMPA [(±)-α-Amino-3-hydroxy-5-methylisoxazole-4-propionic acid hydrate]	Agonist	Agonist	Inactive
NMDA (N-methyl-D-aspartate receptor)	Inactive	Inactive	Agonist
AP-V (DL-2-Amino-5-phosphonopentanoic acid)	Inactive	Inactive	Antagonist
CNQX (6-Cyano-7-nitroquinoxaline-2,3-dione)	Antagonist	Inactive	Inactive
DNQX [6,7-Dinitroquinoxaline-2,3(1H,4H)-dione]	Antagonist	Antagonist	Inactive
Kynurenic acid (4-Hydroxyquinoline-2-carboxylic acid)	Antagonist	Antagonist	Antagonist

a small and less well-defined role in synaptic transmission, and the functional role of delta glutamate receptors remains to be understood. AMPA receptors exhibit both fast activation and deactivation rates and they exhibit rapid and strong *desensitization* (a state where the channel is closed despite ligand being bound to the receptor). The rapid kinetics limit the time course of the AMPA current that is mediated by the aggregate of AMPA channels in a synapse to ~1–2 ms. The kinetics depend on the subunit composition. For example, homomeric GluA1 are particularly fast channels, often expressed on inhibitory interneurons in the cortex. In contrast to AMPA receptors, NMDA receptors have much slower gating kinetics (deactivation can take up to seconds) and mostly lack desensitization. AMPA and NMDA receptors also differ in their respective ion permeabilities (and therefore reversal potentials). Next, we will derive quantitative models of excitatory synaptic currents.

Modeling Excitatory Synaptic Transmission

We will now develop a mathematical framework that enables us to derive the time course of postsynaptic receptor channel activation. This model will guide our discussion of the resulting synaptic currents. Our goal is to know how the conductance G_{syn} evolves as a function of time. Knowing the time course $G_{syn}(t)$, we can then use the framework developed in earlier chapters and discuss how voltage and current relate at the synapse to understand the electrical signaling mediated by synaptic transmission. We will discuss a simple receptor model that represents an AMPA receptor reasonably well [5]. Upon binding of the ligand, the synaptic receptor changes its state (enabling a nonzero postsynaptic conductance). Formally, the receptor is in either of two states: C (closed) or O (open); we model the receptor such that it has a certain probability to be in the closed state $p_C(t)$ and a certain probability to be in the open state $p_O(t)$. In addition, we include transition probabilities that govern the switching between the two states. These transition probabilities are denoted as $p(O \rightarrow C)$ and $p(C \rightarrow O)$, the probability of switching from an open to a closed state and a closed to an open state, respectively. Knowing these transition probabilities enables us to write down a differential equation that describes how $p_O(t)$ changes over time (see toolbox: Differential Equations):

$$\frac{dp_O(t)}{dt} = p_C(t)p(C \rightarrow O) - p_O(t)p(O \rightarrow C) \tag{3.8}$$

Eq. (3.8) is called the *master equation*. The first term on the right side takes into account the probability that the channel was closed and opens with $p(C \rightarrow O)$, thus leading to an increase in $p_O(t)$. The second term models the scenario that the channel was open and then closes with probability $p(O \rightarrow C)$. Furthermore, we assume that the two probabilities are related to each other:

$$p_C(t) = 1 - p_O(t) \tag{3.9}$$

Here, we are less interested in the dynamics of a single channel but rather in the dynamics of an entire assembly of identical channels that together mediate the postsynaptic conductance of a synapse. To this end, we can slightly reinterpret the master equation and introduce state variables $C(t)$ and $O(t)$ that track the fraction of channels in the corresponding state; we also relabel the transition probabilities as transition rates r_{OC}, previously denoted as $p(O \rightarrow C)$, and r_{CO}, previously denoted as $p(C \rightarrow O)$. Making these substitutions, we have:

$$\frac{dO(t)}{dt} = (1 - O(t))r_{CO} - O(t)r_{OC}. \tag{3.10}$$

Rearranging the terms and defining new terms, we can rewrite Eq. (3.10) as:

$$\tau \frac{dO(t)}{dt} = -O(t) + O_\infty \tag{3.11}$$

with time constant:

$$\tau = \frac{1}{r_{CO} + r_{OC}} \tag{3.12}$$

and steady-state value:

$$O_\infty = \frac{r_{CO}}{r_{CO} + r_{OC}}. \tag{3.13}$$

The steady-state term is the value $O(t)$ will converge to once enough time has elapsed and the system has fully settled such that there is no further change and therefore the temporal derivative of $O(t)$ equals zero (see toolbox: Dynamical Systems). Of note, we have assumed that the rates r_{CO} and r_{OC} are constant values. In reality, this assumption is not true. For the synaptic receptor, r_{CO} strongly increases as a function of ligand (ie, neurotransmitter) concentration such that r_{CO} is a function of glutamate concentration ([GLU]). If we assume that the time course of [GLU] in the synaptic cleft is an instantaneous step to a new constant value at synaptic release (a justified simplification), we can solve the differential equation as before and find that:

$$O(t) = O_\infty + (O_\infty - O(t_0))e^{-\frac{(t-t_0)}{\tau}}. \tag{3.14}$$

We will now consider a further simplification where we assume that upon glutamate release, $r_{CO} \gg r_{OC}$ such that we can simplify the equation to:

$$\frac{dO(t)}{dt} = (1 - O(t))r_{CO} \tag{3.15}$$

where we will denote the time scale of opening as $\tau_O = 1/r_{CO}$ such that the fraction of open channels exponentially converges to 1:

$$\tau_O \frac{dO(t)}{dt} = -O(t) + 1 \tag{3.16}$$

As the [GLU] decreases (and the receptors desensitize, not explicitly modeled here), we assume $r_{OC} \gg r_{CO}$ such that:

$$\frac{dO(t)}{dt} = -O(t)r_{OC} \tag{3.17}$$

and the fraction of open channels converges to zero with rate $\tau_C = 1/r_{OC}$:

$$\tau_C \frac{dO(t)}{dt} = -O(t) \qquad (3.18)$$

Knowing the time course of the fraction of open channels, we can easily determine the time course of the synaptic conductance $G_{AMPA}(t)$ by multiplying $O(t)$ with a constant that denotes the maximal conductance G^*_{AMPA} corresponding to the hypothetical state in which all postsynaptic receptors are open:

$$G_{AMPA} = G^*_{AMPA}O(t) \qquad (3.19)$$

We can now determine the AMPA current by using the same model as we used for any other ionic conductance where the current is given by the conductance multiplied with the driving force:

$$I_{AMPA} = G^*_{AMPA}O(t)\left(V_m - V_{eq}^{AMPA}\right) \qquad (3.20)$$

where the equilibrium potential for AMPA receptors is $V_{eq}^{AMPA} = 0$. Fig. 3.3 shows the time course of the excitatory postsynaptic current (EPSC) as it would be measured by voltage clamp (see chapter: Membrane Voltage) for several different command voltages of the postsynaptic neuron.

The negative polarity for all currents with a holding voltage below the equilibrium potential should not cause confusion when we remember that outward currents are positive and

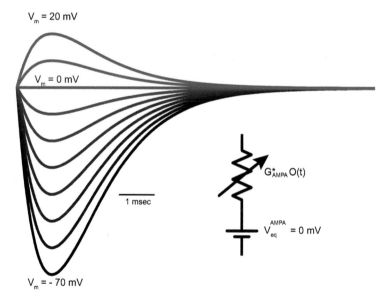

FIGURE 3.3   Time course of the AMPA current (excitatory postsynaptic current, EPSC) for different membrane voltages V_m. *Inset*: Electric circuit equivalent of AMPA-type synapse. The *arrow* superimposed with the resistor indicates that the resistance/conductance changes as a function of neurotransmitter binding.

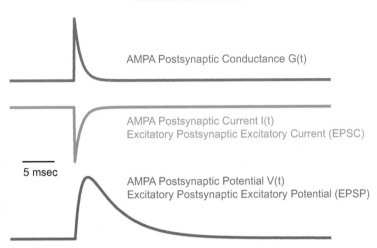

AMPA Postsynaptic Conductance G(t)

AMPA Postsynaptic Current I(t)
Excitatory Postsynaptic Excitatory Current (EPSC)

5 msec

AMPA Postsynaptic Potential V(t)
Excitatory Postsynaptic Excitatory Potential (EPSP)

FIGURE 3.4   Time course excitatory postsynaptic current (EPSC) and excitatory postsynaptic potential (EPSP).

inward currents (such as the AMPA current for $V_m < V_{eq}$) are negative. If we now inject this synaptic current into a cell with only a passive leak current, we can determine the excitatory postsynaptic potential (EPSP). As expected, the EPSP depolarizes the cell (hence "excitatory"). Comparison of the time course of the EPSC and the EPSP reveals that the EPSP is slower than the EPSC, with more sluggish rising and falling flanks (Fig. 3.4).

The reason for this difference in dynamics is that the EPSC needs to charge up the capacitance of the cell membrane (C_m) to change the membrane voltage. The membrane capacitance prevents fast changes of the membrane voltage; in other words, it suppresses high-frequency signals and thus acts as a *low-pass filter*. If we increase the resistance of the cell membrane by reducing the conductance of the leak channel, two things happen. First, the amplitude of the EPSP will increase and the time constant $\tau = R_m C_m$ will increase. Conversely, if we decrease the resistance, the EPSP amplitude is reduced and the time constant is decreased.

Most AMPA receptors are permeable to sodium and potassium ions. However, some AMPA channels are also permeable to calcium ions [6]. Calcium-permeable AMPA channels are additionally gated by a voltage-dependent block mediated by intracellular *polyamines*. Depolarized membrane voltages therefore limit current flux, even in the presence of glutamate bound to the receptors. This preferential inward current (for values of V_m below the equilibrium potential $V_{eq}^{AMPA} = 0$) gives the channel the designation as an *inward-rectifying* channel. These channels do not contain a GluA2 subunit and are limited in their occurrence in the healthy nervous system. Their presence has been associated with central nervous system disorders such as amyotrophic lateral sclerosis and ischemia (stroke).

Unlike AMPA receptors that provide a brief electric postsynaptic signal in response to glutamate released from the presynaptic neuron, NMDA receptors have more complex activation mechanisms [7]. NMDA channels play a crucial role in plasticity, since they can detect both presynaptic (glutamate) and postsynaptic (membrane voltage) activity and these receptors are only activated if both pre- and postsynaptic activity occurs within a relatively narrow time window. This means that NMDA receptors effectively act as coincidence detectors. We will discuss NMDA receptor-mediated plasticity in more detail in the chapter: Synaptic Plasticity. Here, we focus specifically on the biophysical mechanism that mediates the voltage dependence of the NMDA receptor. An extracellular magnesium ion blocks the narrowest part of the NMDA channel, and only depolarization can remove this *magnesium block*. Thus, even in the presence of glutamate, NMDA receptors remain closed unless the postsynaptic cell is depolarized. Therefore the $I-V$ curve is not a straight line anymore but rather is flat and only linearly increases for depolarized voltages. Mathematically, the NMDA conductance is modeled like an AMPA conductance but with different, slower time constants and with an additional multiplicative factor that models the dependence of the state of the channel on the magnesium block. Once the magnesium block is removed, NMDA receptors are permeable to sodium and potassium (similar to the AMPA receptor) but also to calcium ions (which are critical for plasticity). The magnesium block depends on the membrane voltage V_m (in millivolts) and on the extracellular magnesium concentration $[Mg^{2+}]_o$ (in mM) [8]:

$$f_{NMDA} = \frac{1}{1 + (e^{-0.062V_m}) \frac{[Mg^{2+}]_o}{3.57}} \tag{3.21}$$

We can include Eq. (3.21) into an equation that is otherwise identical in structure to the one for the AMPA synaptic current [Eq. (3.20)]:

$$I_{NMDA} = G_{NMDA}(t) f_{NMDA} \left(V_m - V_{eq}^{NMDA} \right), \tag{3.22}$$

where the time course $G_{NMDA}(t)$ models the presence of glutamate bound to the NMDA receptor and f_{NMDA} models the postsynaptic dependence of the channel activation. The equilibrium potential is similar to the one of AMPA receptors, that is, $V_{eq}^{NMDA} = 0$ mV.

Metabotropic Glutamate Receptors

In addition to activating ionotropic glutamate receptors, glutamate also acts on cellular excitability via so-called *second messenger signaling pathways* that are modulated by binding of glutamate to *metabotropic glutamate receptors* (mGluRs) [9]. These receptors belong to the large family of *G-protein-coupled receptors* that are membrane bound and affect the state of a neuron by interaction with G-proteins. Upon binding of glutamate, mGluRs change their conformational state, which activates the G-protein and in turn modulates various effector molecules such as ion channels. The mGluRs are classified into three groups: Group I (mGluR1 and 5), Group II (mGluR2 and 3), and Group III (mGluR4, 6, 7, and 8). Of note, the functional (physiological) roles of these receptor types are very diverse. As a general organizational principle, Group I receptors are often located postsynaptically, and mediate excitation in response to glutamate binding. For example, in the hippocampus, Group I receptors decrease the calcium-activated potassium conductance that mediates spike-frequency adaption (defined as a decrease in spike frequency for sustained constant input, see chapter: Membrane Voltage). This conductance is activated by prolonged superthreshold stimulation, which causes an increase in intracellular calcium concentration. Therefore, in the presence of Group I agonists, spike-frequency adaption is abolished. The same receptors likely have very different properties in other brain areas. In contrast, Groups II and III are often located on presynaptic terminals, where they reduce neurotransmitter release. Therefore those receptors act as a negative feedback system.

INHIBITORY SYNAPTIC TRANSMISSION

We will now turn to the neurotransmitter *GABA*, which is released by *inhibitory neurons*. GABA binds to three types of membrane-bound receptors, denoted as $GABA_A$, $GABA_B$, and $GABA_C$ [10]. $GABA_A$ and $GABA_C$ are ionotropic receptors, whereas $GABA_B$ is a metabotropic receptor. The agonists and antagonists of the GABA receptors are listed in Table 3.2. Ionotropic GABA receptors consist of five protein subunits (ie, they are *pentomeric receptors*) that are arranged to form a pore that is permeable to chloride ions. $GABA_A$ receptors are heteromers; $GABA_C$ receptors are homomers. In theory, $GABA_A$ receptors could be composed of any selection of five of the known 16 subunits (alpha1−6, beta1−3, gamma1−3, delta, epsilon, pi, theta) with any subunit possibly occurring more than once, but in reality there are only a few prevalent configurations. $GABA_A$ receptors are a target for many pharmacological agents. Most of them do not act by binding to the site where GABA would, but rather they bind to a different site that in turn enhances the action of GABA on the receptor (*positive allosteric modulation*). The best-known positive allosteric modulators of $GABA_A$ receptors are *benzodiazepines* and *barbiturates*. Steroids also act on $GABA_A$ receptors, including neurosteroids (synthesized in the brain), sex steroids (gonads), and corticosteroids (adrenal gland). The extent of modulation depends on the concentration of neuroactive steroids, with effects reported for concentrations as low as 1 pM for cortisone. In addition, a large number of naturally occurring compounds interact with $GABA_A$ receptors, including flavonoids (naturally occurring in fruit, vegetables, tea, and red wine), and dietary cholesterol. The specific binding sites and mechanisms of action remain to be fully elucidated for most of these compounds. However, there is growing evidence for changes in behavior (sleep, sedation, memory) associated with $GABA_A$-dependent network activity (see later chapters). "$GABA_C$ receptors" are a historical term as they have been reclassified as a subfamily of the $GABA_A$ receptor, the

TABLE 3.2   Pharmacology of GABA Receptors

Compound	$GABA_A$	$GABA_B$	$GABA_C$
GABA	Agonist	Agonist	Agonist
Muscimol	Agonist	Inactive	Partial agonist
Baclofen	Inactive	Agonist	Inactive
Bicuculline	Antagonist	Inactive	Inactive
Picrotoxin	Antagonist	Inactive	Antagonist
CGP-35348 [3-Aminopropyl (diethoxymethyl)phosphinic acid]	Inactive	Antagonist	Inactive

so-called GABA$_A$ *rho subfamily.* These receptors are uniquely composed of rho subunits, do not respond to GABA$_A$ modulators (barbiturates, benzodiazepines, neuroactive steroids), and are mostly found in the retina.

GABA$_B$ receptors are G-protein-coupled receptors composed of two distinct subunits (GABA$_{B1}$ and GABA$_{B2}$). Presynaptic GABA$_B$ receptors inhibit calcium channels and thereby reduce calcium influx and vesicle release. The terminology makes a distinction between two kinds of presynaptic effects: *autoreceptors* regulate the synthesis and release of their own ligand. In contrast, *heteroreceptors* affect signaling by other than their own ligands. In the case of GABA$_B$ receptors, they are located on presynaptic GABAergic terminals and reduce GABA release (and are therefore autoreceptors) but they also are found on other cell types, where they inhibit the release of other neurotransmitters (they are heteroreceptors). Postsynaptic GABA$_B$ receptors open potassium channels that provide slow hyperpolarization.

Here, we focus on GABA$_A$ receptors, since they provide the majority of the synaptic inhibition that shapes the network activity. GABA$_A$ receptor ion channels exhibit a reversal potential $V_{eq}^{GABA(A)}$ of about -80 mV since they are selectively permeable to chloride. In contrast to glutamate ionotropic receptors, the driving force (difference $V_m - V_{eq}^{GABA(A)}$) is small and inhibitory currents are rather small if the cell is close to its resting potential. However, upon receptor activation the increased conductance reduces the overall resistance of the cell membrane, which in turn reduces the voltage changes induced by other incoming synaptic input. This process, in which excitability of the postsynaptic cell is reduced not by direct hyperpolarization via outward currents but by reduction of the input resistance, is called *shunting inhibition.* An additional consequence of the value of $V_{eq}^{GABA(A)}$ is that relatively small changes in the chloride ion concentration gradient can change the direction of the current flow through GABA$_A$ receptors. Typically, these currents are outward currents (inhibitory) but may become inward currents in case of a decreased chloride concentration gradient. For example, in early development, the chloride concentration gradient is lower because of the lack of KCC2 pumps that extrude chloride from the cytosol to the extracellular space [11]. As a result, GABAergic currents may exert an excitatory effect on neurons during the development of the central nervous system. This switch from excitatory to inhibitory occurs in the second postnatal week. During delivery, there is a transient drop in the intracellular chloride concentration such that GABAergic currents are inhibitory for a short time window around birth, exerting a neuroprotective effect [12]. This process is triggered by the hormone oxytocin, which is involved in triggering labor. In rodent models of autism (valproate and fragile X rodent models), this transient switch is abolished but can be rescued by bumetanide, which decreases intracellular chloride concentration.

ELECTRICAL SYNAPSES

Instead of using long-range communication links enabled by axons and the chemical synapses at their terminals, neurons can also communicate with their direct neighbors via *gap junctions* [13] (often called *electrical synapses*). Gap junctions are channels in the cell membranes of two cells in close proximity that enable direct exchange of intracellular molecules by diffusion. Most cell types in the body (with few exceptions, such as adult skeletal muscle) communicate by exchanging molecules via gap junctions. In the case of neurons, gap junctions not only allow for exchange of molecules such as second messengers (eg, cyclic adenosine monophosphate) but also for direct electric coupling (by ion flux through the gap junctions) such that a change in the membrane voltage of one neuron causes a slight change in membrane voltage in the neurons connected to it by gap junctions. We will later see that such simple and fast bidirectional electric communication plays an important role in synchronizing neuronal network activity in the unit: Cortical Oscillations. Gap junctions are formed by proteins of the family *connexins*. Several members of the connexin family (including Cx26, Cx32, Cx36, Cx43, Cx45, Cx47) have been reported in neurons of adults; probably the best understood is Cx36. Each channel consists of two hemichannels (*connexons*). Each connexon in turn consists of six connexin subunit proteins. Some controversy remains about the presence of gap junctions because of methodological concerns. For example, early studies used antibodies that were not specific enough, and therefore the presence of several connexins in the nervous system was likely overestimated. More definitive evidence comes from electron microscopy, but the presence of gap junctions is often experimentally determined either by dye coupling or electrophysiological measurements of coupling between cells with dual recordings. Interestingly, gap junction coupling of neurons is more prevalent in juvenile animals and decreases during adolescence. In the cortex of adult animals, gap junctions appear to most prominently couple inhibitory interneurons of the same subtype [14]. The direct demonstration of the importance of gap junction coupling in the hippocampus and the neocortex for networks dynamics stems from the study of the $Cx36^{-/-}$ knockout mouse, as discussed in the chapters: Gamma Oscillations and High-Frequency Oscillations. Pharmacological blockade of gap junctions has remained tricky because of the lack of a selective and specific blocker. For example, octanol and heptanol also impair electrical properties of the cell membrane. Therefore a lack of effect when using these compounds is easier to interpret than the presence of an effect.

Electrically, a gap junction is modeled as a resistor that connects the cytosol of two cells. We will first consider a simplified scenario where we ignore the membrane capacitance to understand steady-state coupling between two neurons connected by gap junctions (Fig. 3.5).

We assume that both the presynaptic and the postsynaptic neuron (strictly speaking, this terminology is misleading since gap junctions are symmetric) have an input resistance of $R = R_{pre} = R_{post}$ and are coupled by electric synapses that we model with a resistance R_{gap}. We now use electric circuit modeling to determine the effect of a current injection I into the presynaptic neuron, and, more specifically, what the steady-state changes in the membrane voltages of the two cells are. The injected current will be divided into I_{pre}, the current that depolarizes the presynaptic cell, and I_{post}, the current that depolarizes the postsynaptic cell. According to Kirchhoff's current law (see toolbox: Electrical Circuits):

$$I = I_{pre} + I_{post} \tag{3.23}$$

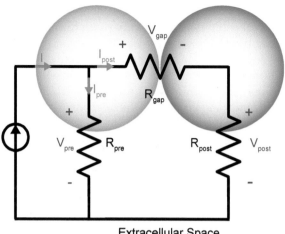

Extracellular Space

FIGURE 3.5   Electric circuit model of an electric synapse (static approximation, no membrane capacitance).

We also know that by Kirchhoff's voltage law, the change in membrane voltage of the presynaptic cell equals the change in postsynaptic membrane voltage plus the voltage drop across the electric synapse modeled by resistance R_{gap}:

$$V_{pre} = V_{gap} + V_{post} \tag{3.24}$$

We can rewrite this equation by using Ohm's law and find:

$$I_{pre}R_{pre} = I_{post}R_{gap} + I_{post}R_{post} \tag{3.25}$$

Together with Eq. (3.24), we now have two linear equations with two unknown variables, I_{pre} and I_{post}, for which we find:

$$I_{post} = I\left(\frac{R}{2R + R_{gap}}\right) \tag{3.26}$$

$$I_{pre} = I\left(\frac{R + R_{gap}}{2R + R_{gap}}\right) \tag{3.27}$$

In agreement with intuition, if there is no gap junction (R_{gap} is infinite), I_{post} equals zero and I_{pre} equals I in this model. Importantly, the smaller the gap junction resistance R_{gap} (and the easier it is for ions to pass between the two cells), the larger the current that depolarizes the postsynaptic cell and the smaller the current that depolarizes the presynaptic cell. For $R_{gap} = 0$, the currents through both cells will be equal.

To understand the temporal dynamics of gap junction coupling, we consider an electric circuit that includes R_{gap} and the postsynaptic passive cell membrane, modeled again with resistance R but also with a capacitance C_m (Fig. 3.6).

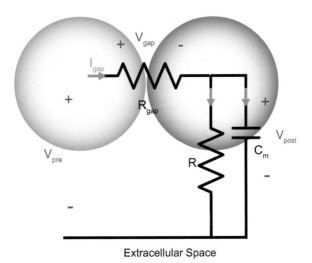

FIGURE 3.6 Dynamics of electric synapse.

We ask how an imagined instantaneous change in presynaptic membrane voltage V_{pre} is reflected in the postsynaptic membrane voltage V_{post}. Initially, all the voltage will drop across the gap junction resistance since the membrane capacitance cannot instantaneously change its voltage (see toolbox: Electrical Circuits). If we wait long enough, the capacitor will be fully charged and can therefore be treated as an open circuit in the steady-state condition (ie, it can be ignored). At this point, the presynaptic voltage V_{pre} is distributed between the gap junction resistance R_{gap} and the input resistance R of the postsynaptic cell. Using Ohm's law, we find:

$$V_{\text{post}} = V_{\text{pre}}\left(\frac{R}{R + R_{\text{gap}}}\right) \tag{3.28}$$

To determine the dynamics, we can again (as previously) use the Kirchhoff voltage rule, which states that the applied voltage in the presynaptic side V_{pre} is equal to the sum of the voltage across the gap junction and the postsynaptic membrane voltage V_{post}:

$$V_{\text{pre}} = V_{\text{gap}} + V_{\text{post}} \tag{3.29}$$

In this case, the current through the gap junction is the sum of the current through the postsynaptic resistance and the current through the postsynaptic capacitance:

$$I_{\text{gap}} = C\frac{dV_{\text{post}}}{dt} + \frac{V_{\text{post}}}{R} \tag{3.30}$$

Using Ohm's law, we can combine Eqs. (3.29) and (3.30) and find:

$$V_{\text{pre}} = R_{\text{gap}}\left(C_{\text{m}}\frac{dV_{\text{post}}}{dt} + \frac{V_{\text{post}}}{R}\right) + V_{\text{post}} \tag{3.31}$$

which after some rearranging ends up looking like our familiar differential equation from previous chapters:

$$\left(\frac{RR_{gap}}{R + R_{gap}}\right)C_m\frac{dV_{post}}{dt} = -V_{post} + V\left(\frac{R}{R + R_{gap}}\right) \tag{3.32}$$

The steady-state solution corresponds to what we have determined previously, and the time constant of the postsynaptic membrane voltage dynamics is:

$$\tau = \frac{RR_{gap}}{R + R_{gap}}C_m \tag{3.33}$$

If there is no gap junction coupling (R_{gap} large), we find $\tau = RC_m$ as determined in earlier chapters (passive cell membrane, no coupling). However, as R_{gap} decreases (more gap junction coupling), τ gets smaller. The key insight from this model is that an instantaneous change in presynaptic voltage V_{pre} does not cause an instantaneous change in the postsynaptic membrane voltage V_{post} because of the membrane capacitance.

SUMMARY AND OUTLOOK

In this chapter, we discussed both chemical and electrical synaptic transmission. For chemical synapses, we reviewed how the probabilistic nature of neurotransmitter release gives rise to the kinetics of postsynaptic currents, learned about receptor composition and pharmacology, and introduced quantitative models. For electrical synapses, we looked at how change in the membrane voltage of one neuron affects the membrane voltage of another neuron that shares a gap junction. Together, this chapter prepares us for the study of networks by introducing the main element by which neurons form networks. Furthermore, the quantitative models derived here are broadly used in a large number of neuronal network simulations. In the next chapter, we will discuss how synapses change their strength through plasticity, the basic mechanism of learning and memory.

References

[1] Simms BA, Zamponi GW. Neuronal voltage-gated calcium channels: structure, function, and dysfunction. Neuron 2014;82(1):24–45.
[2] Cain SM, Snutch TP. T-type calcium channels in burst-firing, network synchrony, and epilepsy. Biochim Biophys Acta 2013;1828(7):1572–8.
[3] Neher E, Sakaba T. Multiple roles of calcium ions in the regulation of neurotransmitter release. Neuron 2008;59(6):861–72.
[4] Traynelis SF, et al. Glutamate receptor ion channels: structure, regulation, and function. Pharmacol Rev 2010;62(3):405–96.
[5] Destexhe A, Mainen ZF, Sejnowski TJ. Synthesis of models for excitable membranes, synaptic transmission and neuromodulation using a common kinetic formalism. J Comput Neurosci 1994;1(3):195–230.
[6] Kwak S, Weiss JH. Calcium-permeable AMPA channels in neurodegenerative disease and ischemia. Curr Opin Neurobiol 2006;16(3):281–7.
[7] Cull-Candy S, Brickley S, Farrant M. NMDA receptor subunits: diversity, development and disease. Curr Opin Neurobiol 2001;11(3):327–35.

[8] Jahr CE, Stevens CF. Voltage dependence of NMDA-activated macroscopic conductances predicted by single-channel kinetics. J Neurosci 1990;10(9):3178—82.

[9] Ferraguti F, Shigemoto R. Metabotropic glutamate receptors. Cell Tissue Res 2006;326(2):483—504.

[10] Bormann J. The 'ABC' of GABA receptors. Trends Pharmacol Sci 2000;21(1):16—9.

[11] Ben-Ari Y, et al. GABA: a pioneer transmitter that excites immature neurons and generates primitive oscillations. Physiol Rev 2007;87(4):1215—84.

[12] Tyzio R, et al. Maternal oxytocin triggers a transient inhibitory switch in GABA signaling in the fetal brain during delivery. Science 2006;314(5806):1788—92.

[13] Sohl G, Maxeiner S, Willecke K. Expression and functions of neuronal gap junctions. Nat Rev Neurosci 2005;6(3):191—200.

[14] Galarreta M, Hestrin S. Electrical synapses between GABA-releasing interneurons. Nat Rev Neurosci 2001;2(6):425—33.

4

Synaptic Plasticity

In the previous chapter, "Synaptic Transmission," we discussed how synapses mediate communication between neurons. The strength of synapses changes over time, a process called *synaptic plasticity*. In this chapter, we will consider two main types of plasticity. First, we will discuss *short-term plasticity*, the modulation of synaptic strength on the time scale of milliseconds to seconds. Second, we will learn about *long-term plasticity*, long-lasting changes to synaptic strength that form the basis of learning and memory. We will learn about the rules that govern synaptic plasticity and introduce intuitive mathematical models of plasticity. The toolboxes "Neurons," "Differential Equations," and "Dynamical Systems" provide helpful background for this chapter.

SHORT-TERM PLASTICITY

Short-term plasticity refers to alteration of synaptic strength on the time scale of milliseconds to seconds. The underlying mechanisms of short-term plasticity are predominantly localized to the presynaptic terminal [1,2]. Broadly speaking, synapses can weaken in an activity-dependent way such that two consecutive presynaptic action potentials cause two postsynaptic excitatory postsynaptic currents (EPSCs), with the second one being weaker than the first one. This process is called *short-term synaptic depression*. Recovery from depression occurs on the time scale of hundreds of milliseconds to several seconds. Synapses can also strengthen in an activity-dependent way, in a process called *synaptic facilitation*. Recovery from facilitation occurs on a similar time scale to that of depression. Both weakening (depression) and strengthening (facilitation) occur as a function of the "use pattern" of the presynaptic release machinery. Short-term plasticity enables the selective enhancement (and suppression) of input as a function of its temporal structure.

Short-Term Depression

Depression is mediated by a decrease in the number of neurotransmitter vesicles that are released because of the depletion of available synaptic resources. Synaptic vesicles in the terminal can be in different states, and only a subset is ready to be released in response to a presynaptic action potential (Fig. 4.1). These vesicles form the so-called *readily releasable*

FIGURE 4.1 Synaptic vesicles are organized into three different pools: reserve pool (largest fraction of vesicles), recycling pool, and readily releasable pool. Vesicles in the reserve pool cannot be directly released.

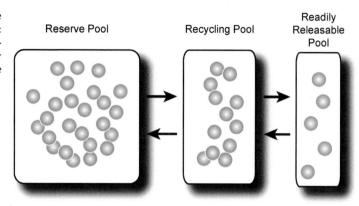

pool, and they represent the minority of all vesicles associated with an active zone. This pool is replenished by the *recycling pool*. The *reserve pool* replenishes the *recycling pool*.

The more vesicles released in response to the first action potential, the fewer there are left to be released in response to the second action potential that quickly follows. Therefore the amplitudes of the first and second EPSCs are negatively correlated. In particular, if extracellular calcium concentration $[Ca]_o$ is high, the release probability and therefore the synaptic strength are high. The price to be paid for such a strong synapse is pronounced depression. In contrast, if $[Ca]_o$ is low and synaptic transmission accordingly weaker, synaptic depression is limited or absent. In in vitro experiments, $[Ca]_o$ is determined by the concentration of calcium in the extracellular medium (*artificial cerebrospinal fluid*); typical values range from 1.0 to 2.0 mM. In vivo, $[Ca]_o$ fluctuates in an activity-dependent way such that elevated neuronal activity reduces $[Ca]_o$.

Synaptic Facilitation

Originally, it was thought that facilitation occurs when presynaptic calcium levels do not fall back to the baseline level after an initial presynaptic action potential (Fig. 4.2A and B). The excess calcium from the first action potential rapidly diffuses in the terminal and leads to a small increase in the new "baseline" calcium level at the time of the next action potential (assuming little time has elapsed between the two action potentials). However, experimental measurements have shown that the residual calcium increase after a presynaptic action potential is too small to explain facilitation. Two possible, alternative mechanisms have been described. The first proposes the presence of two distinct calcium sensors for triggering synaptic release. The main calcium sensor is *synaptotagmin*, which controls release of synaptic vesicles. Synaptotagmin binds calcium with low affinity but fast kinetics (rates) and is therefore the ideal sensor for action potential-induced increases in calcium. In this first model, the residual calcium acts on a different calcium sensor than *synaptotagmin*; higher-affinity calcium sensors could track the residual calcium concentration to mediate synaptic facilitation (Fig. 4.2C). In the second proposed mechanism, a fast buffer binds calcium before it has reached the release sites. Therefore only accumulated intracellular calcium past the saturation of the buffer will succeed in reaching the release sites to enable synaptic transmission (Fig. 4.2D).

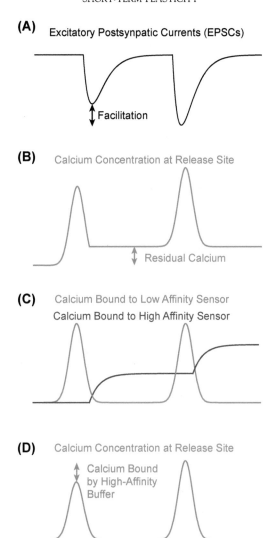

FIGURE 4.2 Mechanisms of synaptic facilitation. (A) Pair of excitatory postsynaptic currents (EPSCs). The amplitude of the second EPSC is larger because of facilitation. (B) Classical model of facilitation. The calcium concentration at the release site does not return to baseline such that the peak calcium concentration is higher in response to the second presynaptic action potential. (C) Facilitation can also be explained by a second, high-affinity sensor to which the low concentration of residual calcium binds. (D) Facilitation can also be explained by the presence of a high-affinity calcium buffer that prevents calcium from reaching the release site. Only when this buffer is saturated can the excess calcium reach the release site [1].

Modeling Short-Term Plasticity

Both depression and facilitation dynamics can be modeled by a system of two state variables [3]. The first state variable, x, denotes the fraction of available resources at the synapse and assumes values between zero and one. If the synapse has not been used in a while, x will have recovered to one since all resources will be available. In contrast, heavy use of the synapses will reduce x to zero. These dynamics can be modeled with the same first-order differential equation that we have encountered throughout the last few chapters (for an introduction to the mathematics see toolbox: Differential Equations).

$$\frac{dx}{dt} = \frac{(1-x)}{\tau_d} \tag{4.1}$$

This equation describes how x converges to $x_\infty = 1$ with time constant τ_d, in effect modeling the recovery from depression. We can extend Eq. (4.1) to reduce x each time there is a presynaptic spike. Mathematically, we denote such events that are limited to an infinitely short duration—in essence the moment of occurrence—with the so-called *Dirac delta function*:

$$\delta(t - t_{sp}) \tag{4.2}$$

which denotes the occurrence of a spike at time point t_{sp}. Each time there is a spike, we spend a certain fraction u of our remaining synaptic resources. Therefore the complete equation reads:

$$\frac{dx}{dt} = \frac{(1-x)}{\tau_d} - ux\delta(t - t_{sp}) \tag{4.3}$$

To model facilitation, we turn u, the fraction of resources used in response to a presynaptic spike, into a state variable. Similar to x, u assumes values between zero and one. Each time a presynaptic spike occurs, we increase state variable u by U_{sp}.

$$\frac{du}{dt} = U_{sp}\delta(t - t_{sp}) \tag{4.4}$$

However, if no presynaptic spike has occurred, u needs to converge to $u_\infty = 0$ such that at the next spike, $u = U_{sp}$ again. We can enforce this by adding a "$-u\tau_f$" term to the right-hand side of the equation; the time constant τ_f defines how fast the synapse recovers from facilitation. We also want to make sure that u cannot grow beyond 1, which we enforce by increasing u not by U_{sp} but rather by U_{sp} multiplied with $(1 - u)$. The final system of equations then reads:

$$\frac{du}{dt} = -\frac{u}{\tau_f} + U_{sp}(1 - u^-)\delta(t - t_{sp})$$
$$\frac{dx}{dt} = \frac{(1-x)}{\tau_d} - u^+x^-\delta(t - t_{sp}) \tag{4.5}$$

We have fine-tuned our notation and added plus and minus signs to the state variable changes induced by action potentials to clarify if the values just before (eg, u^-) or just after (eg, u^+) the occurrence of the spike need to be used.

This model accurately reproduces short-term dynamics of synapses measured in experiments. If the time constant of facilitation is much larger than the time constant of depression, then the synaptic plasticity is dominated by facilitation. In the opposite case, where $\tau_d \gg \tau_f$, the synapse is dominated by short-term depression, since any facilitation quickly recovers because of the comparably fast time constant τ_f. If facilitation dominates, synapses preferentially respond to high-frequency input since under these circumstances facilitation dominates. In contrast, if depression dominates, high-frequency input quickly reduces synaptic strength and thus low-frequency input is preferentially transmitted to the postsynaptic neuron. Therefore, by balancing depression and facilitation, the temporal filtering properties of synapses can be adjusted. Short-term plasticity plays other fundamental roles and provides the substrate for computation and signal processing. For example, synaptic depression enables selective responses to the onset of sensory stimuli since repeated activation of the same synapses will lead to reduced postsynaptic responses because of increasing depression of the afferent synapses.

LONG-TERM PLASTICITY

The basis of learning and memory is the ability of neuronal networks to change in response to external input in a way that the changes are maintained for prolonged periods and previously experienced input can be retrieved. At the cellular and synaptic level, the changes to the networks that enable such memory formation, maintenance, and retrieval are changes in synaptic strength. Excitatory synapses can both become stronger (potentiation) and weaker (depression) on much longer time scales than for the short-term plasticity discussed earlier. The traditional approach to experimentally induce such long-term plasticity is based on temporally patterned stimulation of afferent pathways. The main experimental system for the study of long-term plasticity in mammals is stimulation of the Schaffer collaterals and measurement of the postsynaptic response such as amplitude of the EPSP in hippocampus CA1 (see chapter: Microcircuits of the Hippocampus). High-frequency stimulation induces *long-term potentiation* (LTP, Fig. 4.3, top), whereas low-frequency stimulation induces *long-term depression* (LTD, Fig. 4.3, bottom). Alternatively, both LTP and LTD can be introduced by pairing pre- and postsynaptic action potential firing (*spike-timing-dependent plasticity*, STDP): if a presynaptic spike occurs briefly before a postsynaptic spike, the synapse is strengthened ("pre before post"); if the order is reversed ("post before pre") the synapse is weakened. These mechanisms are forms of *Hebbian plasticity*, a theoretical concept that was postulated before any experimental evidence of long-term synaptic plasticity had been found [4]. According to this theory, if neuron A repeatedly makes neuron B fire an action potential, then the synapse from A to B is strengthened.

Converging evidence shows that the induction of long-term synaptic plasticity critically depends on postsynaptic influx of calcium through *N*-methyl-D-aspartate (NMDA) receptors. In fact, the postsynaptic calcium levels will determine if a synapse is weakened or strengthened, as a function of *how* elevated the calcium concentration is. We will first review several key experimental observations and then proceed to a simple model that elegantly captures the main features that define the dynamics of the plastic changes.

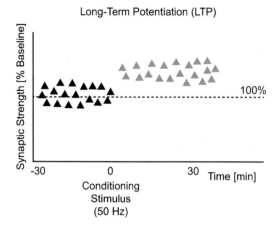

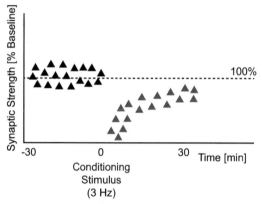

FIGURE 4.3 *Top*: Long-term potentiation (LTP) induced by a 50-Hz conditioning stimulus of the afferent pathway. *Bottom*: Long-term depression (LTD) induced by a 3-Hz conditioning stimulus of the afferent pathway. *Adapted from Dudek SM, Bear MF. Homosynaptic long-term depression in area CA1 of hippocampus and effects of N-methyl-D-aspartate receptor blockade. Proc Natl Acad Sci USA 1992;89(10):4363–7.*

Mechanism of LTP and LTD

The mechanism of LTP has been subject to one of the most intense and protracted debates in neuroscience [6]. The main point of contention was whether LTP is mediated by a pre- or a postsynaptic mechanism. Today, most evidence points toward a postsynaptic localization. However, note that the mechanism depends on the type of synapse. While LTP in CA1 in response to the stimulation of the Schaffer collaterals (see chapter: Microcircuits of the Hippocampus) is driven by a postsynaptic mechanism, other synapses exhibit different forms of LTP. For example, the mossy fiber synapses on CA3 pyramidal cells exhibit LTP that is a presynaptic form of plasticity. This type of LTP is independent of the activation of NMDA receptors. The postsynaptic nature of LTP in CA1 can, for example, be demonstrated by a technique called *glutamate uncaging*. In such an experiment, "caged glutamate" is used that does not bind to glutamate receptors. Two-photon excitation is then used to selectively "uncage" glutamate molecules close to a single dendritic spine to cause a miniature EPSC (mEPSC) that has the same waveform and amplitude as spontaneously occurring mEPSCs. When such activation is combined with postsynaptic depolarization, LTP is induced (in the absence of any presynaptic activation). This process depends on NMDA receptors. The strengthening and weakening of glutamatergic synapses by LTP and LTD, respectively, is driven by changes in the number and type of α-amino-3-hydroxy-5-methyl-4-isoxazolepropionic acid (AMPA) receptors that are located in the synapse. LTP is associated with an increase in the number of receptors; LTD is associated with a decrease in the number of receptors. The process of moving AMPA receptors into and out of synapses is referred to as AMPA *receptor trafficking*. Also LTP is associated with phosphorylation of AMPA receptors, which increases the conductance of the AMPA receptors. The phosphorylation is mediated by *calcium/calmodulin-dependent kinase II* (CaMKII). LTD has been proposed to mirror the mechanism of LTP; indeed, *calcineurin*, a calcium/calmodulin-dependent protein phosphatase, has been implicated in LTD. One attractive feature of these opposing effects of CaMKII and calcineurin is their differential higher calcium affinity. Calcineurin has a higher affinity and is thus activated for lower $[Ca]_i$ than CaMKII. This matches the role of $[Ca]_i$ in LTP and LTD.

Spike-Timing-Dependent Plasticity

STDP is a mechanism by which LTP and LTD can be introduced [7]. STDP changes synaptic strength as a function of the timing between the presynaptic and the postsynaptic action potential (Fig. 4.4). The occurrence of a presynaptic spike a few milliseconds before a postsynaptic spike increases synaptic strength. The shorter the time elapsed between the two spikes, the larger the change in synaptic weight. In contrast, the occurrence of a postsynaptic spike a few milliseconds before a presynaptic spike weakens synaptic strength. Again, the shorter the

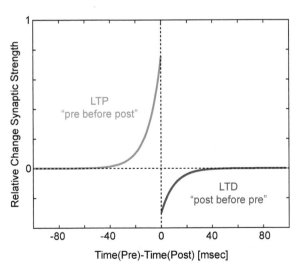

FIGURE 4.4 Spike-timing-dependent plasticity (STDP). If the presynaptic spike occurs before the postsynaptic spike ("pre before post"), the synapse is strengthened (*red*, LTP, long-term potentiation). If the postsynaptic spike occurs before the presynaptic spike, the synapses are weakened (*blue*, LTD, long-term depression). Typically, two action potentials need to occur within at most a few tens of milliseconds for STDP to be recruited.

time interval between the two spikes, the larger the resulting change in synaptic strength. The existence of STDP supports the presence of neural codes that employ precise timing of individual action potentials, because a change in spike time of only a few milliseconds can have opposite effects on the change of synaptic strength. Interestingly, the temporal properties of STDP are altered by neuromodulators such as dopamine (see chapter: Neuromodulators).

Modeling Long-Term Plasticity

In mathematical models [8], the strength of a synapse is called *synaptic weight W*. We start by defining the function $\Omega([Ca]_i)$, which describes how changes in the synaptic weight (strength) W depend on the postsynaptic, intracellular calcium levels $[Ca]_i$ (Fig. 4.5). This function has three distinct regions corresponding to low, intermediate, and high $[Ca]_i$ (measured relative to baseline levels). For low levels, $\Omega = 0$, not enough calcium elevation has occurred to trigger synaptic plasticity. For intermediate levels, $\Omega < 0$, synapses decrease in strength (LTD). For high levels, $\Omega > 0$, synapses are potentiated (LTP).

Since Ω defines the rate of change of synaptic weight W, we can write:

$$\frac{dW}{dt} = \eta \Omega([Ca]_i) \tag{4.6}$$

where η denotes the learning rate, which dictates how fast W changes over time. The problem with this model is that as soon as Ω is nonzero, W will grow without bounds. As an unstable system, it would be a poor model of changes in synaptic strength. To solve this problem, we will now add a decay term, so that the equation reads:

$$\frac{dW}{dt} = \eta\left(-W + \Omega([Ca]_i)\right) \tag{4.7}$$

We recognize that this equation is very similar to the other first-order linear differential equations discussed so far (see toolbox: Differential Equations) with the only difference being that the nonzero steady-state value depends on $[Ca]_i$. Although we introduce a stable

FIGURE 4.5 Omega function defines change rate of synaptic weight as a function of intracellular calcium concentration $[Ca]_i$.

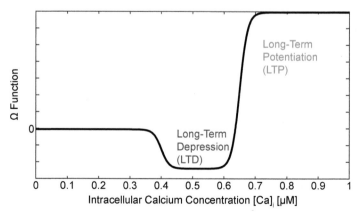

equilibrium with this change, we made things worse in the sense that now changes in W are only transient, since $[Ca]_i$ will return to baseline (and we want the synapse to remain in its altered state). After transient influx of calcium, changes in postsynaptic strength should be maintained. We can make the learning rate η depend on $[Ca]_i$ so that for low $[Ca]_i$ (eg, once $[Ca]_i$ returns to baseline) no more change to W occurs, that is, $dW/dt = 0$. For example, η can exhibit a sigmoidal dependence on $[Ca]_i$ (Fig. 4.6).

The complete model is then given by:

$$\frac{dW}{dt} = \eta([Ca]_i)\left(-W + \Omega([Ca]_i)\right) \tag{4.8}$$

such that if $[Ca]_i$ is low, the rate of change of W is zero and therefore the synapses stay at the augmented level induced by the transient calcium change. It is important to note that this is a phenomenological model that significantly abstracts from the underlying neurobiology.

Next, we need to understand how the calcium current through NMDA receptors (chapter: Synaptic Transmission) affects $[Ca]_i$. We assume first-order kinetics such that:

$$\frac{d[Ca]_i}{dt} = -\frac{1}{\tau_{Ca}}[Ca]_i + I_{NMDA} \tag{4.9}$$

where the clearance of the elevated $[Ca]_i$ occurs with rate τ_{Ca} (typically ~ 20 ms) and the change rate of $[Ca]_i$ is proportional to the NMDA current I_{NMDA}.

Next, we need to investigate under what circumstances the postsynaptic cell depolarizes in a way that enables activation of NMDA receptors. Importantly, the dendrites depolarize not only because of presynaptic input but also because of depolarization induced by action potential firing that propagates back into the dendrites, *backpropagating* action potentials (see toolbox: Neurons). The time course of the backpropagating action potential has a slow and fast decay time constant (τ_{slow} and τ_{fast}, respectively) such that the change in membrane voltage ΔV_m^{BP} induced is:

$$\Delta V_m^{BP} = Ae^{-\frac{t}{\tau_{fast}}} + Be^{-\frac{t}{\tau_{slow}}} \tag{4.10}$$

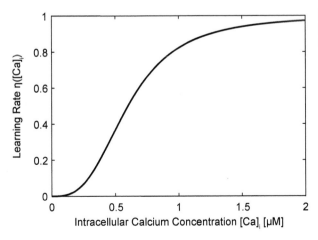

FIGURE 4.6 Learning rate as a function of the intracellular calcium concentration $[Ca]_i$ assumes a sigmoid function.

where *A* and *B* are constants. The depolarization induced by the backpropagating action potential will remove the magnesium block from the NMDA channels and enable sustained calcium influx through these receptors. Because of the slower component of the backpropagating action potential, significant calcium influx can occur. This model can explain STDP (Fig. 4.7). If the postsynaptic spike occurs before the presynaptic spike, a substantial (namely, the fast-decaying) part of the backpropagating action potential is already over before glutamate (released in response to presynaptic action potential) binds to the NMDA receptors. Therefore incomplete synergy between pre- and postsynaptic activity moderately enhances calcium influx through NMDA receptors, leading to long-term depression. In contrast, if the presynaptic spike occurs briefly before the postsynaptic spike, then the backpropagating action potential will induce a postsynaptic depolarization that coincides with the time window in which glutamate is bound to the NMDA receptors. This maximized synergy

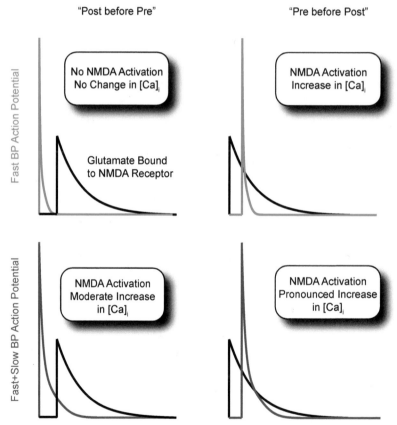

FIGURE 4.7   Role of backpropagating action potential in spike-timing-dependent plasticity (STDP). *Top left*: If the backpropagating (BP) action potential ("post before pre") is short in duration, the resulting postsynaptic depolarization will not overlap with the period during which glutamate is bound to the NMDA receptors and no change in [Ca]$_i$ occurs. *Top right*: For "pre before post," NMDA receptors are activated and [Ca]$_i$ increases. *Bottom*: If the BP action potential exhibits a slow and a fast component, there is moderate NMDA receptor activation for "post before pre" and substantial activation for "pre before post" [8].

causes a more pronounced elevation of intracellular calcium leading to long-term potentiation.

Stability is one of the key challenges that come along with the amazing flexibility of synapses to change their strength through plasticity. As we have already seen, imposing a mechanism that prevents unlimited potentiation was crucial for building a phenomenological model. However, limiting the strengthening of synapses is not only a mathematical issue when building models but also a real issue in biological systems. An early model that brought together activity-dependent strengthening and weakening to keep synaptic strength flexible yet stable is the *Bienenstock—Cooper—Munro model* (often abbreviated as the BCM model of plasticity [9]). Developed based on theoretical considerations, the BCM model made important predictions (particularly about the existence of activity-dependent weakening) that have spurred the discovery and study of LTD. The main features of the model resemble the previously introduced model. We provide a brief summary next.

The BCM model also follows a differential equation that describes the change in synaptic weight W over time:

$$\frac{dW(t)}{dt} = \Phi(c(t))b(t) - \varepsilon W(t) \tag{4.11}$$

where $b(t)$ denotes presynaptic activity and $\Phi(c(t))$ is a function of postsynaptic activity $c(t)$. The function $\Phi(c)$ changes its sign at the so-called modification threshold θ_M such that $\Phi(c) < 0$ for values of $c < \theta_M$ and $\Phi(c) > 0$ for $c > \theta_M$. Furthermore, we recognize the term $-\varepsilon W(t)$ as a stability-inducing decay term. Thus if $c(t)$ is large (above threshold), then synaptic weight W will increase as a function of presynaptic activity. However, if little postsynaptic activity occurs such that $c(t)$ is below threshold, the synaptic weight W will decrease, since the term $\Phi(c(t))b(t)$ will be negative. The modification threshold can be chosen to be the long-term average of postsynaptic activity $c(t)$. Finally, the parameter ε defines how fast W changes.

SUMMARY AND OUTLOOK

This chapter introduced short-term and long-term synaptic plasticity. First, we discussed that changes in presynaptic intracellular calcium concentration can mediate both short-term depression and facilitation of synaptic strength, depending on the activity level of the presynaptic neuron. We then derived a phenomenological model that captures both depression and facilitation. Second, we discussed long-term synaptic plasticity, which again is a function of intracellular calcium concentration, but in this case of the postsynaptic neuron. Finally, we reviewed the required timing of pre- and postsynaptic activity that evokes such longer-lasting changes to synaptic strength. Learning and memory crucially depend on synaptic plasticity. The importance of timing of neuronal activity for inducing synaptic plasticity cannot be overstated. We will revisit this concept in the unit Cortical Oscillations, in which we will discuss the functional role of rhythmically timed activity. We will also encounter plasticity in the chapter "Noninvasive Brain Stimulation," where we will talk about the mechanisms by which the effects of stimulation outlast the application of stimulation.

References

[1] Regehr WG. Short-term presynaptic plasticity. Cold Spring Harb Perspect Biol 2012;4(7):a005702.

[2] Zucker RS, Regehr WG. Short-term synaptic plasticity. Annu Rev Physiol 2002;64:355—405.

[3] Tsodyks MV, Markram H. The neural code between neocortical pyramidal neurons depends on neurotransmitter release probability. Proc Natl Acad Sci USA 1997;94(2):719—23.

[4] Hebb DO. The organization of behavior: a neuropsychological theory. Mahwah, N.J.: L. Erlbaum Associates; 2002.

[5] Dudek SM, Bear MF. Homosynaptic long-term depression in area CA1 of hippocampus and effects of N-methyl-D-aspartate receptor blockade. Proc Natl Acad Sci USA 1992;89(10):4363—7.

[6] Nicoll RA, Roche KW. Long-term potentiation: peeling the onion. Neuropharmacology 2013;74:18—22.

[7] Markram H, et al. Regulation of synaptic efficacy by coincidence of postsynaptic APs and EPSPs. Science 1997;275(5297):213—5.

[8] Shouval HZ, Bear MF, Cooper LN. A unified model of NMDA receptor-dependent bidirectional synaptic plasticity. Proc Natl Acad Sci USA 2002;99(16):10831—6.

[9] Bienenstock EL, Cooper LN, Munro PW. Theory for the development of neuron selectivity: orientation specificity and binocular interaction in visual cortex. J Neurosci 1982;2(1):32—48.

Neuromodulators

Neuronal networks switch among electrical activity patterns as a function of behavioral state. For example, cortical networks exhibit different activity patterns during sleep and wakefulness. During sleep, the brain cycles through distinct stages, and each stage is defined by the temporal structure of cortical activity and by other physiological measures such as muscle tone and eye movements (see chapter: Low-Frequency Oscillations). Behavioral state also fluctuates during waking and is associated with arousal and attention, which correlates with specific patterns of brain activity. In this chapter we will discuss signaling molecules called *neuromodulators*, which play a central role in defining these behavioral and network states. Neuromodulators are endogenous chemicals released by specialized cells that alter the activity of populations of neurons at a large spatial scale. Neuromodulators have received their name in recognition of their difference from classical neurotransmitters that mediate point-to-point communication between individual neurons (see chapter: Synaptic Transmission).

Neuromodulators play a key role in shaping brain activity and take center stage in most models of the biological substrate of mental illnesses. The emphasis on neuromodulators is the result of the success of medications that alter neuromodulatory function in psychiatry (see toolbox: Psychiatry). In addition, neuromodulatory systems are often targeted by drugs of abuse such as cocaine and amphetamines, therefore making them critical to understanding the neurobiological basis of addiction. In this chapter, we will discuss the main neuromodulators acetylcholine (ACh), norepinephrine (NE), dopamine, histamine, and serotonin. With the exception of ACh, these neuromodulators are all *monoamines*, molecules defined by their shared chemical blueprint of an amino group connected to an aromatic ring. Neuromodulators are synthesized in specialized neurons that cluster in specific brain structures associated with a neuromodulator (Fig. 5.1). These neurons exhibit highly divergent projections such that activity in these neurons results in neuromodulators released in large territories. We will focus on the effects of these neuromodulators on the neocortex.

CORTICAL STATE

Neuronal networks are not limited to exhibiting a single pattern of electrical activity. Rather, networks are flexible entities that can rapidly switch among different activity states [1,2]. These transitions can reflect changes in behavioral state that enable animals to quickly

59

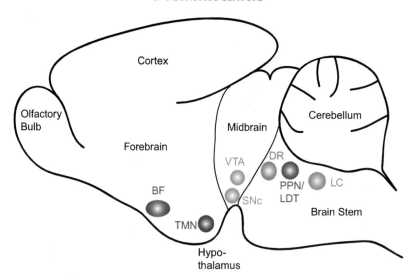

FIGURE 5.1   Location of nuclei that release the neuromodulators discussed in this chapter. Sagittal section of the rodent brain. *BF*, basal forebrain (acetylcholine); *DR*, dorsal raphe nucleus (serotonin); *LC*, nucleus coeruleus (norepinephrine); *PPN* and *LDT*, pedunculopontine tegmental nucleus and lateral dorsal tegmental nucleus (acetylcholine); *SNc*, substantia nigra pars compacta (dopamine); *TMN*, tuberomamillary nucleus (histamine); *VTA*, ventral tegmental area (dopamine).

adapt to events in the environment. For example, the occurrence of an unexpected noise could indicate the presence of a predator, which requires accurate sensory processing and rapid decision making to ensure survival. A fast transition to a state of heightened arousal enables such a behavioral response. In this context, arousal denotes the sensitivity to external input. Cognitive performance exhibits an inverted U-shaped relationship with arousal (Yerkes–Dodson law, Fig. 5.2). Both hypo- and hyperarousal impair cognitive performance.

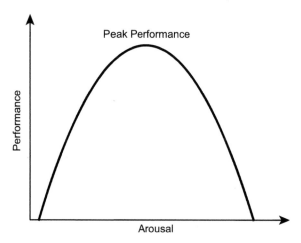

FIGURE 5.2   The Yerkes–Dodson law describes how cognitive performance depends on arousal level. Intermediate arousal levels enable peak performance. The curve assumes an inverted U-shape.

The underlying switch in brain state is controlled by multiple, interacting neuromodulatory systems that we review later.

In addition to changes in brain state during wakefulness, the brain also cycles through distinct stages during sleep. Briefly, sleep is divided into rapid eye movement (REM) sleep and non-REM (NREM) sleep. REM sleep refers to epochs characterized by the lack of low-frequency oscillations (ie, slow waves, see chapter: Low-Frequency Oscillations) in the electroencephalogram (EEG), occurrence of REMs, and reduced muscle tone measured with electromyography. NREM sleep refers to epochs with dominant low-frequency oscillations and normal muscle tone. During a night of sleep, epochs of NREM and REM sleep alternate. The overall functional role of sleep and of the individual sleep stages in particular remains to be fully elucidated. Nevertheless, sleep is an ideal model system to understand state transitions, since the different states are clearly delineated and occur in a prototypical sequence during a night of sleep. This contrasts with a more ambiguous continuum of different states of vigilance and attention in the awake animal. Furthermore, the transitions in the awake animal strongly depend on the external environment, and their occurrence is less predictable. As we will see, the same neuromodulatory systems that regulate cortical state in awake animals are also directly involved in mediating different brain states during sleep.

At the network level, cortical states are primarily defined by the relative presence and absence of different rhythmic activity patterns (see unit: Oscillations in the Brain). If no neuromodulatory inputs are present, the neocortex exhibits continuous, low-frequency, high-amplitude network activity that resembles the network dynamics of NREM sleep. Therefore the cortex relies on neuromodulation for *activation*, which is defined as the switch to high-frequency, low-amplitude activity. Historically, activation is also referred to as *desynchronization*. This term refers to the absence of synchronized activity at low frequencies. However, the activated state is technically also associated with synchronized activity, albeit at higher frequencies that are harder to detect by visual inspection of electrophysiology data such as EEGs. Cortical activation is present in both the awake state and REM sleep. As we will see when we discuss the individual neuromodulators, they are all involved in generating these states.

BASIC PRINCIPLES

All neuromodulators discussed here have been implicated in numerous cellular and synaptic processes, often with conflicting findings. Nevertheless, several basic principles seem to apply to most neuromodulators. First, neurons that synthesize and release neuromodulators exhibit different activity levels and firing patterns as a function of the behavioral state. Second, a small number of neuromodulatory neurons (eg, only a few thousand neurons in the rat) define the neuromodulatory tone of large territories of the CNS. This is accomplished by *volume transmission*—signal transmission mediated by diffusion in the extracellular space. Third, most of what we know about the neuromodulatory systems focuses on the effects of neuromodulators on network dynamics and behavior. However, the neurons that release neuromodulators receive complex and poorly understood glutamatergic and GABAergic input signals from local and distant sources. Fourth, the terms "inhibitory" and "excitatory" do not appropriately describe the complex actions of neuromodulators on the target neurons that express the receptors to bind the neuromodulators. Historically, this notion was developed from experiments that measured changes in firing rates in response to the application to neuromodulators. Fifth, neuromodulation is typically assumed to lack spatial specificity (in comparison to excitatory and inhibitory synaptic transmission) because of the diffuse, divergent projections of the neurons that release neuromodulators. Newer evidence suggests that the spatiotemporal resolution of neuromodulation may be higher than what is commonly assumed [3].

ACETYLCHOLINE

ACh is released by *cholinergic* neurons, which include both projection neurons and local interneurons. In the cortex, the *basal forebrain* (BF) is the major source of ACh. The BF is a diverse set of nuclei composed of the *nucleus accumbens, nucleus basalis, diagonal band of Broca, substantia innominata*, and the *medial septal nuclei*. Specifically, the nucleus basalis provides ACh to the neocortex, whereas the medial septal nuclei provide ACh to the hippocampus. Cholinergic neurons in the *pedunculopontine tegmental nucleus* (PPN) and *lateral dorsal tegmental nucleus* (LDT) project to the thalamus, the hypothalamus, and the BF.

Mechanisms

ACh receptors are divided into *muscarinic receptors* (M1—M5), which are metabotropic receptors, and *nicotinic receptors*, which are ionotropic receptors nonselectively permeable to cations. In the classical model, ACh acts on large numbers of neurons by volume transmission through broad and diffuse release of ACh. Newer evidence suggests more precise patterning of cholinergic projections to the cortex and the presence of synaptic transmission mechanisms that are more specific than volume transmission [3]. Cortical activation by ACh is mediated by muscarinic ACh receptors [4]. In this context, the cellular-level effects of ACh are primarily (1) a depolarization caused by a decrease of several types of potassium conductances, and (2) a reduction in intracortical excitatory synapses. Accordingly, in vitro, the application of ACh has two main effects on the intrinsic excitability of pyramidal cells. First, ACh increases the firing rate of pyramidal cells in response to a constant depolarizing current, which is mediated by an increase in input resistance caused by the decrease in potassium leak current. Second, ACh reduces spike frequency adaption by decreasing M-type and calcium-activated potassium channels [5]. In terms of synaptic transmission, ACh has a layer-specific effect in multiple brain structures, including the hippocampus and the olfactory bulb. Specifically, ACh selectively suppresses excitatory synaptic transmission that mediates recurrent excitatory connectivity (Fig. 5.3). Therefore local interaction between excitatory cells such as between pyramidal cells in the neocortex or in the hippocampus CA3 is reduced (see chapters: Microcircuits of the Hippocampus and Microcircuits of the Neocortex). In contrast, ACh spares or enhances the excitatory synapses that mediate afferent connections, such as for the afferent projections from the thalamus to the cortex [6].

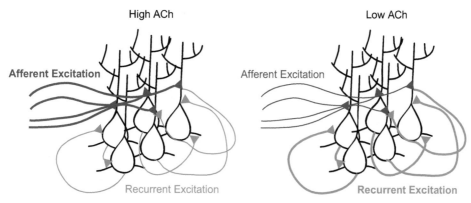

FIGURE 5.3   Balance between afferent (blue) and recurrent (red) excitation is shifted by acetylcholine (ACh). (Left) High ACh levels are associated with a relative strengthening of afferent excitation over recurrent excitation. (Right) Low ACh levels are associated with the opposite, such that recurrent excitation dominates over afferent excitation.

In addition, the response to stimuli can be enhanced by an ACh-mediated increase in firing rate in inhibitory interneurons that reduce background activity levels (ie, noise) in pyramidal cells.

Cholinergic neuromodulation is also of practical interest for in vitro electrophysiology. A large number of studies employ *carbachol*, an ACh equivalent that is not broken down by *acetylcholinesterase*. The application of carbachol activates the network and can induce in vivo-like oscillatory network patterns. For example, hippocampal slices exhibit gamma-frequency oscillations and thalamic slices exhibit alpha-frequency oscillations in response to carbachol application.

Behavior

Cholinergic neurons in both the BF and the PPN/LDT are active during waking and REM sleep but not during NREM sleep. The activity of cholinergic neurons in the BF correlates with cortical activation. During REM sleep, these neurons fire in synchronization with the hippocampal theta oscillation. It is likely that activity in the PPT/LDT follows the same principles. Rats anesthetized with urethane are a frequently used animal model to study cortical activation. In this preparation, the cortex spontaneously switches between periods of awake-like and slow-wave sleep (SWS)-like cortical activity patterns, although the animal is fully anesthetized. In this model, mildly noxious stimuli such as a toe or tail pinch cause a transition to the activated state and an increase in neuronal firing in the brainstem cholinergic nuclei.

In addition to regulation of the sleep—wake cycle, ACh also plays an important role in regulation of attention. The awake state is not a uniform, singular state, but rather is a continuum of behavioral states that span from drowsy but awake to highly attentive. ACh contributes to defining the specific behavioral state in the awake animal. Task-dependent modulation of the cholinergic tone is mediated through top-down feedback projections from the prefrontal cortex (PFC) to both the BF and the brainstem. Importantly, the concepts of *attention*, *vigilance*, and overall *behavioral arousal* are closely related but exhibit differences that may be caused by a difference in underlying mechanisms. Attention is typically conceptualized as a specific enhancement of the processing of certain sensory input at the expense of other sensory signals. It remains to be seen how such localized, specific modulation of neuronal activity can result from a neuromodulatory system that exhibits little spatial specificity. Vigilance, in contrast, is a less selective process than attention. General behavioral arousal differs from vigilance and is associated with running instead of standing still, for example.

DOPAMINE

Dopamine is a metabolite of the amino acid tyrosine and is one of the *catecholamines*, together with *epinephrine* (adrenaline) and *norepinephrine* (noradrenaline). The actions of dopamine are highly complex; dopamine is involved in almost all aspects of behavior ranging from voluntary movement to motivation, mood, attention, memory, and learning. No unifying theory explains either the neurophysiological or behavioral effects of dopamine. Also dopamine has been implicated in a range of neurological and psychiatric disorders, including Parkinson's disease (see chapter: Parkinson's Disease), schizophrenia, attention-deficit hyperactivity disorder, and addiction. The link to psychiatric illnesses is less clear and mostly motivated, at least in the case of schizophrenia, by the clinical success of antipsychotics that block dopamine receptors. There are 10 dopamine-producing nuclei in the mammalian brain, and these nuclei differ from each other. However, most dopaminergic neurons are located in two structures: the *substantia nigra pars compacta* (SNc) and the *ventral tegmental area* (VTA).

Mechanisms

Dopamine binds to G-protein-coupled receptors D1 through D5 [7]. Based on their properties, these receptors are often grouped into two classes: D1-like (D1 and D5) and D2-like receptors (D2 through D4). The two receptor classes employ two different downstream signaling cascades (Gq/s- vs. Gi-dependent signaling pathways, respectively). There is no unifying model of the cellular effects of activating dopamine receptors, and the literature is expansive and highly contradictory. In the cortex, the densest dopaminergic projections are to the PFC. In fact, the presence of dopaminergic projections used to be one of the defining criteria for the PFC until it became clear that the projections are more widespread.

Delineating the synaptic and intrinsic effects of dopamine is challenging given the large number of conflicting studies [8]. Also the effects may be different in various target locations. Here we focus on the effect of dopamine on the PFC neurons [9], which is comparatively well understood. Dopamine increases intrinsic excitability of PFC neurons by activating neurons that express D1 receptors. However, such characterization raises questions, since the effects of dopamine (even a single pulse) often cause a bidirectional or *biphasic* effect, whereby a decrease in activity is followed by a prolonged increase in excitability. Such biphasic action of dopamine is not limited to the PFC but has also been observed in other structures such as the striatum and the hippocampus. Therefore showing the effect at a single time point is insufficient to understand the electrophysiological consequences of dopamine application. Moreover, it is not fully clear which intrinsic ion channels are modulated by dopamine, such that the result is a net increase in excitability. Likely candidates are currents that shape the subthreshold membrane voltage, including the persistent sodium current, the slowly inactivating D-type potassium current, and calcium currents. At the level of synaptic transmission, dopamine appears to reduce evoked glutamate release, leading to a decrease in excitatory postsynaptic current amplitude. Activation of D1 receptors leads to outlasting potentiation of the N-methyl-D-aspartate (NMDA) component of excitatory postsynaptic responses. Yet again, the effect can manifest itself in either direction as a function of dopamine concentration. Dopamine also modulates synaptic plasticity. Interestingly, the effect of dopamine on activity depends on the overall activity levels in the target area. In case of low activity levels, NMDA receptors play a limited role because of their voltage dependence and the enhancement of GABAergic responses by dopamine dominates. In the case of higher activity levels, for example, during periods of persistent activity, the strengthening of the NMDA response by dopamine can lead to an increase in activity levels.

The network-level effects of dopamine have been well characterized in spatial working memory in nonhuman primates, specifically for D1 receptors [10]. Pharmacological blockage or excessive activation of D1 receptors both cause impaired spatial working memory. At the level of network activity, D1 antagonists reduce the signal-to-noise ratio in spatial working memory tasks by unmasking otherwise suppressed responses to the nontarget spatial locations. A moderate agonist dose, however, enhances performance by inhibiting neuronal responses to nonpreferred directions. In contrast, a high agonist dose suppresses all firing activity. Therefore dopaminergic activation of the PFC may optimize the signal-to-noise ratio by suppressing noise levels (ie, activity for nonpreferred input).

Behavior

At the behavioral level, elevated dopaminergic tone is associated with movement and approach motivation (a positive affective state induced by positive stimuli). Traditionally, the SNc and the VTA were associated with two distinct roles: movement and motivation, respectively. The role of dopamine released by cells in the SNc stems from research and treatment development for Parkinson's disease. Motivation is studied in the framework of reward and avoidance of aversive stimuli, and is defined as an induced internal state that produces and reinforces approach behavior. Drugs of abuse such as cocaine and amphetamines increase extracellular dopamine concentration in the ventral striatum, a main target of VTA dopaminergic projections. The role of the dopaminergic system is studied with pharmacological perturbations in animal models, including microinjections of antagonists into the target area of the dopaminergic projections and stimulation experiments. Traditionally, electrical stimulation of the dopaminergic areas was used. One famous example is intracranial self-stimulation, where animals learn to self-administer stimulation of structures that target the VTA. The limitations of electrical stimulation, in particular the uncertainty about which neurons and fibers of passage are stimulated, have been overcome with targeted optogenetic stimulation, which has shown that direction stimulation of the

VTA dopamine neurons alone is sufficient to produce reward-related behavior ([11], see chapter: Optical Measurements and Perturbations).

The role of dopamine in defining cortical and behavioral states is further demonstrated by the powerful, wake-promoting properties of pharmacological agents that interfere with the dopamine transporter (DAT). The DAT is a membrane protein that moves dopamine from the extracellular space back into the cytosol. Cocaine and amphetamines block the DAT and thereby increase the extracellular dopamine concentration.

Dopamine also plays an important role in learning, since dopamine neurons provide a brief (*phasic*) signal that encodes the difference between the actual and predicted rewards [12]. Phasic dopamine release occurs in response to rewarding environmental events; importantly, these signals are unlikely to encode attention since nonrewarding but attention-redirecting stimuli do not trigger phasic dopamine release. In addition, dopamine neurons also fire in response to sensory cues that predict reward such as the sensory signals in classical conditioning. During learning, there is a progressive shift of the dopamine response from the primary reward (eg, food or drink in animal experiments) to the reward-predicting stimulus (called *conditioned stimulus*). Dopamine encodes prediction error in the sense that unexpected reward causes an increase in firing of dopaminergic neurons. Receiving a predicted reward does not alter firing but a smaller than expected reward leads to a decrease in firing.

NOREPINEPHRINE

The *locus coeruleus* (LC) is the main source of NE (also referred to as noradrenaline) in the forebrain. The LC is a small structure (only about 1500 neurons in the rat) with axons projecting widely throughout the CNS.

Mechanisms

NE binds to three main classes of receptors: α_1, α_2, and β NE receptors. At the cellular level, NE reduces spike-frequency adaption by blocking calcium-activated potassium currents (BK channels). NE appears to boost synaptic plasticity, particularly in the context of emotionally relevant stimuli. In a visual discrimination task, LC neurons selectively responded to the target but not the distractor stimulus. Notably, those responses had short latencies ($\sim$100 ms) and preceded the behavioral response. Therefore NE can provide fast and specific enhancement of behaviorally relevant signals [13]. This contrasts with the classical notion of neuromodulators as slow and unspecific mediators of cortical state. Similarly, precise firing of LC neurons has been observed in the context of sleep. The PFC and the LC are reciprocally connected, and during the slow oscillation there is a phase relationship between the brain regions such that a significant fraction of LC cells fires a few tens of milliseconds after the onset of the DOWN state in the frontal cortex [14]. Therefore LC activation may promote the transition from DOWN state to UP state and serve an important role by further boosting the depolarization associated with the UP state (chapter: Low-Frequency Oscillations).

NE also has indirect effects on the neocortex. For example, noradrenergic fibers target the BF and therefore indirectly regulate cholinergic activation. In vitro, application of NE to the BF increased cortical activity [15]; in vivo, application of NE to the BF caused waking [16]. Together, these results suggest that NE signaling boosts cholinergic neuromodulation. However, cortical activation by NE is not solely an indirect consequence of upregulated cholinergic tone, since blocking noradrenergic receptors in the cortex induces a transition to slow rhythmic cortical activity in the awake animal [17].

Behavior

NE release caused by firing of LC neurons is associated with behavioral cues that carry novelty and behavioral relevance. These scenarios span from changes in rule sets in cognitive tasks with reward administration all the way to visceral signals such as pain or signals from the bladder [18]. In its simplest form, the noradrenergic signal can be interpreted as a broadcast signal that indicates a required increase in attention. LC neurons modulate sensory responses; they preferentially fire in response to novel sensory input. NE concentration also follows an inverse U-shape relationship with behavioral performance [19]. In contrast to dopamine, the enhancement of the signal-to-noise ratio is a product of an enhanced response to the preferred stimulus. Stress causes elevated NE levels that bind to the lower affinity α_1 and β receptors, which shut down the functioning of the PFC. In contrast, in the unstressed, alert condition, physiological, intermediate levels of NE activate the high-affinity α_2 receptors.

Similar to cholinergic neurons in the BF, noradrenergic neurons exhibit tonic firing during wakefulness. In fact, it appears that LC activity precedes the behavioral transition from sleep to wakefulness and therefore may play a causal role in the process of waking up. Optogenetic activation of LC causes a switch from sleep to wakefulness accompanied by cortical activation [20]. The activity of noradrenergic neurons is suppressed during SWS. NE neurons are silent during REM sleep—a major difference between noradrenergic and cholinergic neuromodulation.

HISTAMINE

Histamine-releasing neurons are located in the *tuberomammillary nucleus* in the posterior hypothalamus. *Histidine decarboxylase* synthesizes histamine from *histidine*.

Mechanisms

There are four main classes of histamine receptors: H1R, H2R, H3R, and H4R. The first three are widely expressed in the brain. All histamine receptors are metabotropic receptors. H1R and H2R typically contribute to neuronal excitability by—among many other targets—blocking leak and calcium-activated potassium currents leading to increased depolarization. H3R serve as an autoreceptor to regulate histamine release [21]. The role of histamine in promoting wakefulness and activation of cortex has been demonstrated (1) by impairing histaminergic signaling, which decreased waking and increased SWS, and (2) by decreasing histamine degradation that promoted waking. The histaminergic system also indirectly promotes waking through interaction with the cholinergic BF [22].

Behavior

Histamine plays a key role in mediating wakefulness [23]. The effect on wakefulness was demonstrated by the drowsiness caused by first-generation antihistamines prescribed for treating allergies, which cross the blood–brain barrier. Overall, greater histamine concentration increases wakefulness and decreased histamine increases NREM sleep. The histamine levels in the brain fluctuate according to circadian rhythm and correlate with neuronal activity in histamine neurons during waking. The activity of histaminergic neurons is highest during attentive waking, strongly reduced during NREM sleep, and completely suppressed during REM sleep [24].

SEROTONIN

Serotonin (5-hydroxytryptamine, 5-HT) is mostly released by neurons in the dorsal and median raphe nuclei. Again, as for all the other neuromodulatory systems, the axons project widely and thereby likely target the entire neocortex. Serotonergic fibers prefer layer IV of the neocortex as their target.

Mechanisms

Serotonin binds to serotonin receptors. There are seven general serotonin receptor classes with a total of 14 different receptor subtypes: $5\text{-}HT_1$ ($5\text{-}HT_{1A}$, $5\text{-}HT_{1B}$, $5\text{-}HT_{1D}$, $5\text{-}HT_{1E}$, $5\text{-}HT_{1F}$), $5\text{-}HT_2$ ($5\text{-}HT_{2A}$, $5\text{-}HT_{2B}$, $5\text{-}HT_{2C}$), $5\text{-}HT_3$, $5\text{-}HT_4$, $5\text{-}HT_5$ ($5\text{-}HT_{5A}$, $5\text{-}HT_{5B}$), $5\text{-}HT_6$, and $5\text{-}HT_7$. All but one, the $5\text{-}HT_3$ receptor, are metabotropic, G-protein-coupled receptors. Similar to other neuromodulators, the effect of serotonin fails simple classification as either an increase in activity ("excitatory") or decrease in activity ("inhibitory"). The diversity in modulation of neuronal excitability by serotonin stems from the diversity of receptors and the cell-type-specific expression patterns of different receptor types [25]. In the cortex, the most abundantly expressed receptors are $5\text{-}HT_{1A}$, $5\text{-}HT_{2A}$, and $5\text{-}HT_3$. In the PFC, for example, $5\text{-}HT_{1A}$ receptors hyperpolarize pyramidal cells. In contrast, $5\text{-}HT_{2A}$ receptors on the same cells induce a slow membrane depolarization. The balance of these two receptors that have opposing effects appears to be the main effector of serotonin on network activity [26]. It is unclear what determines how a pyramidal cell responds to serotonin, given the opposing effects of the two receptor subtypes. Inhibitory interneurons express $5\text{-}HT_{1A}$, $5\text{-}HT_{2A}$, and $5\text{-}HT_3$ receptors that together again mediate cell-type-specific, complex modulation of the membrane voltage. Likely, parvalbumin- and calbindin-positive inhibitory interneurons express $5\text{-}HT_{1A}$ and $5\text{-}HT_{2A}$ receptors, whereas calretinin-positive interneurons express the $5\text{-}HT_3$ receptor. Overall, the effect of serotonin on network dynamics remains poorly understood. Further complicating the matter is the fact that fibers from the raphe nuclei project directly to the cortex and also indirectly by targeting the nucleus basalis causing a modulation of cholinergic activity in the neocortex. In vitro, the application of serotonin to the nucleus basalis hyperpolarized the cholinergic cells [27]. In agreement with these findings, in vivo serotonin application to the nucleus basalis caused a decrease in gamma oscillations in the neocortex [16]. Serotonin also suppresses hippocampal theta oscillations [28]. Overall, serotonin appears to support a quiet waking state and therefore its role differs from ACh, NE, and histamine.

Behavior

It is difficult to associate serotonin with specific behavioral functions because it appears to be so involved in virtually all behaviors in mammals, including mood, aggression, sexual behavior, and feeding [28]. Serotonin has been implicated in a wide variety of psychiatric disorders—perhaps most prominently depression ("serotonin hypothesis of depression")—that are treated with selective serotonin reuptake inhibitors (see toolbox: Psychiatry). The firing of serotonergic neurons is low during wakefulness and increases phasically in response to an arousal-inducing stimulus. The firing rate is greatly reduced during NREM sleep and suppressed during REM sleep.

MONOAMINE TRANSPORTERS

Monoamine transporters remove excess dopamine, noradrenaline, and serotonin from the extracellular space and are accordingly named DAT, *noradrenaline transporter* (NET), and *serotonin transporter* (SERT). The nomenclature derives from the cell types that the transporters are expressed on and do not indicate selectivity of the transporter. These transporters are targeted both by medications and drugs of abuse. For example, amphetamine and cocaine bind with nanomolar affinity to the DAT and thereby increase extracellular dopamine. Also both drugs interfere with NET and SERT. In contrast, modafinil selectively binds to DAT. Pharmacological agents that block one or multiple neuromodulator transporters are of fundamental importance in psychiatry, particularly for the treatment of depression (see toolbox: Psychiatry).

SUMMARY AND OUTLOOK

Neuromodulators enable neuronal networks to switch between different activity states. These network states can often be mapped onto behavioral states. The most classical role of neuromodulators is their involvement in defining and differentiating wakefulness and sleep. Neuromodulators are also involved in perhaps every behavioral function. In this chapter, we have discussed the neuromodulators ACh, dopamine, NE, histamine, and serotonin. We have seen that the broad range of cellular targets makes explaining the effects of neuromodulators on network dynamics challenging. Nevertheless, we have encountered a number of general properties. A small number of neurons that synthesize and release a given neuromodulator target the entire CNS. Volume transmission by diffusion enables fast and global communication in changes in brain state. A large fraction of the medications for the treatment of mental illnesses targets neuromodulatory tone; therefore bridging the gap between receptor properties and network dynamics will be important for better understanding the actual mechanisms of action.

References

[1] Harris KD, Thiele A. Cortical state and attention. Nat Rev Neurosci 2011;12(9):509—23.
[2] Lee S-H, Dan Y. Neuromodulation of brain states. Neuron 2012;76(1):209—22.
[3] Munoz W, Rudy B. Spatiotemporal specificity in cholinergic control of neocortical function. Curr Opin Neurobiol 2014;26:149—60.
[4] Metherate R, Cox CL, Ashe JH. Cellular bases of neocortical activation: modulation of neural oscillations by the nucleus basalis and endogenous acetylcholine. J Neurosci 1992;12(12):4701—11.
[5] McCormick DA. Cholinergic and noradrenergic modulation of thalamocortical processing. Trends Neurosci 1989;12(6):215—21.
[6] Hasselmo ME, Sarter M. Modes and models of forebrain cholinergic neuromodulation of cognition. Neuropsychopharmacology 2011;36(1):52—73.
[7] Beaulieu JM, Gainetdinov RR. The physiology, signaling, and pharmacology of dopamine receptors. Pharmacol Rev 2011;63(1):182—217.
[8] Tritsch NX, Sabatini BL. Dopaminergic modulation of synaptic transmission in cortex and striatum. Neuron 2012;76(1):33—50.
[9] Seamans JK, Yang CR. The principal features and mechanisms of dopamine modulation in the prefrontal cortex. Prog Neurobiol 2004;74(1):1—58.
[10] Vijayraghavan S, et al. Inverted-U dopamine D1 receptor actions on prefrontal neurons engaged in working memory. Nat Neurosci 2007;10(3):376—84.
[11] Tsai HC, et al. Phasic firing in dopaminergic neurons is sufficient for behavioral conditioning. Science 2009;324(5930):1080—4.
[12] Schultz W. Getting formal with dopamine and reward. Neuron 2002;36(2):241—63.
[13] Aston-Jones G, Rajkowski J, Cohen J. Role of locus coeruleus in attention and behavioral flexibility. Biol Psychiatry 1999;46(9):1309—20.
[14] Eschenko O, et al. Noradrenergic neurons of the locus coeruleus are phase locked to cortical up-down states during sleep. Cereb Cortex 2012;22(2):426—35.
[15] Fort P, et al. Noradrenergic modulation of cholinergic nucleus basalis neurons demonstrated by in vitro pharmacological and immunohistochemical evidence in the guinea-pig brain. Eur J Neurosci 1995;7(7):1502—11.
[16] Cape EG, Jones BE. Differential modulation of high-frequency gamma-electroencephalogram activity and sleep-wake state by noradrenaline and serotonin microinjections into the region of cholinergic basalis neurons. J Neurosci 1998;18(7):2653—66.
[17] Constantinople CM, Bruno RM. Effects and mechanisms of wakefulness on local cortical networks. Neuron 2011;69(6):1061—8.
[18] Sara SJ. The locus coeruleus and noradrenergic modulation of cognition. Nat Rev Neurosci 2009;10(3):211—23.
[19] Arnsten AF. Stress signalling pathways that impair prefrontal cortex structure and function. Nat Rev Neurosci 2009;10(6):410—22.
[20] Carter ME, et al. Tuning arousal with optogenetic modulation of locus coeruleus neurons. Nat Neurosci 2010;13(12):1526—33.
[21] Haas H, Panula P. The role of histamine and the tuberomamillary nucleus in the nervous system. Nat Rev Neurosci 2003;4(2):121—30.
[22] Ramesh V, et al. Wakefulness-inducing effects of histamine in the basal forebrain of freely moving rats. Behav Brain Res 2004;152(2):271—8.
[23] Brown RE, et al. Control of sleep and wakefulness. Physiol Rev 2012;92(3):1087—187.
[24] John J, et al. Cataplexy-active neurons in the hypothalamus: implications for the role of histamine in sleep and waking behavior. Neuron 2004;42(4):619—34.

[25] Andrade R. Serotonergic regulation of neuronal excitability in the prefrontal cortex. Neuropharmacology 2011;61(3):382—6.

[26] Celada P, Puig MV, Artigas F. Serotonin modulation of cortical neurons and networks. Front Integr Neurosci 2013;7:25.

[27] Khateb A, et al. Pharmacological and immunohistochemical evidence for serotonergic modulation of cholinergic nucleus basalis neurons. Eur J Neurosci 1993;5(5):541—7.

[28] Roth BL. Multiple serotonin receptors: clinical and experimental aspects. Ann Clin Psychiatry 1994;6(2):67—78.

Neuronal Communication Beyond Synapses

Neurons mostly communicate by synapses that are specialized contact points between neurons (see chapter: Synaptic Transmission). In this chapter we will discuss other neuronal communication mechanisms that do not require such precise juxtaposition of a presynaptic and postsynaptic cell. We will review communication (1) by messenger molecules other than the classical neurotransmitters (chapter: Synaptic Transmission) and neuromodulators (chapter: Neuromodulators); (2) by fluctuations in extracellular ion concentration, particularly extracellular potassium concentration; and (3) by endogenous electric fields generated by network activity. One common thread is that these mechanisms usually enable communication at a larger scale than just between pairs of individual neurons. Necessarily, this chapter is an eclectic collection of little-studied and poorly understood mechanisms.

CHEMICAL SIGNALING BEYOND SYNAPSES

Chemical signaling between neurons with neurotransmitters can occur at locations other than dedicated synaptic contacts (a mechanism referred to as *nonsynaptic neurotransmission*, Fig. 6.1, [1]). Neurotransmitters can be released from axonal terminals that do not have a postsynaptic match (Fig. 6.1B). Reuptake pumps can work in reverse and pump neurotransmitter molecules into the extracellular space (Fig. 6.1C). Neurotransmitters can spill over from synaptic contacts (Fig. 6.1D), and neurotransmitter receptors can be located at positions other than the postsynaptic site (Fig. 6.1E).

Furthermore, nontraditional mediator molecules that are not classically considered to be neurotransmitters, such as neuropeptides and nitric oxide (NO), also contribute to chemical signaling between neurons. We will focus our discussion on these molecules; their main difference from classical neurotransmitters is that they do not follow the directionality and specificity of synaptic transmission. Release of one of these nontraditional molecules does not have a targeted effect on one specific postsynaptic neuron. Instead, these molecules are released and subsequently present in a volume around the original site of release. The response does not follow the typical input—output relationship of chemical synapses where release on the presynaptic side causes an almost immediate change in the postsynaptic

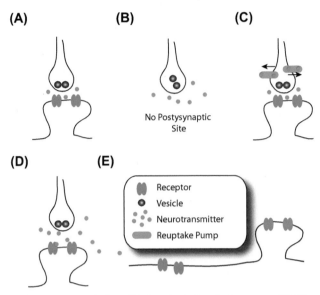

FIGURE 6.1   Nonsynaptic neurotransmission. (A) Standard chemical synapse. (B) Neurotransmitter is released into the extracellular space without the presence of a matched postsynaptic site. (C) Reuptake pumps that act in reverse can "release" neurotransmitter. (D) Neurotransmitter can "spill over" from a synapse and reach receptors other than the ones in the postsynaptic membrane. (E) Extrasynaptic receptors.

neuron (such as a change in membrane voltage). Rather, the action of these molecules occurs through changes in intracellular second-messenger signaling after binding of the molecule to a G-protein-coupled receptor (GPCR) (Fig. 6.2).

Neuropeptides

Neuropeptides are grouped in families based on their amino acid series; more than 100 different neuropeptides are released by neurons in the mammalian brain, and little is known about the functional role of most of them. Neuropeptides assume similar roles to hormones in other organs by providing diffuse chemical signaling. Historically, releasing peptides was considered to be a unique feature of highly specialized neurons in the hypothalamus. More recently, however, it has become clear that most neurons likely employ not only classical synaptic signaling but also mechanisms based on releasing and sensing neuropeptides [2]. In contrast to GABAergic and glutamatergic signaling, neuropeptides are packed in dense core vesicles (named after their appearance in an electron micrograph) that are not necessarily located next to the release zone of a synapse. Studying the electrophysiological consequences of neuropeptide release remains difficult, because in contrast to fast synaptic transmission, the effect of neuropeptides binding to GPCRs can take minutes to manifest. Thus the study of cause and effect, particularly with high temporal resolution, is challenging. One of the better-understood effects of neuropeptides is the modulation of presynaptic GABA and

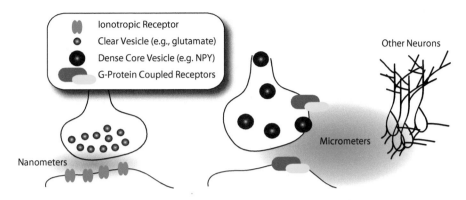

FIGURE 6.2 *Left*: "Classical" synaptic neurotransmission is fast and local. Neurotransmitter is packaged into clear vesicles and released at the synapse. The neurotransmitter molecules reach and bind to the postsynaptic (ionotropic) receptors that are just nanometers away from the release site. *Right*: Neurotransmission with neuropeptides (such as neuropeptide Y, NPY) is slow and less local. Dense core vesicles contain neuropeptides. Once released, the neuropeptides bind to G-protein-coupled receptors on the pre- and postsynaptic sides as well as to other neurons.

glutamate release. For example, neuropeptide Y (NPY) downregulates synaptic release of glutamate in terminals of excitatory hippocampal CA3 neurons that project to CA1 (see chapter: Microcircuits of the Hippocampus, [3]). NPY knockout mice appear to have normal levels of neuronal activity. But experimental induction of seizures causes uncontrollable seizures that lead to death in most NPY knockout animals but not in wild-type control animals [4]. Therefore NPY may provide a slow time-scale mechanism that prevents the genesis of sustained high activity levels, in other words serving as a stabilizing force to put a brake on pathologically high activity levels that occur, for example, during epileptic seizures.

Nitric Oxide

NO provides an additional communication channel between glutamatergic and monoaminergic neurons that works quite differently from classical synaptic transmission [5]. NO is synthesized by nitric oxide synthase (NOS) from the amino acid L-arginine. Specifically, neuronal NOS (nNOS) activity depends on calmodulin, a calcium buffer. This allows NO to be synthesized in response to an increase in intracellular calcium. Interestingly, nNOS is directly linked to the *N*-methyl-D-aspartate (NMDA) receptor at the postsynaptic density (protein-rich specialization at the postsynaptic membrane), and therefore calcium influx predominantly through NMDA receptors enables NO synthesis. As a result, the local NO concentration reflects the overall activation of NMDA receptors. NO is a highly diffusible gas that easily passes through cell membranes and likely reaches cells located hundreds of micrometers away from the synthesizing neuron. This broadcasting communication mode is referred to as *volume transmission*. Because of the free diffusion of NO though both aqueous and lipid environments, NO can act simultaneously on pre- and postsynaptic structures. It is likely that NO signaling plays a broad functional role in the nervous system. Guanylyl cyclase is activated by NO (it is an "NO receptor") and produces the second messenger cyclic guanosine monophosphate (cGMP) from guanosine triphosphate. In turn, cGMP has many downstream targets, including ion channels. Similarly, the effects of NO signaling on synaptic transmission are manifold with up- and downregulation of synaptic function found in different studies. From an electrophysiological standpoint, NO signaling can alter intrinsic excitability through a multitude of pathways. Several in vivo studies found an increase in excitability caused by NO, possibly by a reduction in potassium currents and activation of hyperpolarization-activated depolarizing currents mediated by HCN channels. In further support of NO as a facilitator of neuronal activity, suppression of NO synthesis reduces visual and somatosensory responses. There is intriguing evidence of a cooperative role of NO and acetylcholine in determining the state of the thalamocortical system. Several different experimental manipulations demonstrated that NO contributes to controlling cortical and thalamic excitability. Blocking NOS with NG-nitro-L-arginine reduced action potential firing, but NO *donor* molecules—molecular carriers of NO—*increased* neuronal activity. Both spontaneous and visually evoked activity is subject to such modulation.

EXTRACELLULAR ION CONCENTRATION DYNAMICS

Each time an ion crosses the cell membrane, the corresponding intra- and extracellular ion concentrations change. Several mechanisms are in place to restore and maintain ion concentration gradients. For example, some membrane proteins act as pumps to counteract the changes in ion concentration caused by neuronal activity. However, these mechanisms fail to maintain ion concentration gradients during periods of elevated neuronal activity. Ion concentrations fluctuate with neuronal activity. Any such change in concentration alters the equilibrium potential of the ion channels permeable to the specific ion type (see chapter: Membrane Voltage). As a result, the driving force for the corresponding ion channels is either increased or decreased. Therefore any ion that flows through an ion channel changes the flow of subsequent ions through channels permeable to the same ion type. Notably, such change is not limited to the neuron that caused the original change in the extracellular ion concentration, but also affects neurons in the vicinity that share the same extracellular space. In most cases, such activity-dependent modulation of ion currents by fluctuations in ion concentrations is unlikely to play a major role in electrical signaling because of the extremely small fluctuations in equilibrium potentials. For ion species with high extracellular concentration such as sodium, a small change in extracellular concentration will not have any significant effect on the sodium equilibrium potential or on sodium currents. One noteworthy exception is the extracellular potassium concentration ($[K^+]_o$) [6]. The baseline $[K^+]_o$ is low (3–4 mM) and therefore even small fluctuations represent a significant relative change (Fig. 6.3).

Potassium ions are extruded from neurons not only during rest through leak potassium channels but also during the repolarization phase of the action potential. Indeed, $[K^+]_o$ fluctuates even during physiological activity levels, and therefore the potassium equilibrium potential is not a constant. Despite several mechanisms that clear potassium ions from the extracellular space, such as transporters (most prominently the sodium-potassium ATPase that transports sodium and potassium ions against the concentration gradient), diffusion through the extracellular space, and potentially buffering by glial cells, periods of higher activity levels lead to (local) increase in extracellular potassium concentration. The resulting decrease in potassium driving force makes neurons more excitable because potassium

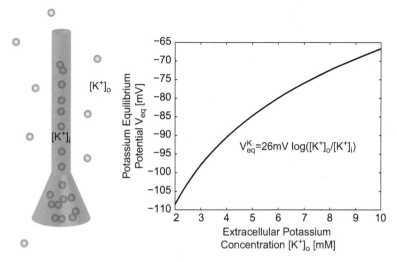

FIGURE 6.3 Increases in extracellular potassium concentration lead to a depolarization of the potassium equilibrium potential.

currents typically repolarize neurons and thereby move them away from the threshold for action potential generation. In the case of elevated $[K^+]_o$, the resting membrane voltage becomes more depolarized (and thus closer to threshold) and neurons can switch from firing individual action potentials to firing bursts of action potentials. The increased activity levels cause a further increase in $[K^+]_o$. Therefore changes in the $[K^+]_o$ may represent a positive feedback mechanism that (pathologically) amplifies activity. Historically, such potassium concentration dynamics used to be a key hypothesis for the mechanism that triggers epileptic seizures. According to this *potassium accumulation hypothesis*, a sufficiently strong increase in $[K^+]_o$ depolarizes neurons and eventually gives rise to a seizure. In this model, once a certain threshold of $[K^+]_o$ is crossed, any further neuronal activity leads to a further activity increase. Once neurons are sufficiently depolarized for sodium channels to inactivate, no more action potentials occur. This state of pathological depolarization is called *depolarization block*. In this model, an increase in $[K^+]_o$ is expected to occur before the onset of pathologically elevated activity and therefore act as a trigger for epileptic seizures (Figs. 6.4).

However, measurements of $[K^+]_o$ during seizure-like events in animal models have shown a delayed onset of the rise in $[K^+]_o$ relative to the increase in neuronal firing. It was also hypothesized that there should be a threshold in $[K^+]_o$ that, once crossed, marks the onset of such pathological activity fueled by $[K^+]_o$. No specific threshold for the occurrence of such "runaway dynamics" was found. Nevertheless, in vitro studies have repeatedly demonstrated that increasing $[K^+]_o$ by changing the amount of potassium in artificial cerebrospinal fluid (ACSF) alters the network dynamics [7]. Elevated $[K^+]_o$ caused the occurrence of network activity patterns that resemble interictal spikes (synchronous firing of populations of neurons), tonic firing (continuous action potential firing), and bursting (rhythmic occurrence of bouts of population activity). Tonic firing and bursting are commonly observed in several types of epileptic seizures (see chapter: Epilepsy). Several caveats need to be considered in the interpretation of these findings. First, these results do not prove a causal

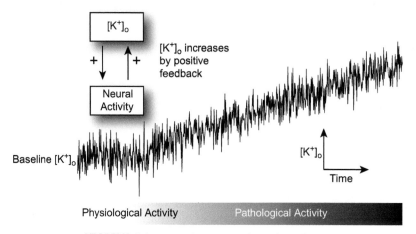

FIGURE 6.4   Potassium accumulation hypothesis.

role of ion concentration fluctuations in the genesis of in vivo activity patterns without an experimentally induced perturbation of $[K^+]_o$. Second, in vitro studies do not address the proposed feedback dynamics since the ACSF acts as a buffer that likely reduces activity-dependent fluctuations of $[K^+]_o$. Dissecting feedback interactions is notoriously difficult in experimental settings. In contrast, computer simulations are an ideal tool, since they allow the direct comparison of network dynamics with and without dynamic $[K^+]_o$. In these simulations, $[K^+]_o$ in a compartment surrounding the individual neurons is treated as an additional state variable. The dynamics depend on the transmembrane potassium currents, potassium pumps, glial buffers, and extracellular diffusion currents. Simulations of networks of neurons that include $[K^+]_o$ dynamics demonstrated that $[K^+]_o$ can indeed cause pathological activity patterns [8,9]. A transient increase in network activity by elevated afferent input to the network can cause an increase in $[K^+]_o$. As a result, activity is increased and sustained after termination of the input that triggered the initial increase in $[K^+]_o$. This positive feedback interaction between increases in activity and $[K^+]_o$ continues until $[K^+]_o$ reaches a level at which neurons switch from tonic firing to bursting because of the impaired repolarization by weakened potassium currents (Fig. 6.5). During bursting, the overall frequency of action potentials is less than during tonic firing and therefore the amount of potassium extruded from the cells is also less. As a result, $[K^+]_o$ decreases during bursting, which will lead to tonic firing again. This differential increase and decrease in $[K^+]_o$ as a function of the activity pattern has been observed in both in vitro and in vivo experimental measures. Note that in this model there is no trajectory to depolarization block except if $[K^+]_o$ is artificially increased to sufficiently depolarize the neurons.

These dynamics differ from the original potassium accumulation hypothesis, which assumed a continued increase in $[K^+]_o$ independent of the activity patterns once these feedback dynamics were triggered. Examining the different activity states a neuron can exhibit as a function of $[K^+]_o$ revealed the existence of a bistable region (Fig. 6.6) in which neurons can exhibit either bursting or tonic firing. This bistability mediates slow transitions between tonic firing and bursting activity.

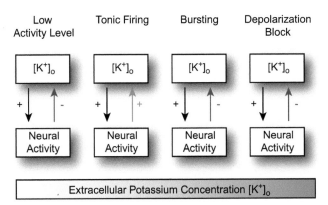

FIGURE 6.5 Interaction between neuronal activity and extracellular potassium concentration is state dependent. Only elevated neuronal activity in the form of tonic firing causes a continued increase in extracellular potassium concentration (positive feedback).

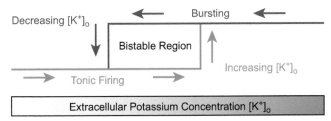

FIGURE 6.6 Bistable regime in which tonic firing and bursting coexist for moderately elevated levels of extracellular potassium concentration. During tonic firing, potassium concentration increases to the point where the neurons switch to bursting. During bursting, potassium concentration decreases down the level where the neurons switch back to tonic firing.

This mechanism explains the transitions between the activity patterns associated with the tonic and clonic phase of epileptic seizures (see chapter: Epilepsy). Elevated $[K^+]_o$ also leads to an increase in activation of the sodium-potassium-chloride cotransporter, which retrieves potassium from the extracellular space at the price of increasing the intracellular chloride concentration. As a result, the reversal potential of $GABA_A$ currents becomes more depolarized and therefore the strength of synaptic inhibition is reduced. In summary, the most simple conceptualization of positive feedback dynamics of $[K^+]_o$, as stated in the potassium accumulation hypothesis, may not be an accurate model of the mechanism that drives epileptic seizures. Nevertheless, computational modeling suggests the presence of more complex interactions between neuronal network activity and $[K^+]_o$. The dynamics may contribute to both physiological and pathological cortical network dynamics.

Strong elevations in $[K^+]_o$ are associated with depolarization block. This regime has been implicated in *spreading depression*, an electrophysiological phenomenon in which cortical activity is transiently suppressed [10]. Spreading depression is easy to detect from the EEG; during spreading depression, the EEG signal is flat and no signs of activity are detected for a period of seconds to minutes. If potassium concentration is sufficiently elevated, neurons are so depolarized that sodium channel inactivation prevents the occurrence of action potentials. This may represent the cellular mechanism of spreading depression. Other processes also likely contribute to spreading depression, such as metabolic rundown phenomena. During metabolic rundown, not enough energy is available to maintain ion concentration gradients by activation of the Na^+/K^+-ATPase. Spreading depression has been associated with migraines, subarachnoid hemorrhage, stroke, and traumatic brain injury. The common feature is a massive disturbance of the ionic microenvironment caused by a lack of oxygen. This leads to a failure of the energy-demanding N^+/K^+-ATPase, which then triggers pathological depolarization because of the failure of maintaining appropriately low $[K^+]_o$. Originally, *cortical spreading depression* (CSD, not to be confused with current source density, also abbreviated as CSD) was characterized as a propagating electric disturbance with a speed of a few millimeters per second induced by electrical or mechanical perturbations in animal models. Clinically, spreading depression can be either benign or detrimental, depending on the underlying disease state. For example, CSD in migraine patients is a benign phenomenon and does not cause permanent damage. On the other hand, CSD in response to acute injury of brain tissue (eg, stroke, traumatic brain injury) may spread to surrounding healthy brain tissue and cause secondary brain damage by neuronal death beyond the location of the original insult.

ENDOGENOUS ELECTRIC FIELDS

Neurophysiology is based on the principle that neuronal activity causes weak but measureable electric fields (see toolbox: Physics of Electric Fields and chapter: LFP and EEG). However, neurons not only generate electric fields, they are also susceptible to electric fields [11]. Neurons can act as antennae if their morphology includes an elongated main axis such as the somatodendritic axis of pyramidal cells [12]. Electric fields impose a gradient in the extracellular potential. Within the neuron, ions are exposed to the same electric field. But those ions are trapped because of the relative impermeability of the cell membrane. As a result, the ions start to rearrange their positions within the neuron until their own charge cancels out the externally imposed electric field. At this point, no net field is left within the neuron and no further charge movement occurs. The membrane voltage is defined as the difference between the intracellular and extracellular potential (see chapter: Membrane Voltage). Therefore an electric field that is parallel to the somatodendritic axis of a pyramidal cell can change its somatic membrane voltage. If the field points from the apical dendrites to the soma, there is an according *neuronal polarization*, with the dendrites hyperpolarized and the soma depolarized. For fields that point in the opposite direction, the effect is opposite, with the dendrites being depolarized and the soma hyperpolarized. Since the membrane acts as a capacitor, the membrane voltage cannot immediately jump to its new value at the onset of an electric field (see toolbox: Electric Circuits). Instead, the membrane capacitance needs to be charged for the membrane voltage to change. Therefore the dynamics of the induced change in membrane voltage are identical to the dynamics observed for a step current injection (see chapter: Membrane Voltage and toolbox: Differential Equations). A (homogeneous) field with a strength of 1 V/m changes the membrane voltage by ~ 0.1 mV if properly aligned.

The key questions we need to ask are: (1) What is the lowest electric field strength that significantly perturbs network activity? and (2) Under what circumstances are fields with such strengths encountered, if at all? The answers to these questions are of interest in several, distinct research fields. First, from a basic science viewpoint, the question arises whether the endogenous electric fields generated by neuronal activity [and routinely measured by electroencephalogram (EEG) and local field potential (LFP)] alter neuronal activity. Second, from a translational neuroscience viewpoint, the question arises whether and how externally applied electric fields can alter brain activity (see chapter: Noninvasive Brain Stimulation). Third, from a biosafety perspective, the question arises whether the electric fields generated by power lines, wireless communication, and other electric devices may alter brain activity. Here we will focus on the first application, the potential role of endogenous electric fields in shaping cortical network dynamics. The second application is discussed later in the book in the context of noninvasive brain stimulation. The third question about biosafety is not further treated here because of the lack of strong evidence that would support such concerns.

Endogenous electric fields in the cortex are very weak and typically measure at most a few V/m. Therefore the resulting change in membrane voltage is less than a millivolt, a very small value in comparison to the amount of depolarization needed to trigger an action potential in a neuron at rest (typically around 20 mV of depolarization). However, in the intact brain, neurons are almost always receiving substantial synaptic input such that the membrane voltage of many neurons hovers around the threshold for action potential firing. Close to threshold, even a small perturbation of the membrane voltage can have a big effect because of the nonlinear, all-or-nothing nature of an action potential. Also synaptic connectivity can amplify and propagate small changes in network activity throughout a network [13]. Weak electric fields have been applied to live tissue slices of the hippocampus and the neocortex [14–16]. In these brain structures, the alignment of the pyramidal cells makes the modulation of the membrane voltage by electric fields effects particularly likely. In these experiments, weak electric fields with amplitudes close to values of the endogenous field changed network activity patterns. Because the perturbation of the membrane voltage by stimulation needs to interact with the depolarization induced by synaptic interaction, timing of the electric field relative to the endogenous activity is crucial. If the small depolarization by stimulation arrives during a period of network hyperpolarization, the effect of stimulation is less than when it arrives during a period of depolarization, when neurons are close to or above threshold. Accordingly, sine-wave electric fields have been demonstrated to enhance intrinsic oscillations, in particular when the stimulation frequency matches the frequency of the endogenous oscillation [16]. Similar results have been found for the application of sine-wave electric fields in animal models [17,18]. The successful modulation of network dynamics by weak exogenous electric fields argues for a similar susceptibility to endogenous electric fields, generated by synchronized network activity. Consequently, the endogenous electric fields generated by oscillatory network activity may play an active role in shaping ongoing oscillation (Fig. 6.7).

Establishing this hypothesized role of *endogenous* electric fields in enhancing neuronal activity patterns is challenging. This hypothesized interaction requires that the electric field caused by neuronal activity would modulate neuronal activity that in turn alters the electric field again. The time constant of this proposed feedback loop is very short, which

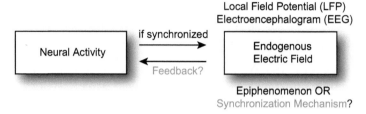

FIGURE 6.7   Feedback interaction between neuronal activity and the accompanying endogenous electric fields. If neuronal activity is synchronized, a pronounced electric field is generated, which is routinely recorded as a readout of neuronal activity [local field potential (LFP) and electroencephalogram (EEG)]. Given the susceptibility of active networks to weak electric fields, those endogenous fields generated by neuronal activity may modulate neuronal activity and thus form a feedback loop between neuronal activity and its endogenous electric field.

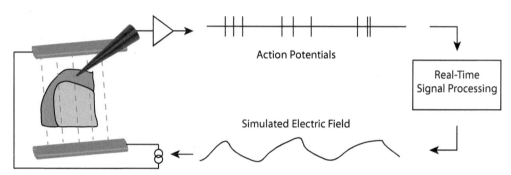

FIGURE 6.8   Hybrid computational–biological system to demonstrate effect of activity-dependent electric field on network dynamics. Action potentials are detected from a cortical slice and processed in real time to compute a "simulated" electric field waveform that is injected back into cortical slice preparation as an electric field generated by a current injection through two parallel, conducting wires. *Red dotted lines*: electric field lines.

makes its dissection in a neurophysiological experiment particularly difficult. However, the use of a biological hybrid system where an electronic circuit generates a "synthetic" electric field waveform in real time enabled direct demonstration that such feedback interaction indeed enhances neuronal synchronization and therefore overall oscillation power (Fig. 6.8) [19].

Similar to the case of the proposed feedback dynamics between neuronal activity and changes in $[K^+]_o$, computer simulations proved helpful to compare the network dynamics of an oscillation cortical network model with and without the proposed feedback by endogenous electric field. In these simulations, a simulated endogenous electric field was computed by averaging the membrane voltages of all pyramidal cells. A scaled version of this field was then injected as a somatic current into the neuron models. These simulations further supported the role of endogenous electric fields in enhancing cortical oscillations. As we will discuss in detail in the unit Cortical Oscillations, rhythmic activity patterns are fundamental to how the brain works, and endogenous electric fields may be a mechanism that enhances and stabilizes such oscillations.

SUMMARY AND OUTLOOK

In this chapter we have covered some of the less conventional and less-discussed modes of how neurons can communicate with each other. First, we looked at two sample signaling molecules, NPY and the gas NO, both of which are implicated in the regulation of overall excitability of neuronal circuits. Second, we discussed dynamic changes in the extracellular ion concentrations, particularly for potassium. Even small changes in the extracellular potassium concentration can change the intrinsic firing pattern of individual neurons. As a result, networks of neurons can exhibit different physiological and pathological activity states, which have been studied mostly in the context of epilepsy and spreading depression. Third, we reviewed how weak endogenous electric fields generated by synchronized brain activity may play an active role in enhancing synchronization and shaping network dynamics. These mechanisms have yet to take center stage in neuroscience research despite their potentially important role in orchestrating network activity. Understanding neuronal networks will likely require taking these mechanisms into account. It may even be that some novel therapeutic opportunities will arise from considering these more "unconventional communication modes" as treatment targets.

References

[1] Vizi ES, Kiss JP, Lendvai B. Nonsynaptic communication in the central nervous system. Neurochem Int 2004;45(4):443–51.
[2] van den Pol AN. Neuropeptide transmission in brain circuits. Neuron 2012;76(1):98–115.
[3] Colmers WF, Lukowiak K, Pittman QJ. Neuropeptide Y action in the rat hippocampal slice: site and mechanism of presynaptic inhibition. J Neurosci 1988;8(10):3827–37.
[4] Baraban SC, et al. Knock-out mice reveal a critical antiepileptic role for neuropeptide Y. J Neurosci 1997;17(23):8927–36.
[5] Kiss JP, Vizi ES. Nitric oxide: a novel link between synaptic and nonsynaptic transmission. Trends Neurosci 2001;24(4):211–5.
[6] Frohlich F, et al. Potassium dynamics in the epileptic cortex: new insights on an old topic. Neuroscientist 2008;14(5):422–33.
[7] Jensen MS, Yaari Y. Role of intrinsic burst firing, potassium accumulation, and electrical coupling in the elevated potassium model of hippocampal epilepsy. J Neurophysiol 1997;77(3):1224–33.
[8] Frohlich F, et al. Slow state transitions of sustained neural oscillations by activity-dependent modulation of intrinsic excitability. J Neurosci 2006;26(23):6153–62.
[9] Frohlich F, Sejnowski TJ, Bazhenov M. Network bistability mediates spontaneous transitions between normal and pathological brain states. J Neurosci 2010;30(32):10734–43.
[10] Pietrobon D, Moskowitz MA. Chaos and commotion in the wake of cortical spreading depression and spreading depolarizations. Nat Rev Neurosci 2014;15(6):379–93.
[11] Tranchina D, Nicholson C. A model for the polarization of neurons by extrinsically applied electric fields. Biophysical J 1986;50(6):1139–56.
[12] Radman T, et al. Role of cortical cell type and morphology in subthreshold and suprathreshold uniform electric field stimulation in vitro. Brain Stimul 2009;2(4):215–228.e3.
[13] Radman T, et al. Spike timing amplifies the effect of electric fields on neurons: implications for endogenous field effects. J Neurosci 2007;27(11):3030–6.
[14] Deans JK, Powell AD, Jefferys JG. Sensitivity of coherent oscillations in rat hippocampus to AC electric fields. J Physiol 2007;583(Pt 2):555–65.
[15] Bikson M, et al. Effects of uniform extracellular DC electric fields on excitability in rat hippocampal slices in vitro. J Physiol 2004;557(Pt 1):175–90.

[16] Fröhlich F, McCormick DA. Endogenous electric fields may guide neocortical network activity. Neuron 2010;67(1):129—43.

[17] Ozen S, et al. Transcranial electric stimulation entrains cortical neuronal populations in rats. J Neurosci 2010;30(34):11476—85.

[18] Ali MM, Sellers KK, Frohlich F. Transcranial alternating current stimulation modulates large-scale cortical network activity by network resonance. J Neurosci 2013;33(27):11262—75.

[19] Frohlich F, McCormick DA. Endogenous electric fields may guide neocortical network activity. Neuron 2010;67(1):129—43.

7

Microcircuits of the Neocortex

In the previous chapters of this unit we focused on the electrical signaling within single neurons, as well as between neurons. What we have not yet discussed in depth is the fact that there is a multitude of different neuron types in the brain, and that the connectivity among them ultimately provides the substrate for the neural circuits that mediate behavior. Interestingly, there are specific connectivity patterns among the neurons in different brain regions, and many of these connectivity patterns are conserved across species. In this and the following chapter we focus on the circuitry in the neocortex and the hippocampus. First, we will discuss the main cell types of the *neocortex* and the circuits they form. In the following chapter, we will look at the hippocampus that is part of the *allocortex*, which is the other type of cerebral cortex aside from the neocortex. The allocortex has a simpler structure than the neocortex, but it shares many of the same cell types. Other brain areas exhibit equally interesting circuits, but we limit our discussion in this book to these two types of cortical circuits. This "corticocentric" view reflects the nature of the later units, in which we discuss mostly neocortical network dynamics.

Neurons can be described and classified in multiple ways using different techniques. The typical classification criteria include morphology (shape), genetic makeup, electrophysiological properties, and connectivity. In both the neocortex and the hippocampus, there are only a few main classes of excitatory cells, but there is substantial diversity of inhibitory (ie, GABA-releasing) cells. The overall goal is to understand individual cell types in terms of their specific roles in shaping overall network dynamics and behavior. We will first discuss the excitatory (glutamatergic) neurons, and then we will tackle the diversity of inhibitory (GABAergic) interneurons. Finally, we will discuss the circuits in the neocortex that use these cell types as building blocks. The neocortex exhibits a *layered architecture* (or *laminar architecture*), with six layers that are stacked on top of each other, with each layer labeled with a roman numeral. Layer I is the top layer, followed by layers II and III (often lumped together as layer II/III), layer IV, layer V, and finally layer VI, which is the bottom layer situated just above the white matter. As we will see in this chapter, these layers differ in terms of their cellular composition, connectivity patterns, and likely also function. The toolbox "Neurons" and the chapter "Synaptic Transmission" provide helpful background with regards to the terminology that describes neurons and synapses.

GLUTAMATERGIC CELLS

Pyramidal Cells

The main class of cortical excitatory neurons is the pyramidal cell (colloquially abbreviated as "pyramids" or PYs, Fig. 7.1). These cells received their name from the pyramidal shape of their cell body. Pyramidal cells in different cortical layers and areas differ in the details of their morphology.

One of the main shared features is the presence of two distinct dendritic trees. The *apical dendritic tree* defines the primary somatodendritic axis of the cell ("vertical" orientation, spanning across cortical layers) and originates at the tip of the pyramid. Typically, the distant end of the apical dendrites is more branched (*apical tuft*). The *basal dendritic tree* originates at the base of the pyramid and is mostly limited to the layer in which the soma is located. Pyramidal cells are the most populous excitatory cell type in the mammalian cortex and receive both excitatory (glutamatergic) and inhibitory (GABAergic) inputs. The majority of excitatory synapses are located on dendritic spines (small protrusions where the synapses are located). In contrast, inhibitory synapses are formed on the soma, axon, and dendritic shafts. If dendritic trees were passive (no voltage-gated ion channels), cable theory (see toolbox: Modeling Neurons) predicts that synaptic potentials would decay in amplitude and would smear out in time as they propagate from their origin toward the soma. To prevent this, dendrites of pyramidal cells include voltage-gated ion channels (referred to as *active* dendrites) that enable more sophisticated signal processing than the passive cell membrane [1]. In fact, the extensive (apical) dendritic tree is sufficiently large that electrical signaling occurs locally and is then modulated by (1) the dendritic structure and (2) the intrinsic, voltage-gated ion channels that together give rise to the global, integrated signal that reaches the soma. Because of the large span of the dendritic trees, synaptic input can reach the axon hillock from a wide range of distances. As a result, both the amplitude and the time course of synaptic inputs can vary substantially when they arrive at the axon hillock. However, voltage-gated ion channels in the dendrites appear to counteract this spatiotemporal distortion of the input, also referred to as *location dependence of synaptic input*.

Active dendrites can also generate action potentials on their own; these so-called *dendritic spikes* are triggered by strong local synaptic input to dendritic branches. Dendritic spikes do

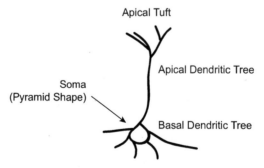

FIGURE 7.1   Schematic representation of pyramidal cells with apical and basal dendritic tree. The name of the cell type refers to the shape of the soma.

not always reach the soma to trigger an action potential at the axon hillock. Also axonal spikes travel into the dendritic tree and are boosted by the voltage-gated ion channels that (partially) prevent the decay of the backpropagating action potential.

Traditionally, pyramidal cells have been classified based on their electrophysiological response to a depolarizing current pulse. For sufficiently strong positive current injections, pyramidal cells fire a series of action potentials. Characterization of the spike frequency as a function of the amplitude of the injected current provides the data necessary for establishing the frequency–current curve (more commonly called *f*–*I* curve, Fig. 7.2). A pyramidal cell fires action potentials at very low rates for an input current that depolarizes the neuron to reach the "threshold." This contrasts with other cells such as fast-spiking inhibitory interneurons (see later) that have a pronounced discontinuity in the *f*–*I* curve, since these cells fire at a much higher frequency once they are pushed past their firing threshold.

Different pyramidal cells, even within the same layer, vary in their spiking response to injected current in terms of (1) the amount of spiking-frequency adaptation (decrease of instantaneous firing rate for step depolarizations) and (2) their ability to fire bursts of action potentials. These differences in spiking behavior are a function of not only the ion channels expressed in the neuron but also the morphology of the neuron.

Spiny Stellate Cells

Spiny stellate cells are excitatory, nonpyramidal neurons predominantly found in layer IV (input layer) of primary sensory areas of the neocortex. Historically, input layer IV in the primary visual cortex (V1) is referred to as the *granular layer* based on its appearance in histological sections caused by the tightly packed stellate cells. Accordingly, the layers above and below layer IV are termed *supragranular* and *infragranular*, respectively. The presence and number of spiny stellate cells differ among both cortical areas and species. For example, in monkeys, layer IV in V1 is divided into sublayers, where sublayer IVc is mostly populated with spiny stellate cells. In contrast, V1 of the mouse is devoid of spiny stellate cells. However, spiny stellate cells are found in the somatosensory cortex of the mouse. Spiny stellate cells received their name from their star-shaped appearance and the presence of a large number of spines on their dendrites. Although spiny stellate cells and pyramidal cells are

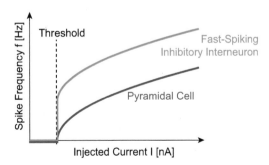

FIGURE 7.2 *f*–*I* curves for pyramidal cells (*blue*) and fast-spiking inhibitory interneurons (*red*). At threshold for action potential generation, there is a pronounced discontinuity in the *f*–*I* curve for fast-spiking inhibitory interneurons.

both glutamatergic cells, the two cell types in layer IV of the cortex greatly differ in their morphology. First, spiny stellate cells do not have dendrites that extend to the cortical surface; their dendrites are confined to layer IV. Second, their axonal projections are limited to the local cortical column (in contrast to pyramidal cells in layer IV). Accordingly, electrophysiological measurements show that incoming synaptic input (in contrast to the pyramidal cells in the same layer) mostly stem from within layer IV of the same cortical column. Only a small fraction of the afferent excitatory synapses onto spiny stellate cells stem from thalamic input (as low as 10%). But this small fraction of overall input is sufficiently strong to control the activity of spiny stellate cells. Spiny stellate cells project to layer II/III cells, but also form collaterals with layer IV and project to the infragranular layers (V and VI).

INHIBITORY INTERNEURONS

Interneurons synapse onto other neurons within spatially confined local circuits, and can be either excitatory or inhibitory. Here we focus on GABA-releasing inhibitory interneurons. These neurons typically have *aspiny* dendrites (ie, they do not have dendritic spines). Interneurons form synaptic contacts with targets both within a cortical column and in other columns. However, they do not typically send projections to the white matter, indicating that they do not send long-range projections to other brain areas. One important feature of the different types of interneurons is their ability to target specific cellular subdomains of postsynaptic neurons such as the axon initial segment. Interneurons can be classified based on their morphology, calcium-binding proteins, or electrophysiological properties. Interneurons can also be functionally classified by their role in circuit dynamics. Given the pronounced diversity of inhibitory interneurons, classification represents a challenge. Classification efforts are ongoing [2,3], and the pendulum has started to swing back in the sense that the number of distinct inhibitory interneurons may be more limited than previous classification efforts have suggested. New tools have aided in narrowing the number of interneuron subtypes. In particular, transgenic mouse lines enable the study of inhibitory interneurons based on their genetic identity. From this viewpoint, there are three major classes of inhibitory interneurons that do not overlap [4]. The three classes are defined by the expression of parvalbumin (PV), somatostatin (SST), and ionotropic serotonin receptor 5HT3a (see chapter: Neuromodulators), which includes interneurons that express vasoactive intestinal peptide (VIP). Despite this emerging classification scheme, there is a large body of literature that classifies neurons by morphology. We will adopt this scheme in the discussion that follows, and will provide information on the molecular classifications that overlap with the morphological classification.

Basket Cells

The most-studied type of interneuron, the *basket cell*, received its name from the fact that the cell forms dense axonal arborization around the soma and the axon hillock of their postsynaptic targets. As a result, the cells exert tight control over the membrane voltage and the spiking behavior of their postsynaptic targets. One main class of basket cells expresses the calcium-binding protein PV (PV-positive, [5]). Basket cells have been extensively studied, and have been found to play a prominent role in physiological network function and in pathological neural dynamics. In terms of their electrophysiology, basket cells are usually of the *fast-spiking* subtype. This label refers to the short duration of the action potential, which is mediated by Kv3 potassium channels that have a high activation threshold and a very fast deactivation time constant. PV-positive cells selectively provide inhibition to pyramidal cells and other PV-positive inhibitory interneurons. It is important to note that the term "basket cell" is the result of a classification by morphology, whereas "fast-spiking" is a classification by electrophysiological properties, and finally "PV-positive" is a classification by biochemical properties. While these three different classification dimensions often overlap, the overlap is not complete and can be the source of confusion. Also most studies do not determine the morphology, the electrophysiology, and the immunohistochemistry of the neurons under investigation. Therefore this cell type is prototypical for the challenges of classifying inhibitory interneurons. We will revisit basket cells in the context of feedforward and feed-back inhibition later in this chapter. Fast-spiking, PV-positive cells are essential for the generation of gamma oscillations (see chapter: Gamma Oscillations) and have been impli-cated in the network-level pathologies of schizophrenia (see chapter: Schizophrenia).

Other Types of Inhibitory Interneurons

Chandelier cells received their name because of their horizontal arbors that give rise to short vertical branches, which carry large *boutons* (ie, presynaptic terminals) that innervate the axon initial segment. Therefore these axo-axonic cells have very strong control over the action potential firing of their postsynaptic targets. Chandelier cells also express PV. *Bipolar cells* express VIP. VIP cells exhibit a unique connectivity pattern—they almost exclusively target other inhibitory interneurons, particularly SST-expressing interneurons, and not pyramidal cells. *Martinotti cells* send axons to layer I in the cortex to inhibit the tufts of pyramidal cells. The axons can spread horizontally in layer I for several millimeters and therefore provide synaptic inhibition across cortical columns. Martinotti cells also target other subcellular compartments. They express SST. SST-positive cells also inhibit other inhibitory interneurons. *Neurogliaform* cells are small neurons with radially organized dendrites and a strongly branched fine web of axonal branches. There are additional, heterogeneous inhibitory inter-neurons in layer I that are poorly understood. However, there is one exception: large *Cajal–Retzius* (CR) cells are confined to layer I and are present mostly during early brain develop-ment [6]. The exact definition of CR cells remains a matter of debate; in most general terms, the name refers to neurons in the *embryonic marginal zone*. This zone of the prenatal cerebral cortex plays an essential role in cellular migration and the formation of the cortical layers. Together with the *subplate*, it sandwiches the cortical plate that gives rise to layers II–VI of the mammalian cortex. In the adult brain, the marginal zone turns into layer I of the neocortex. Layer I is the last cortical layer to develop and has been historically overlooked because of the low number of neurons in adult layer I. CR neurons exhibit ascending pro-cesses that reach to the pial surface and a horizontal axonal plexus. CR cells secrete *Reelin*. Reelin attracts neuronal precursors and may therefore represent an important guidance signal. The general sequence of action for adding neurons to the cortex during development includes (1) recruitment of precursors from the *subventricular zone*, (2) migration to (future) layer I, (3) development of dendritic arborization, (4) establishment of contact with CR cells, and (4) differentiation into an adult neuron. Every migrating neuron bypasses all previously migrated neurons. As layers II–VI develop, all cells need to keep on extending their dendritic arborization. The cells that remain in contact with the CR cells become pyramidal cells (instead of interneurons). Specifically, CR cells target the apical terminal of the dendrites of pyramidal cells. With more and more cells migrating to their final locations, the density of CR cells decreases with development. Not surprisingly, given the role of CR cells in development of the cortex, there are substantial differences in the properties of CR cells across species.

THE CORTICAL MICROCIRCUIT

At the macroscopic level, the cortex is organized into different areas with relatively well-defined functional roles. At the most simplistic level, the anterior half of the cortex is responsible for generating and controlling behavioral *output*, ranging from movement in the motor cortex to higher-order executive function in the prefrontal cortex. The posterior half of the cortex is occupied by sensory areas that process sensory *input*. The cortex is a large sheet that is between 0.5 mm (in small species) and 3 mm (in humans) thick. The brains of mammalian species are either lissencephalic (smooth) or gyrencephalic (folded). The gyrencephalic structure circumvents the size limit of the cortical sheet defined by skull size, or, by extension, by the size of the birth canal. Examples of animals with lissencephalic brains are rat, mouse, mole, and marmoset. Animals with gyrencephalic brains include ferrets, cats, dogs, macaques, and humans. When flattened, the cortical sheet exhibits a two-dimensional, functional organization and can be interpreted as a "computational sheet" (Fig. 7.3, [7]). Within cortical areas, neurons are organized in columns, which group neurons with similar functional properties. These columns are perpendicular to the cortical surface and span all cortical layers. For example, the primate visual system is organized in orientation columns that group neurons that preferentially respond to lines with similar orientation. This functional organization is also reflected in preferentially structural connectivity within columns than across columns. However, any given cortical brain area can have several different columnar organizations for different functional roles. Thus the concept of columns is helpful for the study of the neocortex, but should not be understood as a singular, fully established organizational principle of cortical organization.

The thickness (z-dimension) is referred to as *cortical depth*, and is structured into layers based on cell types and densities (cortical layers I–VI, discussed in detail later). In theory, two fundamentally different organizational principles could be imagined for the blueprint of the circuitry in the neocortex. In the first model, neuronal networks within the cortex are highly specialized and adapted to the specific behavioral role of the corresponding brain area. For example, processing of visual signals requires different computations from processing auditory signals. In this model, the circuits in the visual and auditory cortices are fundamentally different to efficiently perform the different computational tasks. In the second, alternative model, neurons in the cortex form stereotyped circuits essentially identical across cortical areas. Such a general circuit organization in the neocortex is referred to as the *canonical microcircuit*. This model emphasizes the similarities in circuit organization for cortical areas with different behavioral functions [8]. There are several advantages of such a blueprint

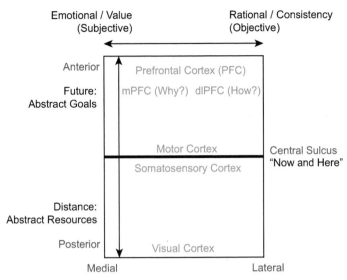

FIGURE 7.3 Cortex can be conceptualized as a "computational sheet" organized along two orthogonal axes that define a behavioral architecture. The anteroposterior axis encodes how close or distant (in time or space) resources and goals are. For example, the prefrontal cortex enables planning for the future. In contrast, the motor cortex, which is more central, is concerned with immediate interactions with the environment. Similarly, the sensory modality that requires direct interaction with the environment, somatosensation, is located closest to the central sulcus. In contrast, vision is located in a most posterior location and enables information about events at a distance to the body. The mediolateral organization separates "objective" and "subjective" processing. One example for this medial–lateral gradient is the different functional roles of the medial prefrontal cortex (mPFC) and the dorsolateral PFC (dlPFC). For example, the mPFC is involved in questions of "why" in terms of affect and motivation, whereas the dlPFC is concerned with the more rational question of "how." *Adapted from Douglas RJ, Martin KA. Behavioral architecture of the cortical sheet. Curr Biol 2012;22(24):R1033-8.*

for the neocortex. For example, it would reduce the number of "developmental programs" to assemble the circuits. Also the shared microcircuit architecture of different regions may enable more flexible compensation for lost function because of injury or illness such as amputation or stroke.

Layered Circuits of the Neocortex

When we consider the cortical microcircuit from a signal processing viewpoint, we can define (1) the input stage, (2) the signal processor, and (3) the output stage. This analogy allows us to classify cortical layers by their role in neuronal signal processing. Cortical layer IV receives the primary inputs from the thalamus and thereby defines our input stage. The superficial layers (I and II/III) represent the signal processor. The deep layers (V and VI) are the output stage. Although this signal processing model is a simplification, it is a useful starting point for the description and study of neocortical circuitry.

Action potentials are the signals of neuronal circuits. The primary effect of an action potential is that postsynaptic neurons are either excited or inhibited, depending on the type of neuron that generated the action potential. Most cells receive an abundance of excitatory and inhibitory inputs that are processed, integrated, and ultimately transformed into an action potential if the net excitation is sufficiently strong. Networks of excitatory neurons typically exhibit some type of *positive feedback* dynamics. In this regime, neuronal firing in the network causes an increase in activity by excitatory synaptic transmission. Therefore activity is amplified and continues to grow until other mechanisms that limit or reduce activity levels are engaged. In contrast, networks with excitatory neurons that excite inhibitory interneurons, which themselves inhibit the same population of excitatory neurons, exhibit *negative feedback* dynamics. In this regime, any increase in activity in the excitatory population is limited by the additional synaptic inhibition. Together, these two opposing mechanisms of neuronal communication need to achieve a delicate balance. Excitatory and inhibitory connections together achieve this balance in brain circuits, including the cortical microcircuit (Fig. 7.4).

Most cortical layers (except layer I) exhibit local excitatory connections within the layer that enable incoming signals to be rapidly amplified. Local synaptic inhibition is driven by the excitatory input it receives and therefore provides an activity-dependent inhibitory signal. Synaptic inhibition not only regulates overall levels of activity, but also shapes the timing of neuronal activity by defining time windows during which excitatory inputs can overcome inhibitory inputs and trigger action potentials in excitatory cells. In addition to these local excitatory and inhibitory loops, the different layers are connected to each other; layer IV

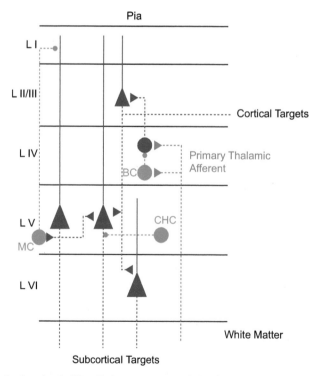

FIGURE 7.4   Cortical microcircuit. *Blue*: Excitatory neurons (*triangle*: pyramidal cells; *circle*: spiny stellate cells). *Red*: Inhibitory interneurons. *Green*: Thalamic afferents. *Solid lines*: Dendrites. *Dashed lines*: Axons. Only a subset of neuron types and connections are shown for clarity. BC, Basket cell; CHC, Chandelier cell; MC, Martinotti cell.

I. NEURONS, SYNAPSES, AND CIRCUITS

projects to layers II/III, which in turn project to the deep layers V and VI. We will now discuss the individual layers and their connectivity in more detail.

Layer IV receives primary thalamic afferents and therefore, in the case of sensory cortices, represents the first location of cortical processing of incoming sensory input. Layer IV contains both spiny stellate and pyramidal cells. Layer IV excitatory cells most prominently project to layers II/III but also send projections to all other cortical layers. However, layer IV pyramidal and spiny stellate cells receive very little top-down input from higher-order cortical areas and therefore are mostly driven by their thalamic inputs. The principle excitatory cells of layer II/III are pyramidal cells that project within the local circuit to layer V and also project to distant targets (other cortical areas, contralateral cortex). Layer V contains two main classes of pyramidal cells that differ in their genetic makeup, morphology, electrophysiological firing patterns, and connectivity patterns. The first class responds to a depolarizing step current with a spike train with spike frequency adaption. These neurons only project to targets in the telencephalon (cortex, striatum). Cortical projections of this class of pyramidal cells are both local (layers II/III and V) and distant. The second class does not exhibit spike frequency adaption and projects to targets outside the telencephalon. The synapses reaching the thalamus are strong. These cells receive input from many cortical layers and are therefore the main output of the cortex because of their projection pattern.

Layer VI of the cortex contains a plethora of different excitatory and inhibitory neural populations [9]. There are two distinct populations of pyramidal cells, referred to as short and tall pyramidal cells. *Short pyramidal cells* form reciprocal circuits with neurons in layer IV and primary (sensory) nuclei in the thalamus. Therefore short pyramidal cells may regulate how information reaches the cortical microcircuit. *Tall pyramidal cells* form reciprocal circuits with layer V and thus are likely to be involved in regulating the output from the cortical microcircuit.

Inhibitory Circuits

Synaptic inhibition plays an important role in sculpting signals within cortical microcircuits. At the most basic level, synaptic inhibition prevents hyperexcitability by providing activity-dependent inhibition. During application of pharmacological blockers of synaptic inhibition (see chapter: Synaptic Transmission), activity levels increase, often exhibiting pathological hypersynchronization that resembles electric seizure activity (see chapter: Epilepsy). Also, and perhaps even more importantly, synaptic inhibition dynamically regulates activity levels and enables advanced processing and routing of information in circuits. In the simplest case, *feedforward inhibition*, afferent excitatory fibers provide input to two populations—excitatory and inhibitory neurons. Neurons in both populations fire action potentials. However, the amount of activity in the excitatory population is limited by the synaptic inhibition it receives from the inhibitory population (Fig. 7.5).

A second, related inhibitory circuit is *feedback inhibition* (Fig. 7.6). Again, an excitatory and an inhibitory population interact. In this case, activity is generated within the excitatory population. The excitatory neurons not only excite other excitatory neurons but also the inhibitory interneurons that in turn provide synaptic inhibition back to them. Therefore the activity level in the excitatory population cannot increase in an unchecked way, since any increase also causes an increase in feedback synaptic inhibition. These inhibitory circuits

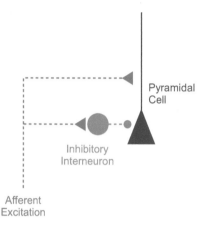

FIGURE 7.5   Feedforward inhibition. Afferent excitation triggers an action potential in the inhibitory interneuron. The pyramidal cell therefore receives a rapid sequence of an excitatory input from the afferent pathway followed by an inhibitory input. This sequence of excitatory followed by inhibitory input defines a narrow time window during which the pyramidal cell can fire an action potential.

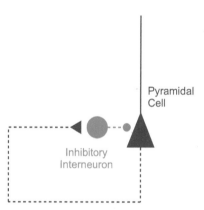

FIGURE 7.6   Feedback inhibition. Action potentials in the pyramidal cell lead to excitation in the inhibitory interneuron, which in turn inhibits the pyramidal cell.

have been extensively studied in the rodent hippocampus and are discussed in more detail in the following chapter "Microcircuits of the Hippocampus."

SUMMARY AND OUTLOOK

In this chapter we discussed the basic elements and organizational principles of neocortical circuits. We reviewed the main excitatory cell types, pyramidal cells and stellate cells. We looked at the staggering complexity of different types of inhibitory interneurons and the different ways of classifying them. Together, these cell types form neocortical circuits that are arranged in a way that different layers of the neocortex play different functional roles. Lastly, we introduced small circuit motifs, feedforward and feedback inhibition, that encapsulate the prototypical interaction schemes between pyramidal excitatory and inhibitory neurons in the neocortex. In the next chapter, we will look at the hippocampus, a simple three-layered form of cortex.

References

[1] Johnston D, Narayanan R. Active dendrites: colorful wings of the mysterious butterflies. Trends Neurosci 2008;31(6):309—16.
[2] Markram H, et al. Interneurons of the neocortical inhibitory system. Nat Rev Neurosci 2004;5(10):793—807.
[3] DeFelipe J, et al. New insights into the classification and nomenclature of cortical GABAergic interneurons. Nat Rev Neurosci 2013;14(3):202—16.
[4] Rudy B, et al. Three groups of interneurons account for nearly 100% of neocortical GABAergic neurons. Dev Neurobiol 2011;71(1):45—61.
[5] Hu H, Gan J, Jonas P. Interneurons. Fast-spiking, parvalbumin(+) GABAergic interneurons: from cellular design to microcircuit function. Science 2014;345(6196):1255263.
[6] Soriano E, Del Rio JA. The cells of cajal-retzius: still a mystery one century after. Neuron 2005;46(3):389—94.
[7] Douglas RJ, Martin KA. Behavioral architecture of the cortical sheet. Curr Biol 2012;22(24):R1033—8.
[8] Douglas RJ, Martin KA. Neuronal circuits of the neocortex. Annu Rev Neurosci 2004;27:419—51.
[9] Briggs F. Organizing principles of cortical layer 6. Front Neural Circuits February 12, 2010;4:3. http://dx.doi.org/10.3389/neuro.04.003.2010. eCollection 2010.

Microcircuits of the Hippocampus

The hippocampus is a cortical structure with a simpler and more orderly organization than the neocortex. Over the last decades, the rodent hippocampus has been one of the most frequently used model systems to study the structure, dynamics, and function of mammalian cortical circuits. Entire subfields of neuroscience rest on discoveries made in the rodent hippocampus, particularly the fields of learning and memory, synaptic plasticity, and spatial navigation. The hippocampus is an ideal model system for basic science studies because of its simple circuit organization. Also, from a translational viewpoint, the hippocampus deserves our attention given its role in Alzheimer's disease and many forms of epilepsy. In this chapter we will first introduce the overall circuit layout of the hippocampus, then we will review its main subdivisions, and finally we will discuss specific local circuits.

CIRCUIT LAYOUT OF THE HIPPOCAMPUS

The human hippocampus (together with the fornix, a C-shaped fiber) resembles a seahorse, which is the origin of the name, which is Greek for horse sea-monster. In the rodent, the hippocampus is a banana-shaped brain structure (Fig. 8.1). Most hippocampal studies use

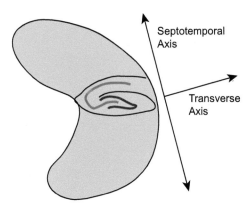

FIGURE 8.1 Schematic representation of the rodent hippocampus. Slice electrophysiology studies typically use transverse sections, which are orthogonal to the septotemporal axis.

Network Neuroscience
http://dx.doi.org/10.1016/B978-0-12-801560-5.00008-2

slices that are in the *transverse* plane, orthogonal to the *septotemporal axis* that connects the two endpoints of the figurative banana. Technically, the term *hippocampus* is reserved for part of the so-called *hippocampal formation*, and is composed of the *CA1, CA2,* and *CA3 subfields*. CA stands for the Latin *cornu ammonis*, or horn of the ancient god Amun. The hippocampal formation also includes the *dentate gyrus, subiculum, presubiculum, parasubiculum,* and the *entorhinal cortex*. Here we will follow the more common use of the term hippocampus to denote the entire hippocampal formation (Fig. 8.2).

Classically, the hippocampus is a *trisynaptic* (ie, three-synapse) circuit (Fig. 8.3). Information comes in through the superficial layers of the entorhinal cortex. Cells in layer II of the entorhinal cortex project to the dentate gyrus and CA3 of the hippocampus. This pathway is called the *perforant path*, since the axons perforate the subiculum, which is located between

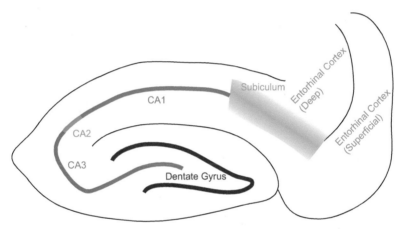

FIGURE 8.2   Overview anatomy of the hippocampal formation in the rodent. The dentate gyrus (dark blue), CA3 (light blue), CA2 (light green), and CA1 (dark green) all exhibit a very thin, densely packed layer of cell bodies (indicated by *thick lines*). In contrast, cells in the subiculum are more spread out. The entorhinal cortex (red) is separated into superficial and deep layers.

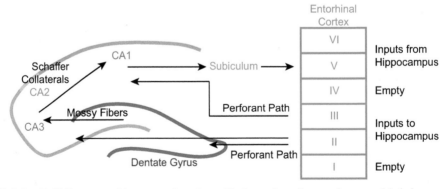

FIGURE 8.3   (Left) Transverse hippocampal section with the main pathways drawn as labeled arrows. (Right) Symbolic representation of entorhinal cortex with its layers I–VI.

the entorhinal cortex and the dentate gyrus. Cells in the dentate gyrus project to CA3 pyramidal cells; the projecting axons form the *mossy fibers*. Pyramidal cells in CA3 form both recurrent connections onto other pyramidal cells in CA3 and project to CA1. This pathway is called the *Schaffer collaterals*. Pyramidal cells in CA1 project to the subiculum and to deep layers of the entorhinal cortex (as does the subiculum).

The dentate gyrus engulfs the CA3 region. The part of the granule cell layer located between CA3 and CA1 is called the *suprapyramidal* layer; the other part is called the *infrapyramidal* layer. The hippocampus itself is divided into subfields CA1, CA2, and CA3. CA2 is a small but distinct subfield in between CA1 and CA3. The subiculum, presubiculum, and parasubiculum are subsumed by the term *subicular complex*. The subiculum starts where CA1 ends (which is defined by the furthest point to which Schaffer collaterals reach) and (in rodents) the dense pyramidal cell layers start to widen. Three major fiber bundles are part of the hippocampal formation. First, the *angular bundle* connects the hippocampus to the entorhinal cortex. Second, the *fimbria–fornix pathway* connects the hippocampus to the basal forebrain, hypothalamus, and brainstem. Third, the hippocampi of the two hemispheres are connected by the *dorsal* and *ventral commissures*. Next, we will discuss the dentate gyrus, hippocampal fields CA1 and CA3, the subiculum, and the entorhinal cortex.

DENTATE GYRUS

The dentate gyrus is a three-layered structure. From the outside there is a relatively cell-free layer called the *molecular layer*, followed by the *principal cell* (or *granule cell*) *layer*, which is densely packed with *granule cells*, and most deeply there is the *polymorphic cell layer*. Granule cells are small cells with elliptical somata that send their dendritic tree into the molecular layer. Granule cells are one of the few cell types in the CNS that undergo adult neurogenesis—new neurons are formed even in adulthood. The axons of the granule cells connect to hippocampal CA3 neurons. Just underneath the granule cell layer are a large number of different types of interneurons. Both the molecular layer (albeit sparsely) and the polymorphic cell layer are occupied by multiple classes of (poorly characterized) neuron types. The dentate gyrus receives its input from the entorhinal cortex (predominantly from layer II); these axons synapse onto the dendrites of the granule cells in the molecular layer. The dentate gyrus in turn projects to CA3 and CA2. The axons from the granule cells are called *mossy fibers* and terminate in CA3 and CA2. Mossy fibers form unique connections with the dendrites of CA3 pyramidal neurons in the form of *en passant* presynaptic terminals called *mossy fiber expansions*, which are a particularly large presynaptic zone with postsynaptic specializations, *complex spines*, on CA3 neurons.

HIPPOCAMPAL FIELDS: CA1, CA2, AND CA3

In the hippocampus itself, the main cell layer is the pyramidal cell layer (Fig. 8.4). This layer is most densely packed in CA1, and is less densely packed in CA2 and CA3.

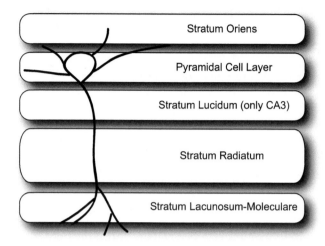

FIGURE 8.4   Layers of the hippocampus shown relative to the position of a pyramidal cell. Note that in contrast to the neocortex, the circuitry for the hippocampus is drawn in a way that the apical dendrites point down, not up.

Above the pyramidal cell layer is the *stratum oriens*, home to the basal dendrites of pyramidal cells and multiple types of interneurons, and is the primary site of input from CA2 neurons. In CA3 and CA2, there is a thin layer just adjacent to the pyramidal cell layer in which the mossy fibers from the dentate gyrus are located, the *stratum lucidum*. Below the pyramidal cell layer (and the stratum lucidum in CA3) is the *stratum radiatum*, in which the recurrent (association) connections within CA3 and the connection to CA1 (Schaffer collaterals) are located. Below the stratum radiatum is the *stratum lacunosum-moleculare*, the recipient zone of input from the entorhinal cortex. Both the stratum radiatum and the stratum lacunosum-moleculare are home to diverse types of interneurons. Pyramidal cells in CA3 are quite variable in their size, while pyramidal cells in CA1 are smaller than the ones in CA3 and are uniform in size. We will focus our discussion on CA1 and CA3 pyramidal cells, which are best understood. Also, to avoid repetition of the previous chapter "Microcircuits of the Neocortex," we will not catalog the different inhibitory interneurons. Instead, we discuss several functional roles of synaptic inhibition in the last section of this chapter.

CA1 Pyramidal Cells

CA1 pyramidal neurons (Fig. 8.5) are among the best-studied neurons in the mammalian nervous system. In the rodent preparation, these neurons are easily accessible for both intracellular and extracellular recordings (in acute slices and intact animals, respectively) and have an afferent pathway, the Schaffer collaterals, that is straightforward to activate and interpret. A further advantage is that it is feasible to record from large apical dendrites using dendritic patch clamp recordings. The dendrites of CA1 pyramidal cells are organized into the *apical* and *basal* dendrites. The basal dendrites are located in the stratum oriens and the apical dendrites are situated in the stratum radiatum (proximal, ie, close, to the cell body) and the stratum lacunosum-moleculare (distal, further away from the cell body).

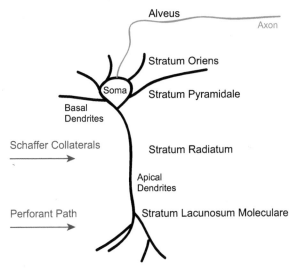

FIGURE 8.5   CA1 pyramidal cell. Blue: Afferent inputs. Red: Axon.

Each CA1 pyramidal cell exhibits around 30,000 spines that receive excitatory input. Direct input from layer III of the entorhinal cortex (perforant path) targets the distal apical dendrites in the stratum lacunosum-moleculare. In contrast, input from CA3 (Schaffer collaterals) forms synapses on the apical dendrites in the stratum radiatum and on the basal dendrites in the striatum oriens. The stratum oriens is the main target of projections from CA2. Also CA1 pyramidal cells are targeted by a diverse set of inhibitory interneurons. For example, the oriens-alveus/lacunosum-moleculare interneurons project to the stratum lacunosum-moleculare and target the apical dendritic branches also targeted by the projections from the entorhinal cortex. The axons of CA1 pyramidal cells project through the striatum oriens in the alveus. In contrast to CA3, almost no collaterals to other CA1 pyramidal neurons are formed.

The most important targets of CA1 pyramidal cell axons within the hippocampus are the pyramidal cells in the subiculum. CA1 pyramidal cells also project to brain areas beyond the hippocampus such as the medial frontal cortex and the olfactory bulb.

The relatively large and complex morphology of CA1 pyramidal cells creates unique challenges for signal propagation [1]. In particular, given the long and elaborate structure of the dendritic tree, distal synaptic input may fail to reach the soma. At least two mechanisms to counteract loss by signal attenuation have been identified. First, the synaptic conductances in the stratum radiatum scale with distance from the cell body such that more distant synapses generate higher-amplitude synaptic currents to compensate for the signal loss caused by propagation to the soma. This mechanism is referred to as *synaptic scaling*. However, synaptic scaling does not apply to the more distal input in the stratum lacunosum-moleculare. Instead, input into the distal dendrites is amplified by voltage-gated ion channels in the dendrites. Simultaneous patch-clamp recordings of dendrites and the soma (see chapter: Membrane Voltage) have shown that injecting current at different distances from the soma always elicited the same response at the soma [2].

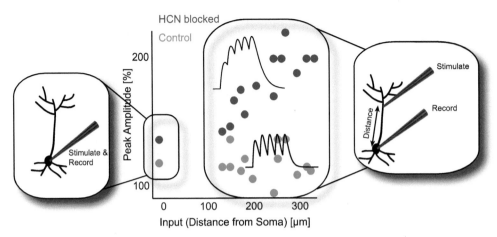

FIGURE 8.6 I_h (mediated by hyperpolarization-activated, cyclic nucleotide-gated, HCN, channels) compensates for spatial filtering by the dendritic tree and enables location-independent synaptic signaling. With HCN channels, the response to stimulation is independent from distance to the soma (control condition, red). Without HCN channels, the amplitude increased with distance because of the temporal smearing of the input, which caused temporal summation of the responses (blue). *Adapted from Magee JC. Dendritic I_h normalizes temporal summation in hippocampal CA1 neurons. Nat Neurosci 1999;2(9):848.*

This finding is in conflict with the cable equation that predicts more temporal smoothing of more distal input. Blocking the hyperpolarization-active depolarizing current (I_h) unmasked such a spatial dependence of the response to the input as a function of distance (Fig. 8.6). I_h deactivates in response to depolarizing (synaptic) input, and therefore it helps shorten excitatory input. In contrast, the amplitude of the incoming excitatory postsynaptic potential (EPSP) is less affected by the relatively slow deactivation time constant of hyperpolarization-activated, cyclic nucleotide-gated (HCN) channels that mediate I_h. There is a strong gradient in the density of HCN channels along the dendritic tree. In agreement with this proposed role of shortening the time course of excitatory events that arrive at distal sites of the dendritic tree, channel density is the highest at the distal end of the dendritic tree.

Voltage-gated ion channels in the dendrite also play a role in the propagation of action potentials from the soma into the dendrites. Specifically, action potentials originate in the initial axonal segment and then backpropagate into the soma and the dendritic tree. This backpropagation is mediated by voltage-gated sodium channels that are uniformly distributed along the somatodendritic axis. Calcium channels are also present in the dendrites, again with approximately constant density. These voltage-gated channels, which enable backpropagating and dendritic action potentials (see toolbox: Neurons), are kept in check by potassium channels in the dendrites. In particular, there is a spatial gradient for the A-type potassium channel that increases nearly linearly with distance on the primary apical dendrite away from the soma [3]. A-type potassium channels exhibit rapid activation and inactivation kinetics. This increase in A-type potassium channels leads to a rapid decrease of the amplitude of the backpropagating action potential. Another important role of backpropagating action potentials is to provide postsynaptic depolarization for long-term

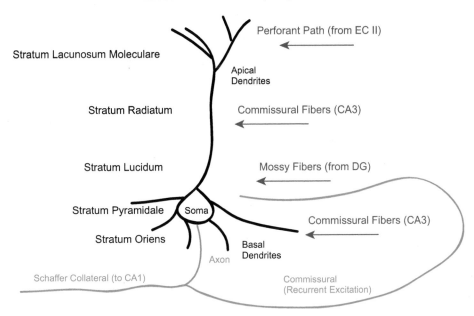

FIGURE 8.7   Pyramidal cell in CA3. *CA*, Cornu ammonis; *DG*, Dentate gyrus; *EC II*, entorhinal cortex layer II.

potentiation. Both backpropagating action potentials (more proximal to the soma) and locally (ie, dendritically) generated action potentials can play this role (see also chapter: Synaptic Plasticity).

CA3 Pyramidal Cells

Pyramidal cells in CA3 (Fig. 8.7) are similar in overall morphology to CA1 pyramidal cells. But in contrast to CA1, CA3 is a highly excitable, recurrently connected network by means of the prominent lateral axon collaterals within CA3. Pyramidal cells in CA3 receive three main classes of excitatory input. First, CA3 pyramidal cells receive input from the entorhinal cortex (layer II) via the perforant path that targets the distal apical dendrites in the stratum lacunosum-moleculare. Second, the input from dentate granule cells is localized to the most proximal part of the apical dendrite (in the stratum lucidum). Third, commissural input is provided by axonal collaterals of other CA3 pyramidal cells that form synapses both in the stratum radiatum and in the stratum oriens. CA3 pyramidal cells connect to CA3, CA2, and CA1 and also the lateral septal nucleus. Unlike CA1 pyramidal cells, CA3 neurons are more prone to intrinsic bursting (firing of several action potentials in an all-or-none fashion). Interestingly, under pathological conditions (experimentally mimicked by suppression of synaptic inhibition or reduction of potassium currents), there are network-wide bursts in CA3. These bursts are referred to as *paroxysmal depolarization shifts*. They are a network phenomenon, since they are mostly mediated by giant excitatory postsynaptic potential, which results from the synchronous activity of large groups of pyramidal cells [4]. This contrasts to physiological bursts in CA3 that are intrinsically generated by voltage-gated ion channels and do not require synaptic transmission to occur.

SUBICULUM

The subiculum represents the major output from the hippocampus. In turn, the subiculum receives input from the superficial layers of the entorhinal cortex and CA1. Therefore subicular neurons (mostly pyramidal cells) receive synaptic input from axons from both areas, with the projections from entorhinal cortex being limited to the more distal dendrites. The subiculum also receives input from many of the brain structures that project to the entorhinal cortex. Pyramidal cells in the subiculum form both local collaterals (targeting other pyramidal cells within the subiculum) and projecting collaterals that target layer V of the entorhinal cortex and also the pre- and parasubiculum. The subiculum also projects to many other brain areas, including the medial prefrontal cortex and many other (sub-) cortical structures. In terms of their electrophysiology, the most prominent features of pyramidal cells in the subiculum are their propensity to fire bursts of action potentials (mediated by a Ca-dependent after-depolarization) and their *subthreshold oscillations*. These oscillations are periodic, small-amplitude fluctuations of the membrane voltage that arise from the dynamic interaction of voltage-gated ion channels that are activated for subthreshold membrane voltages.

ENTORHINAL CORTEX

The entorhinal cortex is the gateway to and from the hippocampus. In essence there are four layers that contain cells (called layers II, III, V, VI according to Cajal; labels are designed to match the neocortex as much as possible) and two layers without cells (I and IV, the latter termed *lamina dissecans*). The superficial, cell-containing layers include *stellate cells* and smaller pyramidal cells. The deep layers contain pyramidal cells (layer V) and cells with a variety of morphologies (layer VI). Stellate cells in layer II have a star-like appearance, with dendrites radiating out from the soma toward both layer I and layer III. Their axons target both granule cells in the dentate gyrus and CA3. Layer II stellate cells exhibit subthreshold oscillations in response to depolarizing current injections in vitro. The frequency of this oscillation ranges from 5 to 15 Hz and corresponds to the theta frequency band as defined in rodents (since there are no thalamocortical alpha rhythms, the theta frequency range is expanded, see chapter: Theta Oscillations). These oscillations are mediated by an interplay of the hyperpolarization-activated depolarizing current I_h and the persistent sodium current I_{NaP}. The pyramidal cells in layer II do not exhibit such subthreshold dynamics, but their projections to the dentate gyrus and CA3 are similar to stellate cells..

The deep layers of entorhinal cortex contain both pyramidal cells and polymorphic neurons (heterogeneous group of neurons). These neurons project to other cortical areas, many of which in turn project to the superficial layers of the entorhinal cortex. In addition, axon collaterals from these neurons also target other cells within the deep layers and the superficial layers. These projections close the loop formed by the superficial layers to the hippocampus and the hippocampal projections back to the deep layers of the entorhinal cortex. Similar to stellate cells in layer II, pyramidal cells in the deep layers exhibit subthreshold oscillations in the theta band. However, these cells exhibit no I_h and so the mechanism for the genesis of subthreshold oscillations may differ between the two cell types. In the presence of activated muscarinic acetylcholine receptors, layer V pyramidal cells exhibit persistent activity after a depolarizing current pulse. Such sustained activity in the absence of input has been speculatively linked to the role of the entorhinal cortex in memory.

FUNDAMENTAL LOCAL CIRCUIT PRINCIPLES

We will now turn our attention to the fundamental principles of synaptic interaction in the hippocampal circuit. Many of the concepts discussed here in the context of the hippocampus also apply to the neocortex (see chapter: Microcircuits of the Neocortex).

The first principle focuses on the number and classification of cell types in the hippocampus. Overall, there are two types of cells: principal cells and interneurons. Strikingly, interneurons constitute only a small fraction of all hippocampal neurons and defy any simple classification approach. For example, although interneurons are classically thought of as inhibitory, this assumption is likely not true in all cases. Cells that release GABA (chapter: Synaptic Transmission) do not necessarily inhibit their postsynaptic target (eg, during development). Furthermore, inhibitory interneurons may not contribute to an overall reduction in activity levels in the circuit since they inhibit other inhibitory interneurons. Also their definition as local interneurons may be inappropriate since there are interneurons that connect to targets outside of the hippocampus. Efforts to classify interneurons are ongoing and are briefly discussed in the chapter "Microcircuits of the Neocortex."

The second principle concerns the stereotyped arrangement of local synaptic connectivity in the circuit. There are several common connectivity patterns, also referred to as *circuit motifs*, that are ubiquitously present (to different extents) across the brain. We briefly mentioned them in the chapter "Microcircuits of the Neocortex," but will review them in detail here since the hippocampus (in particular its acute slice preparation) has been at the center of elucidating the dynamics of these motifs. We will discuss one excitatory connectivity motif (recurrent excitation) and three inhibitory motifs (feedforward inhibition, feedback inhibition, and mutual inhibition).

Excitatory Circuit Motifs

Pyramidal cells in subfield CA3 exhibit much more pronounced lateral (ie, recurrent) connections than the ones in CA1. The majority of excitatory inputs to CA2 and CA3 pyramidal cells stem from their recurrent collateral axons. CA3 pyramidal cells have large axonal arbors. In CA3, the recurrent synapses create EPSPs of around 1−2 mV, and are therefore far from being able to trigger postsynaptic action potentials. However, the effect of a burst of action potentials is quite pronounced because of augmentation of the synaptic response. In contrast, recurrent connectivity of pyramidal cells in CA1 is more limited and the axonal arbors are much smaller in size. This difference in recurrent excitation makes not only the computational properties of the circuits but also their susceptibility to pathological activity patterns (eg, epileptic seizures) very different. Circuits with sufficient recurrent excitation have the important computational property of an *attractor network*. These networks have unique properties that enable the occurrence of distinct activity patterns (multistability) that have been associated with different items stored in memory. Any of these patterns can be activated by a partial input since the recurrent connections amplify and thereby complete the initial activation pattern. Furthermore, the recurrent excitation helps to maintain firing patterns even after withdrawal of the initial input and therefore potentially enables short-term memory capabilities. In terms of pathological activity, the difference in the presence of recurrent excitation leads to differential susceptibility of CA3 and CA1 to epileptic seizures.

Inhibitory Circuit Motifs

Inhibitory circuits perform different functions [5]. Afferent excitatory fibers (from other subfields) often terminate both on excitatory and inhibitory postsynaptic cells. If the inhibitory cells in turn provide inhibition to the excitatory pyramidal cells, the effect of incoming excitatory input on the postsynaptic excitatory cells is both excitatory (monosynaptic, one direct excitatory synapse) and inhibitory (disynaptic, IPSP originates from inhibitory interneuron that receives EPSP from afferent fiber). For this circuit to be effective, the inhibitory interneurons need to receive sufficient input (and be excitable enough) to fire action potentials in response to afferent input. In this case, the postsynaptic excitatory cell receives an EPSP immediately followed by an inhibitory postsynaptic potential (IPSP), with the delay between the two event onsets resulting from the extra synapse in the inhibitory branch of the circuit. This feedforward inhibitory circuit is present in many brain structures and has been extensively studied in the hippocampus. In this circuit, the Schaffer collaterals represent the afferent fibers to both the CA1 pyramidal cell and to soma-targeting, fast-spiking inhibitory interneurons. When activated, these interneurons strongly inhibit the pyramidal cells because of the proximity of their axonal terminals to the action potential initiation site in the pyramidal cell. There is a short delay (a few milliseconds) between arrival of the EPSP and the IPSP in the pyramidal cell (Fig. 8.8). This delay creates a short window of opportunity for the pyramidal cell to fire an action potential in response to the afferent input. Thus feedforward inhibition enforces precise *spike timing*. In the absence of such feedforward inhibition, EPSPs could be integrated (only limited by the relatively long time constant of the postsynaptic pyramidal cell) and produce spiking responses with low temporal precision. Thus feedforward inhibition turns pyramidal cells that are integrators of synaptic input into coincidence detectors that only fire in response to multiple inputs arriving almost at the same time (to sufficiently depolarize the postsynaptic cell before it is subjected to feedforward inhibition).

Feedforward inhibition also mediates *gain control*. Gain denotes the amplification of incoming input to form an output. Neuronal circuits need to be able to respond to input strengths that can span many orders of magnitude. This is most obvious for sensory systems that need to process a range of stimuli from very weak to very strong (eg, in terms of brightness in the visual system). In other words, circuits need to be responsive to extremely weak input, but at the same time not saturate for strong input such that strong inputs can still be distinguished. Feedforward inhibition provides such a mechanism, since the activation of the inhibitory pathway in the circuit depends on afferent input strength. For weak inputs,

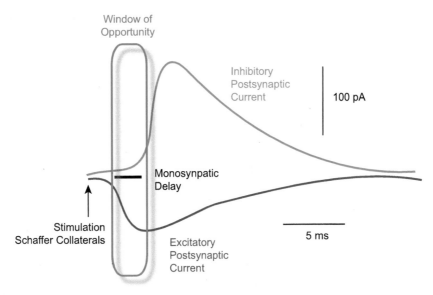

FIGURE 8.8 Feedforward inhibition. Synaptic currents in the pyramidal cell. Red: Inhibitory postsynaptic current (IPSC). Blue: Excitatory postsynaptic current (EPSC). Synaptic inputs are elicited by electrical stimulation of the afferent Schaffer collaterals. Green: Window of opportunity for firing an action potential. *Adapted from Pouille F, Scanziani M. Enforcement of temporal fidelity in pyramidal cells by somatic feed-forward inhibition. Science 2001;293(5532):1159–63.*

inhibition is barely activated and the excitatory input to the postsynaptic pyramidal cells is accordingly unhindered. In the case of strong afferent input, however, inhibition is recruited and makes it harder for the afferent excitatory input to make the postsynaptic pyramidal cells fire action potentials. Together, feedforward inhibition provides both fundamental temporal and spatial processing capabilities to local circuits.

Feedback inhibition denotes a related circuit motif in which activity of pyramidal cells in a given area synapses onto inhibitory interneurons that synapse onto the same pyramidal cell population. The main distinguishing features of feedback inhibition are (1) the population of pyramidal cells needs to be active to trigger and thus receive inhibition and (2) the excitation of the inhibitory cells is locally generated and not provided by a distant cell population. At the most basic level, feedback inhibition limits network activity by recruiting inhibition as a function of the activity of the pyramidal cells. This helps to prevent unwanted, elevated activity.

Inhibitory feedback loops can also contribute to shaping the temporal structure of activity. The dynamics of each loop are a function of the properties of the excitatory synapse onto the inhibitory interneuron, the intrinsic excitability of the interneuron, and the inhibitory synapse back onto the pyramidal cell. For example, in hippocampus CA1 there are two distinct feedback loops mediated by two types of functionally distinct inhibitory interneuron; these loops are recruited in a distinct manner and therefore provide different signal processing capabilities [7]. The first loop is activated by the onset of a train of action potentials in the pyramidal cells and provides perisomatic inhibition. Synaptic depression of the excitatory synapses on these inhibitory interneurons quickly reduces the strength of the inhibition in case of sustained neuronal firing. The second loop is preferentially activated by sustained activity, since the excitatory synapses on this functional class of interneurons exhibit facilitation (see chapter: Synaptic Plasticity). These inhibitory interneurons provide inhibition to the more distal dendrites of the pyramidal cells. The differential targeting of the pyramidal cells in terms of the location of the synapses along the dendritic tree is important because the incoming inhibition interacts with different aspects of the electric signaling in the postsynaptic pyramidal cell. In particular, perisomatic inhibition regulates spike timing (conceptually similar to the feedforward inhibition discussed earlier) and may therefore be important for synchronization and the generation of oscillations. On the other hand, it is likely that inhibition of distal dendrites (electrically far from the soma) preferentially interacts with other excitatory inputs received by that part of the dendritic tree. The third and last inhibitory circuit motif is *mutual inhibition*, in essence the inhibitory counterpart to mutual excitation. Overall (in the hippocampus), inhibitory interneurons are far more likely to target principal cells than other interneurons. However, several classes of interneurons preferentially inhibit interneurons of the same class. Also there are populations of inhibitory interneurons that only inhibit other interneurons and do not target pyramidal cells. Mutual inhibition appears to be an important mechanism that contributes to the genesis of gamma oscillations by synchronized release of inhibition from pyramidal cells (see chapter: Gamma Oscillations).

SUMMARY AND OUTLOOK

In this chapter we reviewed the basic layout of the hippocampal circuits. The entorhinal cortex projects to the dentate gyrus, which connects to CA3, which in turn connects to CA1. The subiculum is the main target of the outputs of CA1 and projects back to the entorhinal cortex. The stereotyped layout of the different pathways within the (rodent) hippocampus greatly facilitates the study of synaptic connections and the microcircuits they form. In particular, the relative lack of recurrent excitation in CA1 makes that hippocampal field an ideal model system for understanding how incoming information interacts and gives rise to output in the absence of endogenous network dynamics.

The previously discussed circuit layout of the hippocampal formation also applies in principle to other species such as monkeys and humans, but major differences in terms of the anatomy have been noted. Most prominently, the CA1 pyramidal cell layer in primates is much thicker than in rodents (up to 30 cell bodies thick in humans). Also, as opposed to rodents, human granule cells have basal dendrites. Although the functional implications of these differences are unknown, these differences are important to note.

NOTES

- An authoritative and comprehensive discussion of the hippocampus is provided in *The Hippocampus Book* edited by Andersen, Morris, Amaral, Bliss, and O'Keefe [8].

References

[1] Spruston N. Pyramidal neurons: dendritic structure and synaptic integration. Nat Rev Neurosci 2008;9(3):206–21.

[2] Magee JC. Dendritic I_h normalizes temporal summation in hippocampal CA1 neurons. Nat Neurosci 1999;2(9):848.

[3] Hoffman DA, et al. K^+ channel regulation of signal propagation in dendrites of hippocampal pyramidal neurons. Nature 1997;387(6636):869–75.

[4] Johnston D, Brown TH. Giant synaptic potential hypothesis for epileptiform activity. Science 1981;211(4479):294–7.

[5] Isaacson JS, Scanziani M. How inhibition shapes cortical activity. Neuron 2011;72(2):231–43.

[6] Pouille F, Scanziani M. Enforcement of temporal fidelity in pyramidal cells by somatic feed-forward inhibition. Science 2001;293(5532):1159–63.

[7] Pouille F, Scanziani M. Routing of spike series by dynamic circuits in the hippocampus. Nature 2004;429(6993):717–23.

[8] Andersen P. The hippocampus book. Oxford; New York: Oxford University Press; 2007. p. 832.

MEASURING, PERTURBING, AND ANALYZING BRAIN NETWORKS

In the previous unit Neurons, Synapses, and Circuits, we discussed individual neurons and how they are connected to form networks in the brain. In this unit, we will learn about the tools that are commonly used to measure, perturb, and analyze signaling in brain networks.

The majority of the methods are typically used either in animal models or in humans, depending on whether or not the methods are *invasive*. The term invasive is defined in this context as a procedure that requires a skin incision or the insertion of an instrument into the body. Classically, animal experiments use invasive electrophysiology, where electrodes are implanted directly into the brain to record action potentials (chapter: Unit Activity) and the local field potential (LFP, chapter: LFP and EEG). More recently, optical methods to record neuronal activity, particularly calcium and voltage imaging, have become a powerful experimental strategy that complements electrophysiology in animals but is also invasive. Light can also be used to perturb neuronal activity by means of optogenetic manipulations (chapter: Optical Measurements and Perturbations).

In humans, magnetic resonance imaging (MRI, chapter: Imaging Structural Networks With MRI) enables detailed, noninvasive imaging of brain structure. The same technology can also be used to measure changes in blood oxygenation as a proxy for neural activity (chapter: Imaging Functional Networks With MRI). Alternatively, the electric fields generated by neuronal activity can be noninvasively measured by electroencephalography (EEG, chapter: LFP and EEG). We will also look at magnetoencephalography and electrocorticography (ECoG), two less commonly used but important methods that also measure signaling in the brain. In addition to recording from the brain, we can also modulate neuronal activity by stimulation with electric and magnetic fields. Both noninvasive brain stimulation such as transcranial magnetic stimulation (chapter: Noninvasive Brain Stimulation) and invasive brain stimulation such as deep brain stimulation (chapter: Deep Brain Stimulation) are used as scientific and clinical tools to perturb brain activity. There are exceptional circumstances in which methods typically reserved for animal experiments, such as invasive electrophysiology, can be ethically used in humans. For example, during certain types of human brain surgery, invasive electrophysiological measurements (eg, ECoG) need to be made to guide clinical decisions. These recordings provide meaningful data for scientific studies in addition to their clinical use. Conversely, the application of tools from human neuroscience and neurology in animal models has helped uncover the underlying mechanisms of these methods. A prime example is the research unraveling the source of the functional MRI (fMRI) signal by combining fMRI with invasive electrophysiological measurement in animal experiments.

In addition to categorizing the discussed methods as either invasive or noninvasive, we will focus on the *temporal* and *spatial resolution* these methods provide. Temporal resolution refers to how frequently in time a measurement is performed, and thus defines the fastest fluctuations in the signal of interest that can be precisely captured. The temporal resolution is specified by the *sampling rate*. The sampling rate denotes the frequency at which measurements are performed. For example, it takes about a second to acquire a single image of activity in a human fMRI scan (sampling rate of 1/s, ie, 1 Hz). Therefore signals that significantly fluctuate within any given 1-s interval (ie, they have a subsecond time scale) are not properly captured. This is in pronounced contrast to the typical sampling rates of EEG devices (1000 Hz and greater, see toolbox: Time and Frequency). The fastest temporal scale we will consider here is the millisecond time scale of action potential firing.

Spatial resolution defines the smallest events in space that can be captured with a given measuring strategy. For example, it is quite common to measure brain activity with a resolution of cubic millimeters in fMRI. In contrast, the raw signals recorded with EEG exhibit very poor spatial resolution since they originate from several square centimeters of brain tissue. Here, we focus on spatial scales ranging from the whole brain to individual neurons. As we will see, methods that optimize spatial and temporal resolution are often more invasive. For example, the spatial resolution of electrophysiological measurements of brain activity can be as small as 100 µm for a recording electrode surgically inserted into the brain. Together with the high temporal resolution of electrophysiological measurements, invasive recordings provide ideal temporal and spatial resolution. This is notable because with few exceptions noninvasive methods with high temporal resolution (eg, EEG) suffer from a poor spatial resolution and vice versa (eg, fMRI). Therefore combining suitable methods has remained one of the most powerful and exciting methodological strategies in network neuroscience.

Collecting data with the methods discussed in this unit is one side of network neuroscience. The other side is to process, analyze, and model the data. In the chapter "Network Interactions," we introduce the main strategies to study the interaction between different neurons or networks of neurons. These interactions are referred to as *functional* or *effective connectivity*.

The chapters in this unit focus on the most common techniques for measuring, perturbing, and analyzing brain activity, while also highlighting a few select applications as examples of the role these techniques can play in network neuroscience. Understanding the advantages and limitations of these tools is vital for the critical analysis and interpretation of data. A further goal of this unit is to foster interdisciplinary collaborations among research groups that use these different techniques.

9

Unit Activity

The brain uses many kinds of signaling systems at different temporal and spatial scales. For the study of networks, we will mostly focus on electrophysiology and related physiological measurements that can be used as surrogate markers to investigate neural activity, such as blood oxygenation levels (chapter: Imaging Functional Networks With MRI) and intracellular calcium concentration (see chapter: Optical Measurements and Perturbations). Electrodes implanted in the brain for the purpose of invasive electrophysiology studies measure both action potentials of neurons close to the tip of the electrode and overall fluctuations in the extracellular voltage. These overall fluctuations reflect the behavior of networks in the vicinity of the electrode. In this chapter, we will focus on extracellular recordings of action potentials. Neurons are referred to in this context as *units*, and we will learn about multiunit (small groups of neurons) and single-unit (individual neuron) recordings. *Extracellular recordings of action potentials* are crucial, as they allow for the simultaneous monitoring of multiple neurons with high temporal resolution. The section marked with an asterisk requires more mathematical background than the remainder of the chapter and can be safely skipped if desired.

TERMINOLOGY AND CONCEPTS

Action potentials are commonly treated as binary (ie, all-or-nothing) events generated by neurons. Typically, action potentials are referred to as *spikes*. While this term is a convenient shorthand, some caution is warranted. To an epileptologist, for example, a spike is a macroscopic electroencephalogram (EEG) signature associated with pathologically synchronized firing of a large number of neurons (see chapter: Epilepsy). When an action potential occurs, the rapid influx of sodium ions and outflow of potassium ions (see chapter: Dynamics of the Action Potential) cause a small, transient deflection in the extracellular voltage that is recorded relative to the reference electric potential of the brain or body of the animal. In practice, these deflections are extracted by band-pass filtering (Fig. 9.1, see also toolbox: Time and Frequency), which removes the slow [ie, low-frequency, local field potential (LFP)] fluctuations and the high-frequency noise from the recorded *broadband* signal (Fig. 9.2). In the remainder of the chapter, we refer to the high-pass filtering although, technically, a band-pass filter is used to not only eliminate the LFP signal but also high-frequency noise.

The amplitude of the *extracellularly (recorded) action potential* (EAP) has an order of magnitude of 100 µV peak to peak, 1000 times smaller than the same action potential measured intracellularly ($\sim$100 mV peak to peak). The duration of the EAP signal ranges from about

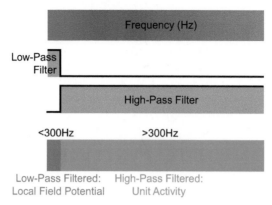

FIGURE 9.1   Separation of extracellular voltage signal into local field potential (LFP) and unit activity by low-pass and high-pass filtering.

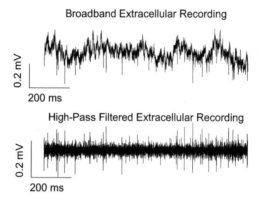

FIGURE 9.2   *Top*: Sample broadband recording of the extracellular voltage in the cortex of an awake animal. *Bottom*: Same trace after high-pass filtering (>300 Hz) to isolate unit activity (appearing as vertical deflections in the trace) and remove the local field potential (LFP).

0.5 to 2 ms. Often multiple active neurons are close to the tip of the electrode, and their EAPs are discernable in the recorded trace. Such activity is referred to as *multiunit activity* (often abbreviated as MU or MUA). With the help of a signal processing method known as *spike sorting*, the various spike waveforms that comprise MUA can be clustered to group together similar-looking waveforms that are assumed to stem from an individual neuron. These clusters are referred to as *single units* (SU). This terminology reflects the fact that we assume that all EAPs in a given cluster originate from a single neuron, although there is no way to verify whether or not that is true (unless simultaneous intracellular recordings are performed). Hence the word "unit," rather than "neuron," is used.

ORIGIN OF THE EAP SIGNAL*

The main challenge present when recording EAPs is the lack of insight into which neurons generated the signal, as there is no visualization of the neurons. Fortunately, computational models of individual cells and the electric properties of their extracellular space allow us to predict both intra- and extracellular waveforms of action potentials [1]. These models can be calibrated with the few available simultaneous intracellular and extracellular recordings of action potentials [2]. The EAP is a measure of the change in extracellular voltage caused by ion current flux across the cell membrane during an action potential. The simplest relationship between voltage and current can be found when the physical medium of interest, here the extracellular space, is purely resistive, that is, the relationship between voltage and current does not depend on the frequency of these signals. The electric field E is a vector field, meaning that every location in space (x, y, z) is associated with a three-dimensional vector (all vectors are denoted in bold print):

$$E(x, y, z) = \begin{pmatrix} E_x(x, y, z) \\ E_y(x, y, z) \\ E_z(x, y, z) \end{pmatrix} \tag{9.1}$$

Fundamentally, Maxwell's equations relate the electric field E to charge density ρ as:

$$\nabla \cdot E = \rho(x, y, z)/\varepsilon \tag{9.2}$$

where the ∇ operator stands for *divergence*, the sum of the spatial derivatives for all three dimensions:

$$\nabla \cdot E = \frac{\partial E_x}{\partial x} + \frac{\partial E_y}{\partial y} + \frac{\partial E_z}{\partial z} \tag{9.3}$$

For example, if at location (x, y, z) the electric field increases along all three axes, the divergence is positive, since all of the derivatives are positive. The *current density* $\rho(x, y, z)$ is a scalar function that depends on location (x, y, z) in space and, in contrast with the electric field, does not have a direction associated with it. The *permittivity* ε is a constant that describes the electrical property of the material under consideration. Since we want to relate the charge that crosses the cell membrane to the extracellular voltage, we need to know how the electric field relates to the electric potential we measure. The electric field is defined as the spatial derivative of the electric potential φ with a minus sign:

$$E = - \begin{pmatrix} \frac{\partial \varphi}{\partial x} \\ \frac{\partial \varphi}{\partial y} \\ \frac{\partial \varphi}{\partial z} \end{pmatrix} \tag{9.4}$$

Therefore the electric potential (our quantity of interest) relates to charge (which we will link to the ionic currents of interest) as:

$$\frac{\partial^2 \varphi}{\partial x^2} + \frac{\partial^2 \varphi}{\partial y^2} + \frac{\partial^2 \varphi}{\partial z^2} = -\rho/\varepsilon \tag{9.5}$$

For a *point source* (charge focused in an infinitesimally small chunk of space), these equations can be solved for the electric potential φ as a function of current I and distance r (derivation not shown):

$$\varphi = \frac{I}{4\pi r \sigma} \tag{9.6}$$

These equations provide us with a framework where we can plug in the ionic currents as calculated from a detailed multicompartment model (see toolbox: Modeling Neurons), then determine the electric potential as a function of the location r (distance of the electrode from the point source) and conductivity of the extracellular space σ. This function will directly provide the EAP waveform for any electrode location by simply summing the potentials that result from all membrane currents in the model (Fig. 9.3).

Such modeling work provides us with several important results [2]. First, the EAP waveform closely matches the total sum of all membrane currents (which consist of a capacitive current and many other types of ionic currents). Of course, the sum of all ion currents in a given compartment does not really represent point sources, but given the relatively large distance from the tip of the electrode to the cell, such an approximation turns out to be well justified. Second, at the onset of the action potential, the sodium current dominates the capacitive current near the soma, leading to the typical negative initial phase of the EAP waveform. If the electrode is closest to the neuron at a more distal site (away from the soma), the capacitive current (generated from passively charging the cell membrane because of voltage changes in the soma) could potentially dominate the recorded signal. However, the overall current flow is low enough that such waveforms (dominated by the capacitive current) are still expected to be below the detection limit of extracellular unit recordings. Therefore EAPs usually exhibit prototypical negative–positive biphasic waveforms, because only the larger signals near the soma, where the ion currents dominate, are detectable. Third, if the electrode is close to the

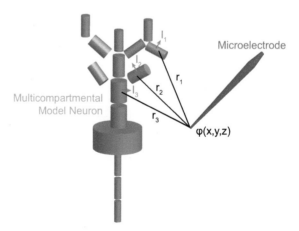

FIGURE 9.3 Multicompartmental neural model reduces the complex neuronal morphology to a set of connected cylinders. Each cylinder is modeled as a point source of current that is the sum of the capacitive and ionic currents. The electric potential φ at the tip of the extracellular microelectrode is the sum of the electric potentials caused by the individual current sources. Three example compartments are labeled here for illustration purposes.

soma and the EAP waveform amplitude is high, the overall EAP signal is essentially dominated by the somatic (and proximal dendritic) currents. In contrast, if the electrode is farther away (lower EAP waveform amplitude), the signals recorded are effectively the sum of backpropagating action potentials (see toolbox: Neurons) along the dendritic arbor that combine to form a broader waveform. Fourth, in addition to distance between the electrode and the cell, the size of the cell body strongly correlates with the amplitude of the EAPs. Together, such detailed biophysical modeling provides strong guidance for the biological interpretation of EAP waveforms.

RECORDING STRATEGIES

Off-the-shelf electrophysiology data acquisition systems allow for the relatively straightforward recording of MUA in animals. Here, we will discuss the main electric properties and behavior of electrodes used to measure MUA [3]. We will dissect the various factors that contribute to the measurement of extracellular voltage, which remain a staple in network neuroscience today.

Overall, the electrical properties of the electrode can be captured by a frequency-dependent *impedance* measure. Impedance allows us to characterize not only resistance, but also capacitance (see toolbox: Electrical Circuits). The impedance of the electrode Z_{el} is in series with the *input resistance* of the amplifier R_{in}. The input resistance is often referred to as *input impedance* because the amplifier technically also has an input capacitance, which we do not discuss here since it does not significantly affect signal collection. Overall, the signal measured, $V_{measured}$, is equal to the electric potential difference present in the tissue (V_{real}), as long as the input impedance of the amplifier is infinite. Otherwise, the recorded signal will be attenuated as follows (Fig. 9.4):

$$V_{measured} = V_{real} \frac{R_{in}}{(R_{in} + Z_{el})} \tag{9.7}$$

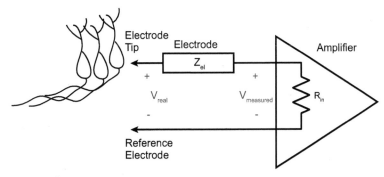

FIGURE 9.4 Electric circuit equivalent of extracellular recording. The amplifier measures the voltage between the electrode tip (positioned close to the neurons of interest) and the reference electrode. The electrode impedance does not matter in this simplified circuit as long as the input resistance of the amplifier is infinite. However, in a nonideal case, the measured voltage is less than the real voltage of interest.

9. UNIT ACTIVITY

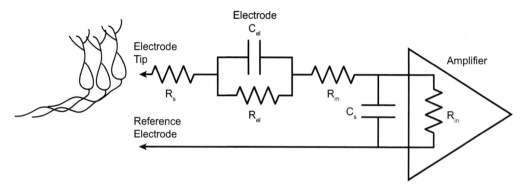

FIGURE 9.5   Electric circuit equivalent of extracellular recording electrode and amplifier.

Ideally, we want R_{in} to be as large as possible and Z_{el} to be as small as possible. We will now discuss the individual components that together give rise to the impedance Z_{el} of a metal electrode (Fig. 9.5).

The electric circuit model [3] includes the extracellular space, the electrochemical interface between the electrode and the extracellular space, the electrode itself (insulated metal shank with exposed tip), and the input resistance R_{in} of the amplifier used to amplify the signal. First, the resistance of the extracellular fluid, R_s, is the *spreading resistance* of the extracellular space, which for convenience we model as a saline bath. For an electrode with a spherical tip, the impedance is determined by the resistance ρ of saline (approximately 70 Ω cm) and the diameter d of the tip:

$$R_s = \frac{\rho}{2\pi d} \tag{9.8}$$

For a typical 1-μm tip, the spreading resistance is small compared to the other resistance values in the electric circuit equivalent. Second, the interface between the electrode and the saline bath is modeled with a capacitor C_{el} and a resistor R_{el}. Without getting sidetracked by the details of electrochemistry, it suffices to know that both partially soluble and inert metals interact with electrolytes. These reactions lead to the exchange of electrons between the electrode and the electrolyte. This process is modeled with an "electrolytic capacitor" C_{el} and a resistor R_{el}. While the electrochemistry is poorly captured by this circuit equivalent, it is important to understand that the main impedance of the electrode is derived from this interface between tissue and the electrode. Empirically, the interface impedance scales by a factor of $1/\sqrt{\omega}$, with ω being the frequency of the measured signal. As a result, for any electrode impedance measurement, the frequency at which the measurement was performed should be specified. Typically, a current with a 1-kHz sine-wave current waveform is passed through the electrode, and the resulting voltage across the electrode is measured to determine the impedance. It is incorrect to attempt to measure the resistance of an electrode with an ohmmeter, as it will find an incredibly high, meaningless value. Similarly, very slow fluctuations in the extracellular signals cannot be detected. The resistance of the metal of the electrode itself, R_m, is computed as:

$$R_m = \frac{4\rho L}{\pi d^2} \tag{9.9}$$

In this equation, ρ denotes the specific resistivity of the electrode metal (typically at least 10^{-5} Ω cm), L represents the length, and d represents the diameter of the electrode. The resistance values are very small compared to the impedance caused by the interface between the electrode and the extracellular space. Therefore the choice of material for the electrode is guided by its electrochemical interactions with electrolytes and not by the resistivity of the material. Lastly, we need to consider the stray capacitance C_s between the conductor (the electrode) and the surrounding material, that is, the saline bath. Typically, C_s is very small and can be neglected. However, if the electrode must be deeply inserted into the brain to record from areas that are distant from the surface of the brain, significant stray capacitances can arise. This equivalent circuit model demonstrates the importance of reducing the overall impedance of the electrode as much as possible while also maintaining a reasonable electrode size that can be inserted into the brain with minimal damage.

SPIKE SORTING

EAPs are detected from raw recordings of extracellular voltage by initially applying a high-pass filter to the signal to remove lower-frequency LFP components. Action potentials are then extracted by applying a threshold to the data (typically -3 to -5 times the standard deviation of the band-pass-filtered signal). Next, putative EAPs are collected, and their times of occurrence (ie, time stamps) and waveforms are recorded. Such MUA can be further processed into SU by *spike sorting* (Fig. 9.6, which shows examples of well-isolated SU).

Spike sorting refers to the process of separating recorded EAPs into different clusters such that EAPs with similar waveforms constitute individual clusters. EAP waveforms are determined by distance of the neuron from the electrode, cell morphology (particularly size, as discussed earlier), and ion channel complements. The concept of spike sorting is based on the assumption that all of these properties do not change (ie, they are *stable*) throughout the duration of a recording and are at least reasonably distinct for individual neurons. Therefore the more similar two EAP waveforms are, the more likely it is that they have originated from the same unit (ie, cell). Although this process of clustering EPA waveforms seems relatively straightforward, the realities of spike sorting are far more complex and unclear. We will begin

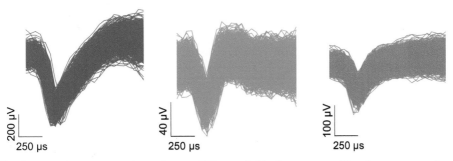

FIGURE 9.6   Waveforms for sample single units (SU) recorded in the neocortex. Waveforms are superimposed for each unit.

by discussing the fundamental challenges of spike sorting. Then, we will examine some commonly used approaches to spike sorting and how these approaches address the introduced challenges.

The first challenge is that there is no direct way to validate the output from a spike-sorting procedure (lack of *ground truth*). Do all the spikes from one cluster (unit) really correspond to a single neuron? The only way to answer this question is to perform difficult, simultaneous intra- and extracellular recordings. Even this method, if performed, would only provide validation for the one unit that corresponds to the cell targeted by the intracellular recording. Both false positives (spikes that were erroneously included in an SU) and false negatives (spikes that were erroneously excluded from an SU) represent a challenge to the quality of the sorted spike data. The second challenge is that because of the lack of a universal strategy to evaluate the performance of spike-sorting algorithms, there are many different spike-sorting approaches used with little cross-validation. The third challenge is that the waveform from a given neuron is not always stable over time. If a neuron fires a *burst* (ie, a rapid sequence) of action potentials, the waveforms of the individual spikes that form the burst are different in their time course. It is likely that spikes from a bursting neuron are assigned to multiple SU clusters. Also the level of underlying depolarization just before spike initiation alters the spike waveform because of the different levels of available deactivated sodium channels (see chapter: Dynamics of the Action Potential). Additionally, the distance between the neuron and the electrode tip can change during an experiment, since pulsations introduced by breathing and the heartbeat cause the brain to move.

Spike sorting is a two-step process. First, features of the spike waveform that can be used to distinguish the individual EAPs must be identified. Second, those features are used to cluster (ie, group) the EAPs into units. The most straightforward (but not the best) approach would be to simply sort spikes via their peak amplitude, disregarding the actual waveform and other related features. Alternatively, algorithms can be used to find complex features that capture more of the overall variance of the waveforms and therefore provide a better starting point for subsequent clustering. Principal component analysis (PCA) is one frequently used algorithm. Conceptually, PCA identifies the features that drive the overall variance in the data set; these features are often helpful for identifying clusters of similar waveforms. Whatever process is used to identify features, they ideally exhibit a multimodal distribution that can then be separated into individual clusters (Fig. 9.7). In reality, building spike clusters will always lead to false positives and false negatives, which impair the overall performance of the spike sorting (Fig. 9.8). Often, an actual human being performs the separation into clusters rather than an automated clustering algorithm. Despite the issues that arise with manual separation into clusters, this still represents one of the most prevalent methods of spike sorting.

The primary disadvantage of automated clustering algorithms is that they commonly assume a normal (ie, Gaussian) distribution of features for individual units. Simultaneous intra- and extracellular recordings have demonstrated that this assumption does not hold [4].

A strategy crucial to improving the isolation quality of SU is to use electrodes that have multiple contact sites that are close enough to each other to detect activity from the same neuron (Fig. 9.9). This approach provides more information about every spike. For example, two units may be equally distant from one recording electrode and therefore generate a very

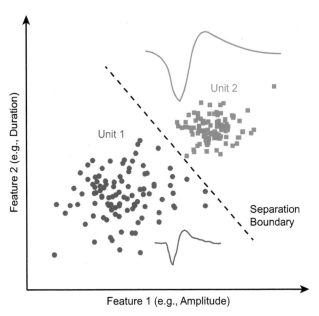

FIGURE 9.7 Ideal case for spike sorting, where the waveforms of two units nicely separate in a two-dimensional feature space. Every symbol corresponds to a single action potential. In this illustration, the separation boundary was determined by hand to cluster the waveforms into Unit 1 (*blue*) and Unit 2 (*red*).

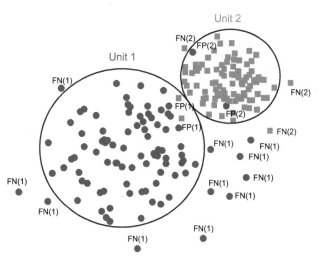

FIGURE 9.8 Example of a less-than-ideal spike-sorting result, where both units have false positives (FP) and false negatives (FN). Individual spikes correspond to *blue circles* and *red squares* in an arbitrary feature space.

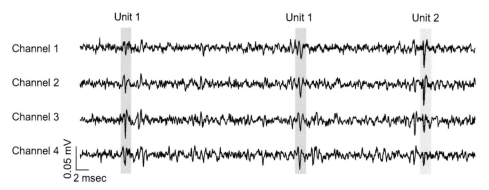

FIGURE 9.9 Simultaneous recording of extracellular spiking activity on four neighboring channels of a depth electrode in the neocortex. Spacing of electrode sites was 50 μm. Spikes from two units are highlighted; spike waveform is detectable on multiple channels.

similar signal on that electrode. But if there is a second electrode at a different location, one of the two neurons will be closer to it than the other one. Therefore considering the EAP detected by the second electrode will enable improved discrimination between EAPs from the two neurons. This strategy is typically used in the hippocampus of the rodent, where high cell density makes spike sorting particularly difficult since so many similar waveforms are picked up by each electrode. The most frequently used electrode configurations are *stereotrodes* (two electrodes closely spaced) or *tetrodes* (four electrodes closely spaced).

To a certain extent, the waveform of the EAP can be used to make an inference about the type of cell from which the EAP was recorded. In many recordings, the duration of the EAP (or of its negative-going initial phase) clusters into two distinct groups. The units with shorter EAPs likely correspond to fast-spiking neurons with short intracellular action potentials. Although there are several types of neurons that generate fast (ie, short) action potentials, the most prevalent are fast-spiking inhibitory interneurons (chapter: Microcircuits of the Neocortex). Thus such spike waveform metrics can be used to distinguish between excitatory (more correctly: regular-spiking) and inhibitory (more correctly: fast-spiking) cells.

Despite these limitations, spike sorting provides a way to greatly increase the resolution of extracellular unit activity recordings. Despite the caveats discussed, spike sorting remains an important data analysis procedure in network neuroscience.

ANALYZING AND VISUALIZING UNIT ACTIVITY

In this section, we will begin with the output of any spike-sorting procedure, the time stamps of when individual neurons fired action potentials, and then review the basic visualization and analysis strategies applied to such "spiking data." The analyses used for spiking data greatly differ from those used for continuous data such as LFP and EEG recordings. Spiking data is often conceptualized as the result of so-called *point processes*, which are random processes for which any realization is set of points in time (ie, spike times). The most basic representation is the so-called raster plot, where a small tick mark is drawn at all spike times (Fig. 9.10, top). These plots are typically used to visualize data that were

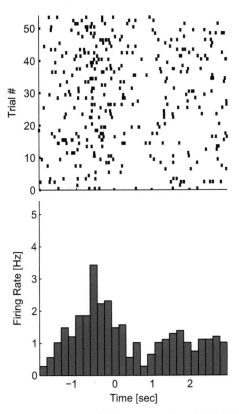

FIGURE 9.10 Example of single unit (SU) recorded from more than 50 trials during a behavioral task. Time-locked modulation of firing rate is visible both in the raster plot (*top*) and peristimulus time histogram (PSTH) (*bottom*). Time zero denotes the onset the behavioral response of the animal to a visual stimulus. This SU increased its firing rate before the behavioral response and subsequently suppressed its activity.

collected for multiple trials, that is, repeats of the same stimulus or behavioral task. Each row in a raster plot corresponds to an individual trial. This approach enables visual appreciation of individual neuron activity levels, as well as how activity fluctuated as a function of time within the trials. Raster plots are often reduced by averaging across trials. To do this, the frequency of spike occurrence (*firing rate*, measured in Hz) is determined for small, consecutive windows of time (called *bins*). The firing rate is modulated in time by exogenous (eg, sensory input) and endogenous (eg, intrinsic network oscillations) factors. The binned firing rate is determined and plotted as a bar graph, in which each bar corresponds to the firing rate for a short, nonoverlapping window in time. These plots are referred to as *peristimulus time histograms* (PSTHs) or sometimes *perievent time histograms*. Such analysis and visualization (Fig. 9.10, bottom) is built on the assumption that neurons employ *rate codes*, where the firing rate encodes the relevant information. A classic example of a cell that exhibits a rate code is a cell in the visual cortex with a firing rate proportionate to the brightness of the visual stimulus. In contrast, the precise timing of individual spikes may also carry relevant information (*spike-time code*), and such information is lost by constructing the PSTHs. It remains unclear when, why, and to what extent neurons employ rate codes or spike-time codes.

In agreement with the rate-code model, spike trains often appear random in terms of the occurrence of individual spikes. The simplest random structure defines the *Poisson spike train*, for which the following assumptions are true: (1) all spikes are generated randomly, (2) every spike occurs independently of every other spike, and (3) spikes have a uniform probability of occurrence over time. For such Poisson spike trains, the distribution of *interspike intervals* (ISIs, time elapsed between consecutive spikes) follows an exponential distribution $p(t)$:

$$p(t) = \frac{1}{\mu} e^{-\frac{t}{\mu}} \tag{9.10}$$

where μ is the average firing rate and t is the ISI. In general, the variability of a spike train is defined as the coefficient of variation C_V of the ISIs, which is defined as the ratio of the standard deviation σ of the ISIs and the mean μ:

$$C_V = \frac{\sigma}{\mu} \tag{9.11}$$

Alternatively, the variability of spiking response can be determined by counting the number of spikes per trial. The *Fano factor F* describes variability in the spike count and is defined by the ratio of the variance V to the mean of the spike count N across trials:

$$F = \frac{V}{N} \tag{9.12}$$

In the case of a Poisson process giving rise to the spike train, the Fano factor F has a value of one.

SUMMARY AND OUTLOOK

In this chapter, we have reviewed the principles of unit recordings from the viewpoints of data acquisition and signal processing. We discussed how ionic currents give rise to brief deflections in the extracellular electric fields during action potentials. Spike sorting is a commonly used data processing strategy that—perhaps surprisingly—remained a mostly manual process with no common agreement on how to best perform it. Despite this limitation, the recording of extracellular action potentials has been around for decades and still represents the staple electrophysiology technique for in vivo studies in the intact animal. The number of recording electrodes used in a given experiment has been steadily increasing, so the development of more advanced algorithms for spike sorting and subsequent analysis strategies is a priority area in network neuroscience.

References

[1] Gold C, Henze DA, Koch C, Buzsáki G. On the origin of the extracellular action potential waveform: a modeling study. J Neurophysiol May 2006;95(5):3113–28 [Epub 2006 Feb 8].
[2] Henze DA, et al. Intracellular features predicted by extracellular recordings in the hippocampus in vivo. J Neurophysiol 2000;84(1):390–400.
[3] Robinson DA. The electrical properties of metal microelectrodes. Proc IEEE 1968;56(6):1065–71.
[4] Fee MS, Mitra PP, Kleinfeld D. Automatic sorting of multiple unit neuronal signals in the presence of anisotropic and non-Gaussian variability. J Neurosci Methods 1996;69(2):175–88.

10

LFP and EEG

One of the greatest discoveries in neuroscience was that electric activity of the human brain can be measured noninvasively with electrodes attached to the scalp. The *electroencephalogram* (EEG) has become an invaluable tool in both research laboratories and clinical practices. Similarly, invasive extracellular recordings of the *local field potential* (LFP) are a routine technique to measure network dynamics in animal experiments. While both the LFP and the EEG share many features, the main difference is that the LFP refers to signals from an invasive recording of extracellular voltage that is more local in origin compared to EEG signals. In contrast to extracellular recordings of action potential firing (see chapter: Unit Activity), both LFP and EEG signals exhibit quite limited spatial resolution. In this chapter, we will focus on how these signals are generated, and on the most commonly used recording configurations and data analysis strategies. Understanding these mesoscale (LFP, reflecting activity from at most a few hundred micrometers from the electrode tip) and macroscale (EEG, reflecting activity from up to tens of square centimeters of cortex) signals is a prerequisite for understanding the remainder of the book.

ELECTRIC FIELDS CAUSED BY NETWORK ACTIVITY

Extracellular electric fields caused by neuronal activity can be observed in all neuronal structures [1]. Transmembrane ion currents (both synaptic and intrinsic currents) and capacitive currents generate electric fields that are recorded as EEG and LFP signals (toolbox: Physics of Electric Fields). An influx of cations (eg, sodium and calcium ions) into the cells (ie, inward current) creates a *current sink* in the extracellular space. Current sinks locally reduce the density of positive charge in the extracellular space. At a larger spatial scale, however, the amount of overall charge always remains the same, that is, no charge can magically appear or disappear in a closed system (here the macroscopic tissue volume). This principle is called *electroneutrality*. Because of electroneutrality, a current sink is matched by a current that leaves the cell (outward current, typically at a different location on the cell) and thereby creates a *current source*. The pair of current source and sink forms a *dipole*, which is defined as a pair of charges separated by a distance. The more spatially separated the source and sink are, the larger the contribution of the corresponding dipole to the LFP or EEG signal. This is typically the case for larger cells, such as neocortical layer V pyramidal cells, where the distance between the soma and the apical dendrites is significant. A large number of such current sources need to sum to generate electric fields that are detectable by EEG or LFP. This makes the parallel alignment of neurons in the cortex ideal for performing these measurements because of the spatial summation of signals from individual cells. More tightly packed cells cause higher LFP amplitudes because of the relatively lower conductivity of the extracellular space (see toolbox: Physics of Electric Fields). For example, cell density in the mouse cortex is higher than in the rat cortex. Accordingly, the LFP measured in the mouse will exhibit higher amplitude than the one measured in the rat cortex. Theoretically, a broad range of electrical processes could contribute to the LFP and EEG signals. But in reality, synaptic currents are the most prominent contributors to the LFP and EEG signals [2]. Synaptic currents last longer than action potentials, and therefore synaptic currents from multiple neurons are more likely than EAPs to overlap in time and thus sum. Therefore synaptic currents add up to form the meso- or macroscale signals. In the theoretical case of completely asynchronous activity (no temporal organization of activity across neurons), the temporal summation would average out the individual events and no LFP or EEG would be generated. In other words, electric fields from the activity of individual neurons only add up to a meso- or macroscopic signal if the neurons are active in synchrony, exhibiting a shared temporal patterning of their activity. Therefore neither EEG and LFP signals measure overall activity levels but rather measure the temporal structure of the overall network activity. As a consequence, the scientific and clinical questions that can be answered with such meso- and macroscopic electrophysiological signals and the associated data analysis strategies focus on dissecting the temporal structure of neuronal network activity.

ELECTROENCEPHALOGRAM

The EEG is a noninvasive measure of macroscopic brain activity recorded by electrodes attached to the scalp [1]. The measured scalp potentials are caused by current sinks and sources in the brain, other physiological current sources that are not part of the brain (such as muscle activity), and environmental electrical noise. The collection and analysis of EEG data is optimized to maximize the contribution of current sources in the brain to the measured signal. Simultaneously, the aim is to minimize the signal contributors (eg, electrical activity in muscle tissue) that are unrelated to brain activity. EEG recordings are *differential recordings*, where the difference between voltages at two points is measured. Typically, the pair of electrodes used for a single-channel EEG recording is referred to as a *recording* and a *reference electrode*. All EEG recordings follow this *bipolar* structure, which selectively eliminates signal components that are identical at the locations of the recording and the reference electrode (rejection of the so-called *common-mode signal*). More formally, we can assume that the voltages V_1 and V_2 measured at two points on the head are both a sum of a voltage signal specific to that location (U_1 and U_2, respectively) and a common-mode signal U_{cmd}, which is identical at the two locations:

$$V_1 = U_1 + U_{cmd}$$

$$V_2 = U_2 + U_{cmd} \tag{10.1}$$

In the most simplified case, voltages U_1 and U_2 are caused by current sources in the brain (ie, the neuronal signal of interest), and U_{cmd} is an electrical potential caused by electrical noise in the room where the recording is performed. To simplify matters, we can assume that this exogenous noise is identical at different locations on the scalp, since the distance between electrodes is relatively small compared to the distance between the electrodes and the noise source. Differential amplification scales the difference between the two measured signals with factor (*gain*) A_V:

$$A_V(V_2 - V_1) = A_V(U_2 + U_{cmd} - U_1 - U_{cmd}) = A_V(U_2 - U_1) \qquad (10.2)$$

thereby eliminating the common mode component. Interpretation of the resulting signal is not straightforward, because it is the difference between the neuronal activities measured from two different electrode locations. The choice of where to position the reference electrode can address this concern to some extent. The reference electrode provides the second input to the differential amplifier that measures the EEG signal as the difference between the electrode positioned over the area of interest and the reference electrode at a distant location. Ideally, the distant location is not associated with an electric field caused by neuronal activity. The goal is to position the reference electrode so that it picks up the same overall electrical potential and noise as the recording electrode. If the reference electrode is too close to the recording electrode, it picks up the neuronal signal of interest. This signal would then be subtracted and cause the resulting recording to be an underestimation of the real neuronal activity underneath the recording electrode. Typical choices for the position of the reference electrode are therefore on the head but relatively distant from the brain, such as earlobes, nose, and mastoids (referring to a location over the mastoid bone, which is situated behind the earlobe). Nevertheless, finding a truly "quiet" reference on the head is impossible for most types of activity patterns measured in EEG. Alternatively, *bipolar recordings* can be performed. Bipolar recordings are a misnomer, because all EEG recordings are bipolar by nature (see earlier). Rather, bipolar recordings in the context of reference electrode location denote recordings where the reference electrode is very close to the recording electrode such that a significant amount of neuronal signals are suppressed by the differential amplification. However, the advantage is that the bipolar signal is an actual local signal; therefore this strategy is commonly used in the clinic to localize pathological signals.

Ideally, we can determine the true scalp potentials (here denoted as U_1 and U_2) without contamination by signals from the reference electrode (issue in unipolar recordings) and without signal loss (issue in bipolar recordings). The signal from the reference electrode (V_R) can be removed by rereferencing. In this process, signals from two electrodes V_1 and V_2 are subtracted:

$$V_1 - V_2 = (U_1 - V_R) - (U_2 - V_R) = U_1 - U_2 \qquad (10.3)$$

An alternative strategy for rereferencing is to use the *average reference*. This approach is based on averaging the measured scalp potentials and subtracting this average potential from individual scalp potentials. In case of a large number of electrodes (eg, 128 electrodes), the averaged scalp potential V_{ref}^{avg} is approximately equal and of opposite sign to the potential of the reference electrode since the sum of all N scalp potentials V_i is approximately zero because of the sphere-like nature of the head (derivation not shown):

$$V_{ref}^{avg} = \frac{1}{N} \sum_{i=1}^{N} (V_i - V_R) = \frac{1}{N} \sum_{i=1}^{N} V_i - \frac{1}{N} \sum_{i=1}^{N} V_R = \frac{1}{N} \sum_{i=1}^{N} V_i - V_R \approx 0 - V_R \qquad (10.4)$$

Therefore subtraction of the average scalp potential approximately determines the reference-independent potentials:

$$V_1 - V_{\text{ref}}^{\text{avg}} = (U_1 - V_R) - (-V_R) = U_1 \tag{10.5}$$

In addition to the recording and reference electrodes, there is a ground electrode that connects to the internal ground of the amplifier, which is required to keep the amplifier hardware at an appropriate electrical potential relative to the signals of interest. The ground electrode prevents contamination by line noise that will inevitably occur when the participant and the device are not at the same relative electrical level. Of practical importance, if two amplifiers are connected to a participant at the same time, it is vital to have one shared ground. Otherwise, if the two pieces of equipment are at different voltages relative to ground (eg, earth ground has failed on one of them), the potential difference between the two machines can cause a dangerous amount of current to pass through the participant.

EEG electrodes are made of different materials with different properties, and there are several different strategies to ensure good electrical contact between the skin and the electrodes. Silver-chloride-coated electrodes are used for recordings of very slow signals such as shifts in DC potentials. In contrast, gold electrodes exhibit a significant capacitance (unlike the silver-chloride electrodes); therefore DC potentials cannot be measured (see toolbox: Electrical Circuits). Gold electrodes cannot pick up slow electrical signals, but they also do not exhibit the problem of slow drift associated with changes in the electrode interface with the scalp. Electrodes are applied to the scalp after skin preparation by using conductive gel or paste to ensure good contact with the skin. The electrode impedance serves as a measure of the quality of the electrical contact and under normal circumstances impedances with values below 50 kΩ are deemed acceptable. Electrode impedance is measured by application of a weak electrical current through the electrodes. The higher the input impedance of the amplifier, the less critical a low impedance of the electrode is (the same principle as for extracellular unit recordings, see chapter: Unit Activity). Typically, the skin is simply cleaned of oily residues. In the past, the skin was abraded to remove the epidermis (the outermost layer of the skin with highest electrical resistance). Since modern EEG amplifiers have sufficiently high input impedance, skin abrasion is no longer necessary.

Positioning of individual recording electrodes is performed using standard coordinates that are determined relative to three skull landmarks: (1) *nasion*: the deepest point between nose and forehead, (2) *inion*: the lowest rear point of the skull above where the neck starts, and (3) *preauricular points*: the intersection of the cheek bone in front of the ear and the *tragus*, a small backward-pointing structure in front of the ear canal. Placement of the electrodes is then determined by equidistant meridians between these reference points, expressed as percentages of the full distance between the skull landmarks. The most commonly used electrode placement strategy based on these landmarks is the 10−20 *system* (Fig. 10.1). The numbers 10 and 20 refer to the spacing of the electrodes in percent of the measured head size.

The individual EEG electrode locations are denoted by a combined letter and number. The letter indicates the corresponding cortical lobe (*F*: frontal, *T*: temporal, *P*: parietal, *O*: occipital) and the letter *C* is used to denote the center (50% of both the nasion−inion and the intraaural distances). The number indicates the relative position; even and odd numbers correspond to the right and left hemisphere, respectively. For example, electrode location O2 refers to

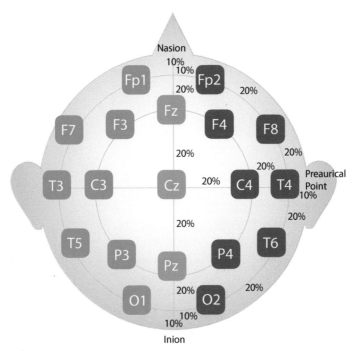

FIGURE 10.1 The standard 10–20 system designates electrode locations based on head measurements. The letters denote which lobe the electrode is placed above. The numbers identify individual electrode locations. Odd numbers are on the left, even numbers on the right.

location over the right occipital lobe. Electrodes along the midline from nasion to inion are designated with a second letter (z) instead of a number. Research EEG systems often have larger electrode counts, with up to hundreds of electrodes typically organized in a cap or net (Fig. 10.2). This helps to reduce the time of electrode placement since not all electrode locations need to be individually determined.

Next, we will review how electrical signals from the EEG electrodes are measured. The EEG signal is first amplified; the typical gain of EEG amplifiers is at least 1000, meaning that the signal at the output is equal to the input signal multiplied by a factor of 1000. As we will see, amplification is important for *digitizing* the signal. *Analog-to-digital converters* read analog signals such as the amplified EEG signal and digitize the signal for computer storage. These devices have a resolution defined by the number of bits used to encode the analog input signal. If this resolution is low, the digitized signal will exhibit visible steps in amplitude. A usual resolution is 16 bits; in this case the system can represent 2^{16} (65,536) different amplitude values. For an operating range of ±100 mV, the resolution of the system is 200 mV/65,536 = 3.1 μV. Also computers cannot store truly continuous signals but instead sample the signal at periodic intervals (sampling rate, see toolbox: Time and Frequency). Theoretically, the sampling frequency needs to be twice the highest frequency present in a signal. In practice, signals are collected with sampling rates ~10 times as high as the highest biological frequency of interest, making this a nonissue.

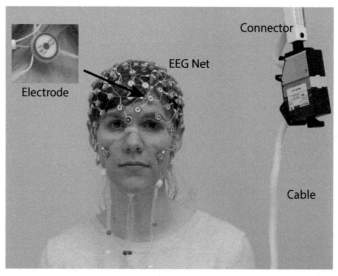

FIGURE 10.2   High-density electroencephalogram (EEG) net. *Inset:* Individual electrode (ie, sensor). The cable connects the EEG sensors to the EEG amplifier.

EEG recordings are prone to electrical artifacts that contaminate the recorded signal, some of which are of physiological but not neurophysiological origin (Fig. 10.3). Muscle artifacts (*myogenic potentials*) are a very common EEG artifact, often observed in frontal and temporal muscles (they are caused by activity in the frontalis and temporalis muscles). Frontal

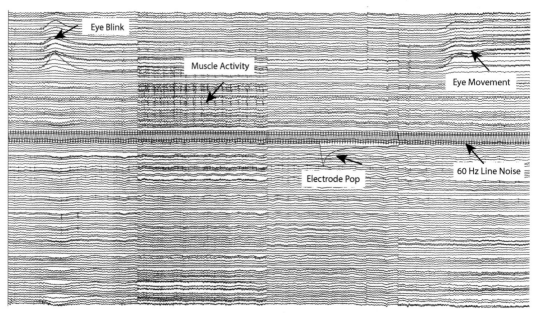

FIGURE 10.3   Examples of common electroencephalogram (EEG) artifacts in a 128-channel EEG recording (eye blink, muscle activity, electrode pop, eye movement, 60-Hz line noise). Four different 1-s segments are shown.

electrodes are often contaminated by eye movements and blinking. Eye movements cause electrical transients since the eye acts as a dipole; the cornea is positively charged and the retina is negatively charged. Activity of muscles in and around the orbit can cause additional artifacts. EEG channels can be contaminated by electrocardiogram (EKG) artifacts that can easily be identified by simultaneous recording of the EKG from the chest. In addition to these physiological artifacts, recordings can be contaminated by nonphysiological noise sources. Most typically, line noise (50 Hz in Europe, 60 Hz in the United States) from ambient electromagnetic radiation contaminates EEG signals. Furthermore, any rapid change in the electrical interface between the recording electrode and the scalp (eg, an abrupt change in impedance) also causes a recording artifact that is limited to the single, affected electrode ("electrode pop").

LOCAL FIELD POTENTIAL

Compared to an EEG, recording of the LFP is less susceptible to artifacts, since the electrode is placed inside the brain. As a result, however, LFP recordings are mostly limited to animal studies (except for the invasive electrophysiological recordings in human patients discussed later). A range of different materials can be used for LFP electrodes (such as tungsten, platinum, and iridium). The material properties and geometry of the exposed electrode tip are typically optimized for picking up extracellular action potentials, since the LFP is

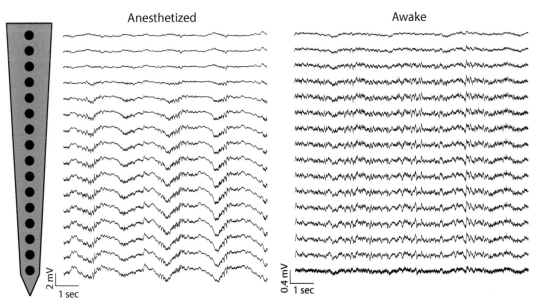

FIGURE 10.4 Recording of the local field potential (LFP) in the prefrontal cortex of an animal with a 16-channel silicon depth probe. *Left*: Low-frequency, high-amplitude fluctuations dominate the LFP during anesthesia. *Right*: Recordings from the same location in the awake animal. Changes in signal as a function of depth (top to bottom) are less obvious. The LFP signal exhibits lower amplitude and higher-frequency content.

comparably easy to record and presents fewer requirements in terms of electrode properties. Electrode impedance is mostly determined by the exposed area at the tip of the electrode. LFP recordings are not very sensitive to the size (and thus the impedance) of the electrode (in contrast to single- and multiunit recordings). The other, more important consideration regarding choice of electrode geometry is the amount of tissue damage caused, which is primarily a function of size of the electrode. For the study of LFP signals, depth probes with regularly spaced contact sites are particularly helpful (Fig. 10.4), since they enable the determination of activity as a function of depth with current source analysis (see later). The recording hardware requirements closely mirror those for EEG. The reference electrode is typically positioned nearby the LFP electrodes, but it has lower electrical impedance than the other contact sites. This reduces the likelihood of action potentials being picked up by the reference electrode. However, such close proximity of the reference electrode to the recording electrodes leads to an underestimation of the global LFP signals. This is a reasonable sacrifice to make, since it avoids the recording of signals from other brain areas that are detected at a distance (referred to as volume conduction). The ground electrode is often a silver wire that is wrapped around a skull screw and inserted through a small craniotomy.

DATA ANALYSIS

The most common analysis strategy to characterize the temporal structure of LFP and EEG signals is *frequency decomposition*, which determines the presence of rhythmic structure at different frequencies (see toolbox: Time and Frequency). These frequencies of interest range from ~1 to 300 Hz. The lower cut-off frequency is often chosen to exclude slow drifts that are artifacts and not neuronal signals. The upper cut-off frequency is designed so as to prevent inclusion of spiking activity, since individual spikes have a 1–2-ms time scale (see chapter: Unit Activity). The spectra of neuronal activity are often displayed on a logarithmic scale, since the overall structure roughly follows a so-called $1/f$ distribution (f stands for frequency), meaning that there is more power in lower frequencies. The physiological interpretation of oscillations at different frequencies is discussed in the unit: Cortical Oscillations. One of the main features of network dynamics is their nonstationary nature, which means that the brain rapidly switches between different activity states as a function of internal and external demands. Therefore the spectrogram, defined as the time-dependent representation of the spectrum, is useful to delineate changes in the relative presence of different oscillations over time (an example of an episode of alpha activity is shown in Fig. 10.5, toolbox: Time and Frequency).

Such time courses are particularly important for studying network dynamics relative to specific events that occur at well-defined time points, such as the presentation of a sensory stimulus or a behavioral response. In such studies, data are typically collected from a relatively large number of trials (ie, repeats of the same stimulus). There are two fundamentally different strategies to analyze EEG and LFP signals that are structured into such a trial format: *evoked* and *induced* responses. The classical method is to average the time course

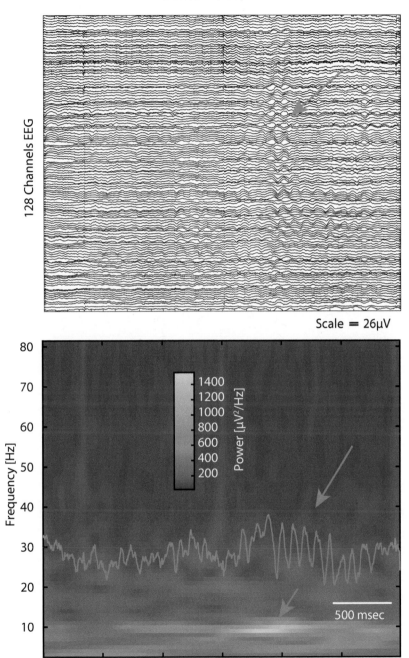

FIGURE 10.5 Electroencephalogram (EEG) recording (128 channels) showing alpha oscillations (*red arrows*). *Top*: Raw data. *Bottom*: Spectrogram with sample single channel trace drawn on top (*red*).

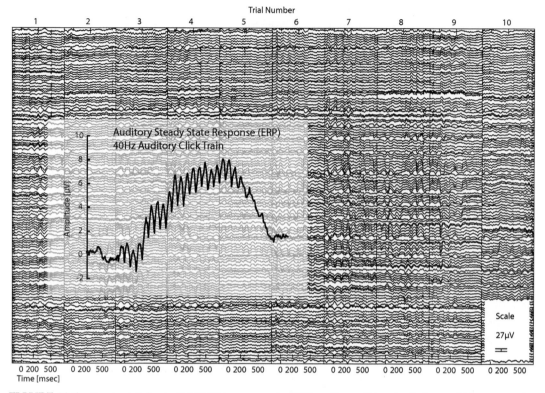

FIGURE 10.6   Event-related potential (ERP) in response to 40-Hz auditory click train. *Background*: Ten consecutive trials for 128 electroencephalogram (EEG) channels. *Front*: Single-channel ERP (auditory steady-state response, ASSR).

across trials to determine an averaged response. This strategy used to be very common, especially in EEG studies. Averaging reduces noise (which is inherently a major concern in EEG signals because of the small signal amplitudes) and isolates the network response that is time locked to the stimulation or the behavioral event used as a marker to align trials. In the case of EEG signals where the signal-to-noise ratio (SNR) is often low, such averaging is a convenient way to enhance SNR by reducing noise. Typically, these so-called *evoked potentials* (original terminology) or *event-related potentials* (ERPs, today's terminology, a broader term) exhibit a sequence of positive and negative peaks designated as "P" or "N" followed by a number. The number denotes the typical delay of occurrence (in milliseconds) after the trigger signal (Fig. 10.6 shows an ERP evoked by auditory stimulation; Fig. 10.7 shows a "cognitive" ERP evoked by an auditory oddball paradigm). The downside of this method is that by definition no trial-by-trial analysis can be performed such as correlating the network response with behavioral responses across trials. Also oscillatory network dynamics are underestimated by this approach, because oscillations are often *induced* and not *evoked*. Induced responses refer to activity patterns that are triggered by an event but that do not occur with the same timing across trials. Averaging the raw LFP or EEG signals across trials will suppress induced responses. For example, a visual stimulus can reliably turn on an oscillatory activity

Standard Tone (ERP)

Deviant Tone (ERP)

Anterior (Nasion)

1 s

Posterior (Inion)

FIGURE 10.7   Event-related potential (ERP) from auditory oddball paradigm, in which 80% of the trials were one tone (standard tone, *blue*) and 20% were another, different tone (deviant tone, *red*). The participant is asked to count the number of deviant tones. ERPs are arranged according to the position on the scalp from where they were recorded. The ERPs differ most prominently at the central/posterior location. The positive peak at about 300 ms is referred to as P300.

pattern such as a burst of gamma oscillations. But the phase of the oscillation is not synchronous across trials. Therefore at any given time point the raw signal can either assume a positive or negative value, depending on the onset phase. On average, the signals cancel each other out. To study induced oscillations, the spectrogram is first computed for each trial and then averaged across trials.

The unit Cortical Oscillations reviews the cellular mechanisms and functional roles of oscillations. In the subsequent unit Network Disorders, we will learn that changes in oscillations are a hallmark of a wide range of neurological and psychiatric illnesses. Before we conclude this chapter, we turn our attention to signal processing strategies that aim to overcome the low spatial resolution of EEG and LFP.

SOURCE LOCALIZATION

The high temporal resolution of LFP and EEG recordings comes at a price of low spatial resolution. Just because a voltage fluctuation is recorded from a given location does not mean that the source of the signal is localized to the same place, since electric fields can be picked up far from the actual source. Fortunately, by making certain assumptions about the electromagnetic properties of biological tissue, electromagnetic theory (Maxwell equations) can be used to determine the underlying current distribution. These methods require simultaneous recordings at multiple sites with known locations.

For localization of the LFP sources, the *current source density* (CSD) is determined [3]. LFP signals are typically recorded in the cortex, where the layered structure favors the genesis of such signals. This structure also simplifies the analysis, because homogeneity of properties in the transverse (x and y) plane can be assumed for a reasonably small part of a cortical sheet. Depth electrodes with equally spaced contacts are used to simultaneously record the LFP across cortical layers, by inserting them perpendicular to the cortical surface. The current density is determined as the second spatial derivative of the measured extracellular electrical potential $V(X)$ (toolbox: Physics of Electric Fields). Briefly, the first spatial derivative denotes the electric field measured in V/m (volts per meter). The electric field then relates to the charge density (the first derivative of the electric field, thus the second spatial derivative of the measured voltage). The spatial derivative is computed by determining the difference between voltages at neighboring sites $V(X)$ and $V(X + \Delta X)$ (and normalizing by the distance ΔX

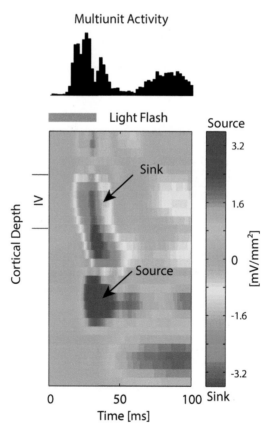

FIGURE 10.8   Multiunit activity (*top*) and current source density (*bottom*) in the primary visual cortex of a ferret in response to light flashes. The initial current sink (*red*) delineates cortical layer IV.

between the probes). The process of taking the spatial derivative is performed twice to calculate the second spatial derivative:

$$\text{CSD} = \frac{1}{\Delta X}\left(\frac{V(X+\Delta X)-V(X)}{\Delta X} - \frac{V(X)-V(X-\Delta X)}{\Delta X}\right)$$

$$= \frac{V(X+\Delta X)-2V(X)+V(X-\Delta X)}{\Delta X^2} \tag{10.6}$$

The issue with this approach is that taking the second derivative of an already noisy signal further exacerbates noise. Thus typical CSDs are computed from a large number of trials such as the repeated presentation of simple, bright visual stimuli. In this framework, electrical signaling can be tracked through cortical layers, at least for the first few synapses. Sensory input arrives first in layer IV of the cortex through afferent projections from the thalamus. Therefore the first current sink/source pair in the cortex can be used to identify and localize layer IV (see also chapter: Microcircuits of the Neocortex). Results are usually shown as color-coded CSD (Fig. 10.8) as a function of time (x-axis) and cortical depth (y-axis). Positively charged ions, such as sodium ions, flowing into cells during the rising phase of the action potential, correspond to a current sink (negative CSD values, often drawn in *red*). Positively charged ions flowing out of cells accordingly correspond to a current source (positive CSD values, often drawn in *blue*).

A conceptually similar approach can be used to localize sources that contribute to the signals measured by EEG electrodes, referred to as the *Laplacian*, which is proportional to the cortical surface potential in the case of localized sources. This procedure estimates the dipole

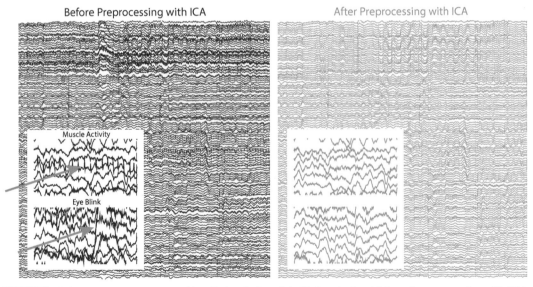

FIGURE 10.9 Electroencephalogram (EEG) data before (*left*, *blue*) and after (*right*, *red*) preprocessing with ICA. Each line represents the signal from one of 128 EEG electrodes. *Insets:* Example of EEG signals contaminated with muscle activity and eye blinks (*left*, *red arrows*) and remaining activity after removal (*right*).

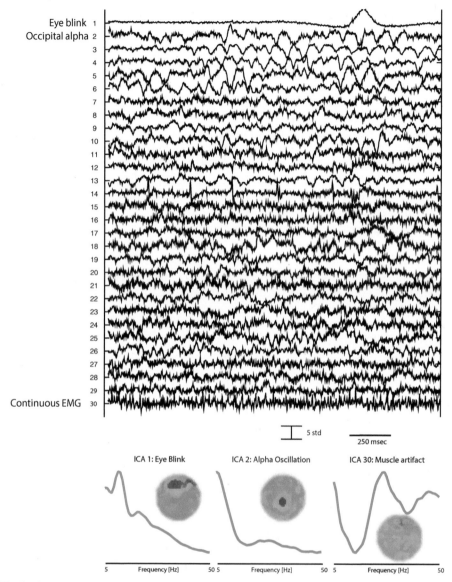

FIGURE 10.10   Independent component analysis (ICA) components. Three components are highlighted (spectra and spatial distribution shown in the following). ICA 1: Isolated eye blinks. ICA 2: Posterior midline alpha oscillations. ICA 3: Electrical activity (electromyographic signal, *EMG*) from muscles in the forehead.

location. An alternative, data-driven approach uses blind source separation strategies such as independent component analysis (ICA). In this framework, the sensor signals are considered to be the result of linear superposition (ie, sum) of statistically independent source signals. ICA is used to find the different source signals that are "mixed" together at the scalp, where they are picked up. ICA analyzes the statistical dependence of the simultaneously recorded sensory signals, which is caused by the fact that any given source signal is likely to contribute to multiple sensory signals. ICA determines the source signals in an iterative way by maximizing their statistical independence. This is also referred to as solving the *inverse problem*. ICA can also be used to "clean" the EEG signal by removing nonneurophysiological signals such as eye blinks or muscle artifacts that appear as localized sources to specific locations such as frontal electrodes for blink artifacts or temporal electrode sites for muscle artifacts (Figs. 10.9 and 10.10).

INVASIVE RECORDING IN HUMANS

Direct recordings from the human brain are performed in a small number of patients who undergo brain surgery. These neurophysiological recordings are used to localize tissue with pathological electrical signaling. In addition, electrical stimulation through the recording electrodes is used to cause transient, reversible *virtual lesions* by temporarily disrupting neuronal signaling. Stimulation locations where negative side effects are observed (eg, impaired speech) are not resected. The most frequent application of this approach is in patients with pharmacoresistant epilepsy. As a last resort, some patients are candidates for resective surgery of the brain tissue from which epileptic seizures originate. Noninvasive EEG recordings do not provide sufficient localization ability (at least not with the standard 10–20 electrode location system with only 19 electrodes) for surgical planning. Also seizure onset zones are often deep in the temporal lobe. As a result, the pathological activity is not only spatially blurred when measured at the scalp, but is also incompletely captured by standard EEG. Invasive electrophysiological monitoring is typically performed with a subdural grid (Figs. 10.11 and 10.12) or with depth electrodes.

The terminology is not uniform across research groups; commonly used terms include ECoG, intracranial electroencephalogram (iEEG), and depth or stereo EEG. Invasive recordings provide unique access to human network electrophysiology and provide a great window into brain function. Most of the electrodes used are only able to pick up a local field potential (LFP)-like signal, since the size of the contact sites is too large to pick up action potentials.

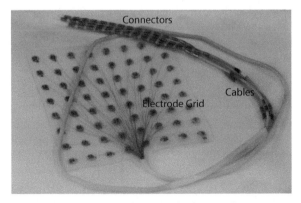

FIGURE 10.11   Subdural electrode grid with 64 electrodes for human electrocorticography (ECoG).

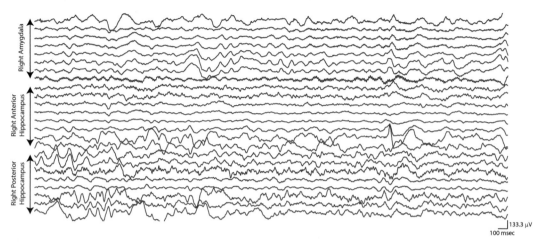

FIGURE 10.12   Sample electrocorticography (ECoG) recording from patient with depth electrodes implanted into the right amygdala, and the right anterior and posterior hippocampus.

However, there are also depth probes available that have a set of protruding microwires that can record action potentials. As with any technique, invasive electrophysiological recordings in human patients come with their own distinct drawbacks when used to perform scientific experiments. First, study design is difficult because of the heterogeneity of patients, diagnoses, and electrode placement (which is exclusively dictated by clinical needs). Second, patients are typically heavily medicated at the time they are enrolled for scientific recordings (typically a few days before surgery). Most of the medications target the excitability of the central nervous system since they are used to treat epilepsy. Despite these drawbacks, ECoG studies have significantly contributed to unraveling the relationship between network activity patterns and behavior with high spatial resolution.

MAGNETOENCEPHALOGRAPHY

Neuronal activity not only causes an electric field, but also a magnetic field that reflects the direction of the transmembrane ion currents. Magnetoencephalography (MEG) measures this magnetic field. Thus, in contrast to EEG, MEG is most sensitive to currents tangential to the skull. The magnetic fields are incredibly small (typically 10—100 fT), and measuring them is a challenge that has been overcome by the invention of so-called *superconducting quantum interference devices* (abbreviated as SQUIDs). Today's state-of-the-art MEG devices use a helmet with several hundred SQUID-based sensors to generate maps of neuronal activity with reasonable spatial resolution that is typically enhanced with source localization algorithms. In contrast to EEG, there is less spatial smearing of the source signals. Thus MEG has both an impressive temporal (submillisecond) and a relatively high spatial (a few square millimeters) resolution. For studies that depend on high-quality source localization, MEG is preferable to EEG. Nevertheless, MEG is not as widely adopted as EEG, mostly because of the expense of the MEG device and the complexity of its operation.

SUMMARY AND OUTLOOK

In this chapter, we discussed the main theoretical and practical aspects of measuring meso- and macroscopic electrical signals that reflect the activity of a large number of neurons. Tiny changes in the extracellular voltage from the activity of individual neurons sum to signals strong enough to be noninvasively recorded from the scalp in the form of the EEG. LFP measurements are the equivalent invasive measurement method, with better but not ideal spatial resolution in comparison to EEG. We learned about the different analysis strategies, and discussed the difference between evoked and induced (oscillatory) activity. Finally, we looked at more exotic approaches to record similar signals: invasive recordings in human epilepsy patients and MEG that measure the magnetic instead of the electric field caused by neural activity. Most of what we will discuss in the units Cortical Oscillations and Network Disorders is based on the methods introduced in this chapter.

References

[1] Nunez PL, Srinivasan R. Electric fields of the brain: the neurophysics of EEG. 2nd ed. Oxford, New York: Oxford University Press; 2006. xvi, p. 611.
[2] Buzsaki G, Anastassiou CA, Koch C. The origin of extracellular fields and currents—EEG, ECoG, LFP and spikes. Nat Rev Neurosci 2012;13(6):407—20.
[3] Mitzdorf U. Current source-density method and application in cat cerebral cortex: investigation of evoked potentials and EEG phenomena. Physiol Rev 1985;65(1):37—100.

Optical Measurements and Perturbations

Electrical measurements represent the most widely used strategy for the study of neuronal activity in animal models. The temporal resolution of such electrophysiological measures matches the millisecond timescale of neuronal spiking activity. But this advantage comes at the expense of information about the location and identity of the cells that contribute to the recorded network activity. Several methods have been developed to recover this information from electrophysiological recordings, but they remain imperfect. Another limitation of using electrophysiology to record action potentials of single neurons is that recording probes must be inserted into the brain, an approach that physically disturbs the tissue in the area of interest. In contrast, imaging strategies that visualize neuronal activity can to a certain extent overcome these limitations, and so can provide an important alternative strategy for recording neuronal activity. In this chapter, we will first discuss two main imaging strategies with the potential to reshape the landscape of network neuroscience studies in animal models: *calcium imaging* and *voltage imaging*. Then we will focus on *optogenetics*, a strategy for cell-type-specific modulation of neuronal activity with light-activated transmembrane proteins.

CALCIUM IMAGING

Calcium imaging is based on fluorescence microscopy. Fluorescence is a process in which a fluorophore, the molecule that exhibits fluorescence, absorbs a photon of one wavelength and emits a photon of another wavelength, which is detected as the readout signal. The absorption of a high-energy photon moves an electron to a higher energy state. In the case of fluorophores, this process can lead to *excitation*. After some time, the electron switches back to its *ground state* and loses the additional energy by releasing a lower-energy photon. This process is called *emission* (Fig. 11.1). The excitation and emission spectra (corresponding to wavelengths, ie, color) indicate to what extent a fluorophore is excited by light at a given wavelength and the resulting wavelength of emitted light, respectively. For the purpose of this explanation, we ignore the difference between absorption (ie, switching to a high-energy state) and excitation (switching to a high-energy state that causes the release of a photon). Ideally, the excitation and emission spectra do not overlap such that the light can be

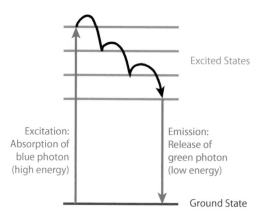

FIGURE 11.1   Mechanism of fluorescence for green fluorescent protein (GFP). Fluorophores are excited with a high-energy photon (shorter wavelength, *blue arrow*) and emit a lower-energy photon (longer wavelength, *green arrow*). Transitions between excited states release energy as heat rather than as photons (*black arrow*).

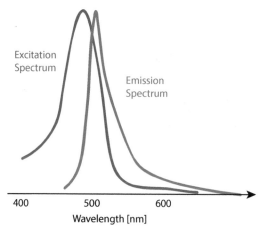

FIGURE 11.2   Excitation and emission spectra for green fluorescent protein (GFP). The two spectra overlap but are sufficiently separated to allow fluorescence microscopy. For example, a color filter used for GFP could have a 450–490-nm reflection band, which prevents the light used for excitation to reach the image acquisition system, and a 505–800-nm transmission band to capture as much fluorescence as possible.

separated by color filters. The most famous example in biology is the green fluorescent protein (GFP), which is excited by blue light and emits green light as it switches back to the ground state (Fig. 11.2).

Any change in membrane voltage is associated with transmembrane ion currents that alter intracellular and extracellular ion concentrations. The intracellular calcium concentration ($[Ca]_i$) is a particularly attractive quantity that reflects neuronal activity because of its important role as an intracellular messenger (see chapter: Synaptic Transmission). Upon depolarization, calcium ions enter the cell through voltage-gated calcium channels. In a first approximation, fluctuations in [Ca]$_i$ reflect neuronal activity. Calcium-sensitive indicators bind to intracellular calcium and, in response, change how they respond to light in terms of their fluorescence. These changes in fluorescence correlate with neuronal activity and are measured by a combination of a microscope and an image-acquisition system.

As we will see later, one way to build a *calcium indicator* (ie, a molecule that measures calcium concentration) is to modify the GFP such that in the absence of calcium binding its absorption of excitation light is minimal, but fluorescence is enabled upon binding of calcium. This strategy provides a relative measure of the intracellular calcium concentration. Calcium indicators are excited with a light source. Resulting calcium-dependent changes in fluorescence of the indicator are then measured with a microscope combined with a light-detecting device, such as a camera or a photomultiplier tube (PMT) that measures fluorescence emission. The time course of the change in fluorescence is then used to track the dynamics of neuronal activity.

Calcium Indicators

Two main types of calcium indicators can be distinguished: *small molecule indicators* and *genetically encoded calcium indicators* (GECIs). Small molecule indicators are directly applied to the surface of the brain in preparation for a single imaging session. In contrast, GECIs are delivered by the creation of transgenic animal lines or by infection of the target brain tissue with a virus that carries the GECI gene as a payload. GECIs have the advantage of enabling long-term imaging, and the development and optimization of GECIs is an active area of research. The general blueprint of calcium indicators consists of a calcium-binding protein and one or two fluorescent proteins. In the case of one fluorescent protein, calcium binding changes the fluorescence intensity. When two fluorescent proteins are attached, calcium binding alters fluorescence through Förster resonance energy transfer (FRET). In FRET, a high-energy donor chromophore transfers energy to an accepting lower-energy chromophore such that the brightness of the latter fluorescent protein increases at the expense of the former fluorescent protein.

The choice of calcium indicator depends on the research question of interest. Ideally, calcium indicators report changes in activity level without perturbing the endogenous function of neurons. However, calcium indicators are by design calcium buffers that bind to calcium and therefore alter the concentration of free calcium in the cytosol. For a given calcium concentration, low-affinity buffers are less likely to bind to calcium than high-affinity buffers. Low-affinity buffers or calcium indicators are ideal in this regard since they only minimally alter the endogenous calcium-buffering capabilities of the neurons. However, to detect calcium signals from individual action potentials, the indicator must have a high affinity for calcium to exhibit sufficient sensitivity. Such indicators can significantly alter endogenous calcium dynamics and will also saturate more easily at higher neuronal activity levels. So the tradeoff between calcium affinity and alteration of endogenous calcium dynamics needs to be managed as a function of the specific requirements of a given experiment. Today, GCaMPs [1] are the most frequently used GECIs. The GCaMP molecule includes GFP and calmodulin (calcium buffer).

Calcium Imaging Setup

The basic imaging setup for calcium imaging consists of a light source, a microscope, and a detector system. Also the excitation light needs to be separated from the emission light before it reaches the detector. This separation of the two light paths is based on the different peak wavelengths of the excitation and emission signals, and is achieved by a *dichroic mirror* that reflects the light from the light source (for excitation) and lets the light emitted from the calcium indicator pass through (Fig. 11.3). Calcium imaging can be performed both in vitro and in vivo. For in vitro applications where the imaged neurons are close to the surface of the sample, a camera or (historically) an array of photodiodes can be used to detect the fluorescence emission by the calcium indicator. Alternatively, laser-scanning microscopy, in particular *confocal* or *two-photon microscopy*, can be used to achieve higher image resolution. Confocal microscopy combines laser scanning of the sample with the selective collection of photons from the in-focus plane (eg, by a pinhole). This process is called *optical sectioning*. However, in-plane photons from deeper in the tissue can fail to reach the typically used PMT because of the light-scattering properties of tissue. As a result, application of confocal microscopy is mostly limited to in vitro approaches. Two-photon microscopy overcomes this limitation by using near-infrared photons that need to interact to excite the fluorophore. Specifically, two photons with exactly twice the wavelength of the excitation wavelength (specific to individual calcium indicators) need to be delivered to the sample in very rapid succession so that they act like one unified photon to excite the fluorophore (Fig. 11.4, [2]). The excitation is thereby limited to a very small volume where such photon density can be achieved. Such spatially limited excitation prevents excitation of the calcium indicator in other locations that are not in focus, which would contribute to blurring the acquired image. The lower energy of the individual photons that do not result in two-photon activation limits *bleaching*, an interaction between indicator and other endogenous molecules that leads to signal degradation. To achieve sufficient excitation for the two-photon effect, lasers that

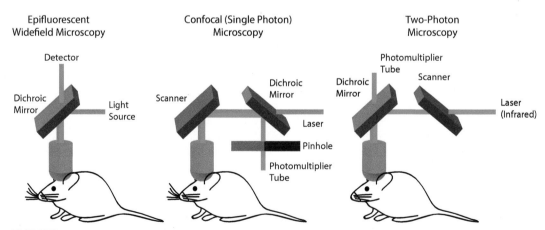

FIGURE 11.3　Fluorescence microscopy strategies. *Left*: Epifluorescent widefield microscopy. *Middle*: Confocal microscopy, which provides better spatial resolution. *Right*: Two-photon microscopy, which provides better tissue penetration.

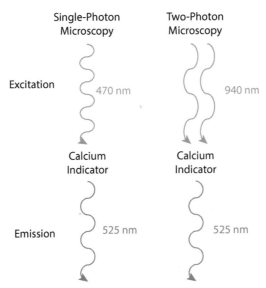

FIGURE 11.4 Single-photon and two-photon excitation of a calcium indicator. In two-photon microscopy, a pair of low-energy photons excites the calcium indicator to emit a single high-energy photon.

generate brief pulses (typically around 100 fs) are required. Because excitation occurs only in a very small volume, all detected emission photons originate from the region of interest, and no pinhole is required for optical sectioning. A further advantage of two-photon microscopy is the reduced scattering of the longer-wavelength light within the brain tissue, as opposed to the shorter-wavelength light used in single-photon microscopy strategies such as epifluorescence and confocal microscopy.

Two major developments in in vivo calcium imaging are the use of two-photon imaging in the head-fixed animal (Fig. 11.5) and the use of miniaturized epifluorescence microscopes to record calcium transients from awake, freely moving animals. Simultaneous imaging of a large number of neurons in the intact animal enables the study of spatiotemporal activity patterns and their role in behavior (Figs. 11.6 and 11.7).

Applications

Calcium imaging can be used to investigate phenomena across multiple spatial scales, from subcellular components such as dendritic spines all the way to populations of thousands of neurons. Recording population dynamics with cellular resolution is a unique strength of calcium imaging. In particular, the individual cells can be visually identified, indicating that no error-prone procedures (such as spike sorting in the case of electrophysiological unit recordings) are required to identify individual cells (see chapter: Unit Activity). Visualization of the cells also provides the opportunity to describe the spatial relationship of the individual cells to each other, and thus can help to elucidate the fine spatial structure of a circuit. The main downside of calcium imaging in comparison to electrophysiology lies in the temporal domain; today's calcium indicators have rise and decay rates of tens to

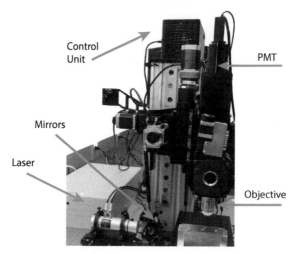

FIGURE 11.5   Typical two-photon microscope for in vivo imaging. The laser to excite the calcium indicator is deflected by a series of mirrors to reach the objective. Photomultiplier tubes (PMTs) measure the emitted photons. The control unit guides the scanning system (not visible in the photograph), which moves the laser beam to sample the image.

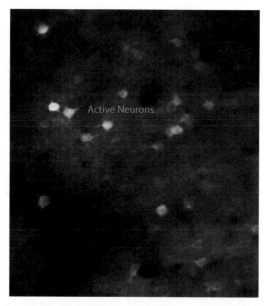

FIGURE 11.6   Snapshot of a movie collected by two-photon calcium imagining in mouse frontal cortex. Bright objects are active neurons. Pyramidal shape of the soma indicates that the cells are pyramidal cells (see chapter: Microcircuits of the Neocortex). *Image with permission from Joshua Trachtenberg.*

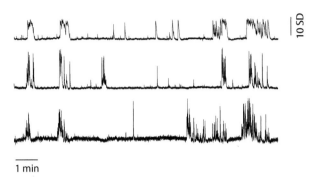

FIGURE 11.7 Sample traces of calcium indicator signals from three pyramidal cells. These signals are determined by computing the brightness over time of individual neurons in movies such as the one shown in Fig. 11.6. *Image with permission from Joshua Trachtenberg.*

hundreds of milliseconds. Moreover, the performance of algorithms to extract individual action potentials from calcium signals depends on image quality and neuronal firing patterns. In particular, high firing rates make the process of extracting the calcium transients that correspond to individual action potentials more difficult because of the nonlinear dynamics of calcium indicators. Additionally, calcium imaging studies typically focus on the more superficial layers of the cortex, although it should be noted that two-photon imaging allows acquisition of data from deeper cortical layers.

VOLTAGE-SENSITIVE DYE IMAGING

The ability to simultaneously measure the membrane voltage fluctuations of a large number of neurons in a network is highly desirable. Unlike electrophysiological or calcium-imaging measurements of neuronal activity, measurements of subthreshold fluctuations of the membrane voltage enable the study of how incoming signals are processed by neurons. Although the local field potential and the electroencephalogram (EEG) provide a signal that predominantly reflects subthreshold activity (see chapter: LFP and EEG), the spatial resolution of those techniques is extremely poor. In contrast, voltage imaging enables such measurements of fluctuations in the membrane voltage by imaging with much higher spatial resolution. *Voltage-sensitive dye imaging* (VSDI) is based on the staining of the brain with a *voltage-sensitive dye* (VSD) that attaches to the external aspect of cell membranes and undergoes changes in optical properties as a function of the membrane voltage [3]. We introduce the basic technical setup used for VSDI and review its strengths and limitations. As we will see, VSDI plays an important role in the study of spontaneous and sensory-evoked network dynamics in the neocortex.

VSDI Setup

The basic VSDI approach requires the administration of the dye, illumination of the dye with a light source, a macroscope (a large field of view contrasted to microscopes) or microscope, a fast camera, and a computer-based acquisition system. An in vivo VSDI experiment starts with opening the skull and dura to allow the administration of the VSD to the surface of the brain. There is no cell-type specificity in terms of which cells the VSD stains, so the resulting signal will stem from all types of cells. Since the dye passes through the brain by diffusion, superficial cortical layers are more heavily stained. VSDs are very fast (microsecond response times) and theoretically permit a spatial resolution in the range of micrometers. To leverage these advantages, a very fast, high-resolution camera is needed. Individual pixels of the camera reflect the blurred average of the actual VSD signal (in both space and time), and therefore VSDI does not report the activity of individual neurons but rather of a local neuronal population. Practically, spatial resolutions in the tens of micrometers have been achieved with such setups. Overall, VSDI has an outstanding combination of temporal and spatial resolution (compared to EEG, for example, which has great temporal resolution but very poor spatial resolution). The signal is proportional to the amount of stained membrane. Therefore it stems mostly from dendrites and unmyelinated axons and not somata.

The development of new VSDs is an ongoing enterprise. Overall, one of the goals is to shift the excitation toward longer wavelengths (toward red). The two main advantages are reduced overlap in light absorption by hemoglobin (leading to heartbeat and breathing artifacts) and reduced light scattering and thus better depth penetration. Adoption of VSDs has been limited because of their toxic (pharmacological) side effects and photobleaching (decreasing signal strength with continued illumination during imaging) in the past. Recent improvements have mitigated some but not all of these concerns.

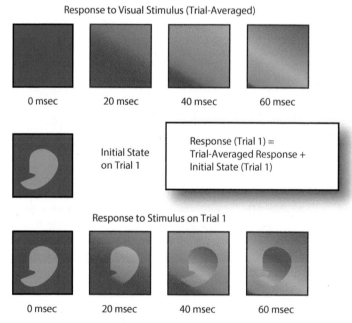

FIGURE 11.8 VSDI revealed that spatiotemporal response patterns in the visual cortex (shown here as a simplified colored square for consecutive acquisition time points) are the sum of the trial-average visual response (*top row*) and the initial state at trial onset (*middle row*). Brain state therefore explains the trial-to-trial variability of the visual response (*bottom row*). Details in Ref. [4].

Applications

VSDI is particularly suited for the study of large-scale activity patterns with high temporal resolution. For example, VSDI can be used to efficiently map orientation tuning of the visual cortex by presentation of visual stimuli with different orientations and imaging the entire surface of the visual cortex. VSDI is also instrumental in unraveling spontaneous, large-scale network dynamics in the cortex and has demonstrated that the spatial patterns of spontaneous activity before stimulus onset predict sensory response after stimulus onset (Fig. 11.8).

Another application of VSDI is the simultaneous recording (ie, imaging) of the membrane voltage in a single neuron at a large number of points in vitro with a microscope instead of a macroscope. This technique enables measurement of the membrane voltage in subcellular locations (eg, axons) that are not accessible to other methods such as patch-clamp recordings (see chapter: Membrane Voltage).

Together calcium and voltage imaging have rapidly evolved and have become staple techniques in network neuroscience. Their combination with electrophysiological measurements will allow unprecedented, simultaneous measurements of neuronal activity at multiple spatial and temporal scales.

OPTOGENETICS

In addition to using imaging to measure brain activity, light application can also be used to perturb neuronal activity. *Optogenetics* is a stimulation strategy based on the expression of light-sensitive proteins in the neuronal cell membrane that, upon illumination, alter the electric state of the neuron [4a]. That alteration can be assessed by whole-cell patch-clamp recording (Fig. 11.9) or extracellular unit recordings (Fig. 11.10). One key advantage of optogenetics is that genetic strategies can be used to achieve targeted expression and ultimately

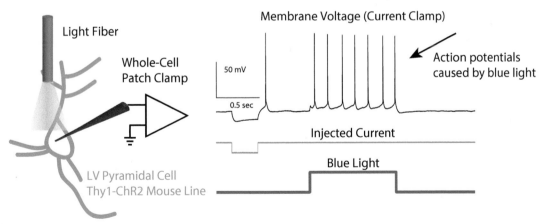

FIGURE 11.9 Optogenetic activation of a layer V pyramidal cell in an acute slice of a Thy1-ChR2 mouse. Blue light delivered via a light fiber causes a train of action potentials measured in current clamp of the whole-cell patch-clamp configuration.

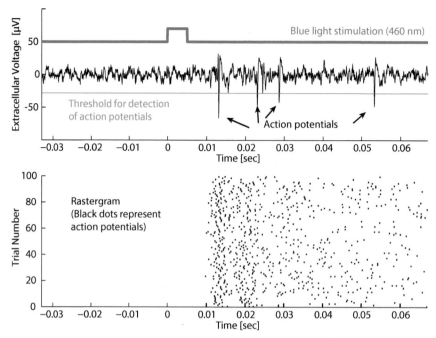

FIGURE 11.10   Optogenetic activation of a cortical network with a brief pulse of blue light in vitro (Thy1-ChR2 mouse). *Top*: The sample trace of extracellular voltage shows clear action potential response. *Bottom*: 100 trials of the same stimulation show stereotyped initial response followed by continued firing of the neurons in the absence of light stimulation.

activation of these light-sensitive proteins in specific cell types. Therefore the role of activity in a given cell population can be determined without activating other cells and fibers of passage, as would occur with electric stimulation. A second advantage of stimulation with light is the ability to simultaneously stimulate and record from the same cellular population. In contrast, electrical stimulation causes an electric artifact that prevents the detection of neuronal activity by electrophysiological means for at least several milliseconds.

Optogenetic Probes

The light-sensitive proteins used as optogenetic probes (*opsins*) were first discovered in algae and have been subsequently expressed in neurons. The most commonly used optogenetic probe is *channelrhodopsin*, which depolarizes neurons upon application of blue light. Channelrhodopsin is a light-sensitive ion channel that is permeable for cations (positively charged ions). The most commonly used variant of channelrhodopsin is channelrhodopsin-2 (ChR2), which has a peak excitation wavelength of $\sim$480 nm. The closing time constant is about 10 ms, which defines how much time needs to elapse before the channelrhodopsin can be activated again. There are numerous variants of channelrhodopsins with improved properties adapted for specific applications. Optimization of channelrhodopsins typically attempts to (1) increase the peak and steady-state photocurrent that flows through the channel (to gain control over more neurons with the same light intensity), (2) alter the duration of the open state (ie, closing time constant) after discontinuation of light application (to switch neurons on or off for longer periods of time), and (3) shift the peak sensitivity to longer wavelengths (so-called red-shifted opsins, to reduce overlap with excitation for calcium imaging). As opposed to the depolarizing action of ChR2, both halorhodopsin (a chloride pump) and archaerhodopsin (Arch, a proton pump) are proteins that hyperpolarize the membrane voltage in response to light application.

Targeting: Achieving Specificity

The most common approach for expressing optogenetic probes in the neuronal membrane is delivery of the encoding gene by viral injection (Fig. 11.11). Specifically, most applications use the adeno-associated virus (AAV) for delivery. AAV is nonpathogenic, and is commonly found in humans and other species. A large number of different AAV versions with different coat proteins can also be artificially modified through mutagenesis. Some versions of AAV preferentially infect certain cell types, whereas others preferentially enter via axon terminals. Through genetic engineering, viruses can be assembled to have a set of desired properties for targeting specific cell populations.

The key strength of optogenetics is the ability to specifically target certain cell types (Fig. 11.12, [5]). First, the injection site of the virus provides an initial level of targeting; this only requires a stereotactic strategy (ie, use of a three-dimensional coordinate system) to precisely and reliably target a given area with the viral injection. Second, conditional

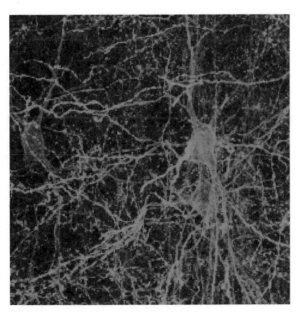

FIGURE 11.11 Example confocal image of a virally delivered construct that combines optogenetic actuators channelrhodopsin-2 with yellow fluorescent protein (YFP).

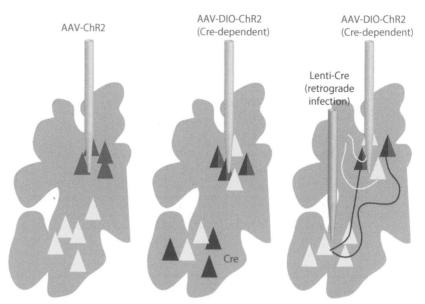

FIGURE 11.12   Optogenetic targeting strategies. In all three cases, the light fiber is positioned over the somata of the cell population that expresses ChR2. *Left*: When AAV-ChR2 is injected into the target area, ChR2 will be expressed in cells near the injection site. *Middle*: Cell-type-specific expression by injection of ChR2 in a Cre-dependent construct (*red*) into a transgenic animal that expresses Cre (*blue*) in a specific cell class. ChR2 will be expressed only in Cre-expressing cells near the injection site. *Right*: Injection of a retrograde virus such as the lentivirus with Cre as payload into the area where targeted cell types project to (*blue*). Cre will be expressed in neurons that project to the injection site. This is followed by injection of an anterograde virus, typically AAV, with Cre-dependent ChR2 as payload into area of interest. Only the neurons that project to the target area of interest will express ChR2.

expression systems can be used (such as site-specific recombinase, eg, Cre). Cre enables the expression of a specially prepared genetic sequence that remains otherwise inaccessible and thus not expressed. With this tool, cells with a given genetic makeup can be targeted so that only the Cre-expressing cells will end up expressing the optogenetic protein. The downside of this method is that it typically requires a genetically modified animal that expresses Cre in a certain cell population; the creation of such an animal is a time-consuming and expensive enterprise. Third, connectivity patterns can be used to achieve expression in restricted cell populations. This strategy requires a specific connectivity pattern together with a virus that infects neurons in a specific way. For example, by using a virus that selectively infects neurons via axons (retrograde), illumination of the soma (if distant) can then selectively activate cells that project to the area of the injection. Certain types of virus are effective at retrograde labeling (eg, rabies-based viruses), but cytotoxicity often limits their use. Less-toxic viruses (eg, lentivirus and certain AAV serotypes) are preferred, but suffer from limited efficiency in retrograde labeling.

In addition to these strategies for expressing opsins in specific cell populations, targeting the light for stimulation can also be used to achieve specificity (Fig. 11.13). Both LEDs and lasers are typically used for stimulation. For in vitro studies, light delivery for stimulation can be integrated into the microscope. For in vivo studies, light fibers are implanted at the

Targeting Brain Area

Targeting Projection

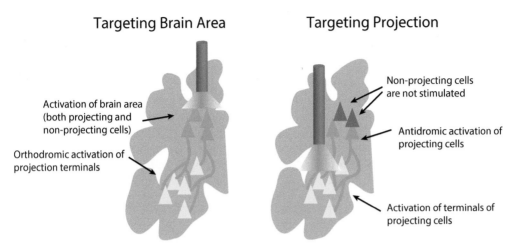

Activation of brain area
(both projecting and
non-projecting cells)

Orthodromic activation of
projection terminals

Non-projecting cells
are not stimulated

Antidromic activation of
projecting cells

Activation of terminals of
projecting cells

FIGURE 11.13   A light fiber (*shown in blue*) can be positioned over cell bodies to activate an area of interest. The stimulation causes action potentials that propagate to the axonal terminals in the target areas (*left*, orthodromic activation). Positioning of the fiber over the axons of the ChR2-expressing cells (*right*) enables selective activation of a given input to the target area (here, input to yellow cells). Stimulation of the axons elicits action potentials that propagate back to the soma (antidromic stimulation). Cells that do not project to the target (*shown in red*) are not directly stimulated.

target site. Spatial control of the light can be achieved by various strategies such as scanning of the light path over the sample or simultaneous projection of several light beams with digitally controlled optical mirrors.

Applications

Optogenetic perturbations are being used in a rapidly growing number of applications. Most fundamentally, the functional (and causal) role of specific neuronal populations during behavior can be determined by either activating or silencing the neurons and documenting them according to change in behavior. In such studies, it is important to perform simultaneous electrophysiological recordings to verify the desired effect of the optogenetic perturbation. The majority of optogenetic studies have focused on dissecting the complex subcortical circuitry that has been linked to psychiatric diseases such as anxiety and depression [6]. Typical behavioral readouts of such studies are anxiogenic (causing behavior associated with anxiety) and anxiolytic (reducing anxiety-related behavior) outcomes. Such anxiety-related behavior is assessed in rodents by the elevated plus maze and the open-field test. Other behavioral dimensions that have been successfully modulated by optogenetic perturbations are appetitive and aversive behaviors closely related to the reward system. Together, a growing number of studies have demonstrated that activation and inactivation of specific subcortical circuits can profoundly alter behavioral state. However, it has emerged that activating or silencing a specific area is often not sufficient to elucidate its functional role. In those cases, a projection-specific approach is necessary to uncover the complex interaction of the different subpopulations of neurons within a given structure. For example, projections from the medial prefrontal cortex (mPFC) can either induce an active-coping or a passive-coping response in the forced swim test, depending on whether projections from the mPFC to the dorsal raphe nucleus or projections to the lateral habenula were stimulated [7]. Critically, stimulation of the somata in the corresponding target areas does not lead to the same specific behavioral responses. Together, such studies emphasize that targeting of projections (not just of cell types or brain areas) is feasible and crucial to dissect the role of neuronal circuit activation in behavior. Nevertheless, the actual neural code (electrophysiological signals) in these pathways remains poorly understood. The application of optogenetic stimulation to modulate sensory processing and cognitive function is in its infancy. It is likely that the complexity of the spatiotemporal code in these circuits will require even more sophisticated tools than those used for turning on and off pathways. For example, optogenetic perturbations can be used to induce specific oscillatory activity patterns. Such experimental paradigms not only demonstrate the causal role of specific cell types in a given oscillation pattern (such as parvalbumin-positive cells in the genesis of gamma oscillations, Fig. 11.14, [8]), but also the behavioral consequences of such temporal patterning of network activity [9].

Light activation can also be used to "optically tag" specific neurons when combined with electrophysiological recordings. In such an experiment, channelrhodopsin is expressed in a specific cell type by one of the targeting strategies discussed earlier. Light activation is used to make cells fire, and their action potential waveforms are then used to identify these cells in large-scale electrophysiological recordings. This strategy can be used to tag neurons by shining light on somata (same location of light fiber and recording electrode) or axonal terminals in the projection zone (light fiber close to axonal terminals, recording electrode close to somata). In the latter approach, the evoked retrograde action potentials are recorded in the area where the somata of the activated, projecting neurons are. Furthermore, channelrhodopsin can be used to map connectivity at a range of spatial scales [10]. At the scale of a small circuit, ChR-expressing neurons can be activated by light and the postsynaptic responses can be measured by whole-cell patch-clamp recordings. This type of experiment is

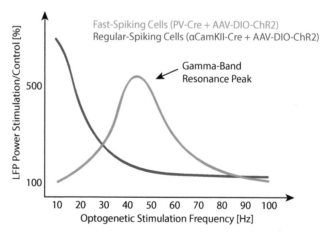

FIGURE 11.14 Optogenetic stimulation at a range of frequencies revealed that fast-spiking inhibitory interneurons exhibit a response peak in the gamma frequency band (*red*), indicative of resonance and a causal role of this cell type in the genesis of gamma oscillations. Regular spiking (presumably pyramidal cells) did not show this resonance peak (*blue*). Details in Ref. [8]. *LFP*, Local field potential.

typically performed in vitro. Of note, a ChR-expressing cell does not need to be present in its entirety since cut axons also can be stimulated with light application in vitro. Thus the post-synaptic effect of brain structure "A" on brain structure "B" can be assessed by expressing ChR in neurons of structure "A" and preparing live brain slices of area "B." Light application to these slices will trigger action potentials in the projecting fibers from area "A" and thereby trigger postsynaptic responses in cells in area "B" that can be recorded with patch-clamp electrophysiology. At larger spatial scales, network connectivity can be probed by optogenetic activation of an area and monitoring of a large potential target area with VSDs. Optogenetics can also be used to probe whole-brain functional connectivity, by combining light stimulation with functional magnetic resonance imaging (opto-fMRI, see chapter: Imaging Functional Networks with MRI).

SUMMARY AND OUTLOOK

In this chapter, we discussed optical methods to both record and stimulate neurons. The main strength of these methods over electrophysiological measurements is the fact that both the calcium indicators and the opsins for optogenetic stimulation can be targeted to genetically defined cell populations. Also calcium imaging of populations of neurons circumvents the problem of spike sorting, since the individual neurons can be directly visualized. Nevertheless, the optical methods reviewed in this chapter are complementary to electrophysiological strategies, which are still unsurpassed in their temporal resolution and their relative ease of use in the freely moving animal.

References

[1] Chen TW, et al. Ultrasensitive fluorescent proteins for imaging neuronal activity. Nature 2013;499(7458):295—300.
[2] Svoboda K, Yasuda R. Principles of two-photon excitation microscopy and its applications to neuroscience. Neuron 2006;50(6):823—39.
[3] Chemla S, Chavane F. Voltage-sensitive dye imaging: technique review and models. J Physiol Paris 2010;104(1—2):40—50.
[4] Arieli A, et al. Dynamics of ongoing activity: explanation of the large variability in evoked cortical responses. Science 1996;273(5283):1868—71.
[4a] Fenno LE, Davidson TJ, Mogri M, Deisseroth K. Optogenetics in neural systems. Neuron 2011. http://dx.doi.org/10.1016/j.neuron.2011.06.004.
[5] Packer AM, Roska B, Hausser M. Targeting neurons and photons for optogenetics. Nat Neurosci 2013;16(7):805—15.
[6] Deisseroth K. Circuit dynamics of adaptive and maladaptive behaviour. Nature 2014;505(7483):309—17.
[7] Warden MR, et al. A prefrontal cortex-brainstem neuronal projection that controls response to behavioural challenge. Nature 2012;492(7429):428—32.
[8] Cardin JA, et al. Driving fast-spiking cells induces gamma rhythm and controls sensory responses. Nature 2009;459(7247):663—7.
[9] Sohal VS. Insights into cortical oscillations arising from optogenetic studies. Biol Psychiatry 2012;71(12):1039—45.
[10] Lim DH, et al. Optogenetic approaches for functional mouse brain mapping. Front Neurosci 2013;7:54.

12

Imaging Structural Networks With MRI

One of the most impressive breakthroughs in the application of physics to medicine and biology has been the advent of magnetic resonance imaging (MRI), which allows for the noninvasive, detailed visualization of biological matter (Fig. 12.1). MRI has revolutionized neurology and neuroscience. In this chapter, we will first look at the physics of MRI to gain a conceptual understanding of the inner workings of this essential tool. Then we will review the two main applications of anatomical MR scans: imaging of brain structure and imaging of network connectivity. The latter is achieved by an MRI-based technique called *diffusion tensor imaging* (DTI) that visualizes white matter tracts. This chapter will equip us with the knowledge to critically interpret structural neuroimaging studies while also preparing us for the next chapter, "Imaging Functional Networks With MRI," where we will look at how MRI can be used to measure proxies of brain activity. The section labeled with an asterisk requires more mathematical background than the remainder of the chapter.

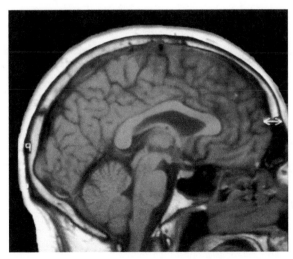

FIGURE 12.1　Magnetic resonance imaging (MRI) of the human head. *Tero Sivula/shutterstock.com*

PHYSICS OF MRI

Nuclear Magnetic Resonance

MRI is based on *nuclear magnetic resonance*, a physical property of nuclei that varies depending on the composition of protons and neutrons within each nucleus. In the case of MRI, we focus on the nucleus of the hydrogen atom (a single proton, often labeled with its abbreviation H for hydrogen). The most abundant source of hydrogen in the human body is water. Protons have a *spin*, which is a concept from quantum mechanics. Here, we will examine a simplified explanation of the phenomenon as a necessary step to understand how the MR signal is generated [1]. Visualize the spin as a rotation of the proton around some axis, which gives the proton an angular momentum. The electric charge of the proton, together with this rotation, gives the proton magnetic properties. Protons can therefore be conceptualized as small magnets (Fig. 12.2).

If an external magnetic field B_0 is applied, the rotation of the magnetic protons will align according to the field (Fig. 12.3). Specifically, the individual protons either align *with* or *against* the field (we will skip over the quantum mechanics that explain this phenomenon) such that the individual magnetic fields mostly cancel each other out. However, slightly more protons are aligned with than against the field. The number of excess protons aligned with the field is proportional to the field strength B_0. These excess protons form the basis of the MR signal.

Next, let us visualize a proton with a spin as a spinning top. If an external perturbation is applied, the spinning top will wobble. In other words, it exhibits *(gyroscopic) precession*. This phenomenon is defined as the change of rotational axis of a rotating body. In the case of a spinning top, gravity is the external force causing precession (Fig. 12.4).

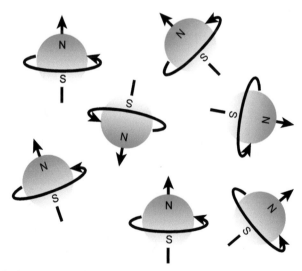

FIGURE 12.2   Individual protons rotate around their own axes and thereby act as magnets (indicated by the labels for the north and south poles). In the absence of an external magnetic field, these magnets are not aligned.

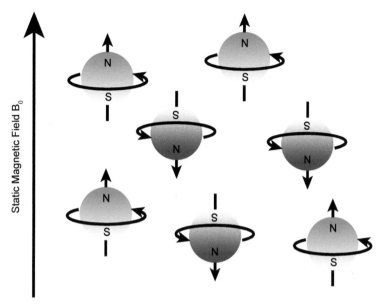

FIGURE 12.3 Protons align with or against the static external magnetic field with strength B_0. There are more protons aligned with the field (*red*) than protons aligned against the field (*blue*). This preference for alignment with the external field leads to magnetization of the tissue.

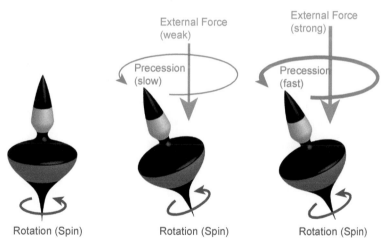

FIGURE 12.4 Precession as a function of an external force, illustrated by spinning tops. In a case without external force, no precession occurs (*left*). Precession is caused by an external force, and the oscillation frequency of the precession increases with the strength of that external force (*middle and right*).

The same mechanism of precession applies to protons. In the presence of an external perturbation, the rotating protons will wobble or *precess* about the axis of the applied magnetic field. The wobble itself has a frequency called *precessional frequency* (also referred to as the *Larmor frequency* ω). The value of the Larmor frequency is determined by the strength of the static magnetic field B_0 and the properties of the hydrogen nucleus. For a field strength of 1 T, the Larmor frequency for hydrogen is 42.57 MHz. An external radiofrequency (RF) electromagnetic pulse with the same frequency can interact with the proton. Specifically, a frequency-tuned RF pulse (at the Larmor frequency) targets the *magnetization M_0* of the tissue. This magnetization is caused by the small difference between the number of protons with spins aligned with the field and the number of protons with spins aligned against the field. In the absence of an RF pulse, M_0 is aligned with B_0 (Fig. 12.5, top). Note that we draw M_0 as aligned with the z-axis such that the magnetization vector only has a z-component. An orthogonal RF pulse B_1 tips the magnetization away from the axis of the static field B_0 such that the magnetization vector is aligned with the x−y plane (M_{xy}, Fig. 12.5, middle). Once the pulse is no longer applied, the magnetization will gradually move back toward the z-axis (Fig. 12.5, bottom) while it exhibits precession at the Larmor frequency around the z-axis.

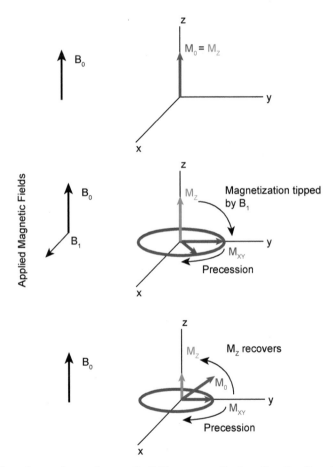

FIGURE 12.5   Effect of an orthogonal magnetic field on magnetization. *Top*: Baseline. *Middle*: Application of transient field B_1 tips magnetization away from the z-axis toward the x−y plane, where precession occurs. *Bottom*: After completion of the pulse, magnetization reverts back to its initial alignment with the z-axis.

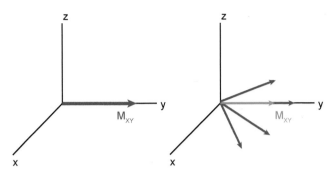

FIGURE 12.6 *Left*: Initially all of the spins are aligned and give rise to a maximal value of M_{xy}. *Right*: With time, the individual spins start to diphase and the vector sum of all magnetization vectors M_{xy} is reduced in length.

Application of a uniform magnetic field does not necessarily lead to a uniform magnetic field within the tissue. Small variations in the composition of the tissue can distort the field. Since the spin precession (and its Larmor frequency) is a function of the static magnetic field, precession will occur at different frequencies for different spins (Fig. 12.6). Note that the RF pulse is required to tip the magnetization away from the direction of the static magnetic field, but the strength of the static field defines the precession frequency in response to the RF pulse. The more time elapses, the less synchronized the precession of all the spins, and thus the smaller their sum will be. As a result, the overall magnetization in the $x-y$ plane exhibits a signal loss over time, characterized by the time constant T_2^*. This so-called *spin dephasing* is the result of two factors.

The first factor is the inhomogeneity in the tissue, which leads to different magnetic fields and thus Larmor frequencies and thereby desynchronizes the system. The second factor is the dynamic inhomogeneity caused by the magnetic interaction of the protons themselves (referred to as spin–spin interactions), which also desynchronizes the spin. Thus the overall decay time constant T_2^* is determined both by the time constant T_2' of static inhomogeneity and the time constant T_2 of dynamic inhomogeneity. Mathematically, the three time constants relate to each other as follows:

$$\frac{1}{T_2^*} = \frac{1}{T_2'} + \frac{1}{T_2} \tag{12.1}$$

Over time, the signal in the $x-y$ plane is lost because of the diminishing spin phase (as determined by the T_2 value). A second, different process also contributes to the gradual reduction of the signal in the $x-y$ plane: as time elapses after the RF pulse, the direction of the magnetization returns to the direction of the static magnetic field. The time constant of this process is referred to as T_1.

Localizing the MR Signal

The next step is to understand how these nuclear resonance properties can be used to generate an MR image based on the signal from the rotating $x-y$ component of the magnetization vector. Stated simply, the challenge is that the signal measured in response to the RF pulse, that is, the electromagnetic radiation caused by the spin precession, is a single time series, and therefore no localization information is encoded in the signal. To select a specific "slice" of tissue, this composite signal must be separated into three dimensions for localization. This is achieved by the application of static magnetic field gradients such that the strength of the magnetic field is a function of all three dimensions, x, y, and z. As a result, not all tissue is exposed to the same magnetic field, and precession occurs at different Larmor frequencies. As a first step, the three-dimensional localization problem is reduced to a two-dimensional problem by introducing a gradient of the magnetic field along the z-axis such that different locations along the z-axis exhibit different magnetic field strengths and thus different Larmor frequencies. Spins can only be tipped by RF pulses that match their Larmor frequencies. By using RF pulses with different frequencies (matched to the range defined by the gradient of the static magnetic field), signals can be generated that originate from specific z-locations. Because of frequency mismatch, all other locations would not exhibit precession. Of note, this gradient only needs to be implemented during the application of the RF pulse. This process is called *slice selection*. Accordingly, the gradient of the magnetic field is referred to as the *slice-selection gradient*: the stronger the gradient is, the more different is the precession frequency between two points on the z-axis. Any RF pulse has not only a single frequency but rather a range of frequencies (referred to as bandwidth). For weak gradients, two quite distant points may still be affected by the same pulse, since they exhibit similar precession frequencies. This results in limited resolution of the data along the z-axis. In the case of a strong gradient, only points close to each other have sufficiently similar precession frequencies to be affected by the same RF pulse, resulting in higher z-axis resolution (Fig. 12.7). At this point, for a given frequency of the RF pulse, we are left with a single time series that corresponds to the entirety of a two-dimensional image of the selected slice.

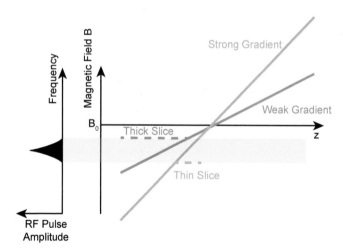

FIGURE 12.7   Slice selection. The frequency at which the spin magnetization can be tipped to elicit an electromagnetic signal from the tissue is a function of the static magnetic field. If the strength is a linear function of the z-location (gradient), pulses at varying frequencies selectively cause spin precession in a slice of tissue. Thus such a slice-selection gradient enables the targeting of specific tissue slices. The *blue-shaded area* delimits the range of magnetic field strengths for which a sample radiofrequency (RF) pulse (*black, left*) can interact with the spin.

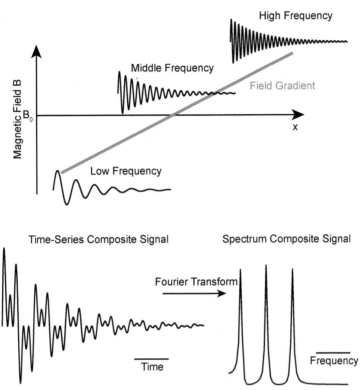

FIGURE 12.8   Frequency encoding of x-axis position with gradient that causes precession at different frequencies as a function of x-axis position during signal readout. The signal is decomposed using the Fourier transformation to identify the signal for the individual columns of the image.

With the slice-selection gradient, we can now acquire a signal that corresponds to a given $x-y$ plane. We still need additional steps to address specific locations within a given slice (columns and rows). Note that there is no consecutive sampling of different spatial locations. Rather, additional gradients in the magnetic field in the x- and y-directions are employed to enable reconstruction of slice images. These gradients exploit the fact that the frequency of the precession triggered by the RF pulse is a function of the magnetic field strength after the RF pulse is over. Therefore a magnetic field gradient during the readout, that is, measurement of the RF radiation generated by the spin precession, will generate signal components that exhibit different frequencies as a function of the local field strength (Fig. 12.8). This mechanism is used to encode the spatial location along the x-axis of the slice image (columns). The resulting composite signal is in essence the sum of different sine waves with varying frequencies determined by the corresponding locations on the x-axis. Frequency decomposition, such as through the Fourier transformation (see toolbox: Time and Frequency), can be used to extract the amplitude of the different frequency components that correspond to the signal strength of the individual image columns.

Finally, to decompose individual columns by row (ie, pixels), a different approach is used. The application of RF pulses followed by readout is performed multiple times. At each iteration (sweep) a different gradient in the y-direction is briefly added to static magnetic field. This brief gradient application causes precession at different frequencies as a function of the y-position. Because the duration for which the gradient is applied is the same for all locations, differing amounts of phase advances of the precession signal will occur as a function of the y-position (Fig. 12.9). This process is repeated for every row of the acquired image. The resulting set of numbers is not a time series, but nevertheless will exhibit "fast" and "slow" components. At the center, the field strength will be the same for all gradients, and there will be no change in signal as a function of the different gradients. In contrast, the signal for the points at the extreme ends of the gradients (the lowest and highest field strength) will assume very different values for each gradient, since the magnetic field strength (and therefore the precession frequency) differ substantially across sweeps. When the data across sweeps are assembled, large changes in phase across sweeps correspond to the edges of the image. Analysis of the frequency structure of the data points collected for different gradients will reveal the y-position. Together with the slice selection and the frequency encoding scheme discussed earlier, this process enables spatial localization in three dimensions.

In summary, a typical MR sequence consists first of an RF pulse for excitation, during which the slice-selection gradient is turned on. Shortly thereafter, a phase-encoding field is applied, followed by the frequency-encoding gradient. It is during the application of the frequency-encoding gradient that the signal is recorded (Fig. 12.10). The one aspect we have omitted for clarity is that there is a second RF pulse used to undo the dephasing caused by tissue inhomogeneity such that the signal decays with time constant.

Now that we have established how the MR signal is localized to form an image, we will return to the question of what exactly the MR signal measures. The measured signal depends

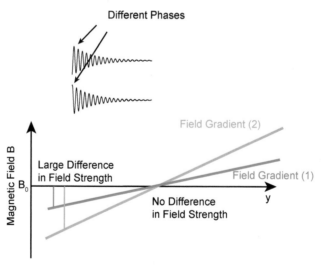

FIGURE 12.9   Different y-positions are encoded by different phase shifts caused by magnetic field gradients along the y-axis. Responses for different gradients are collected (here: *red and orange*). Proceeding through gradients, the phase shifts rapidly change for locations far from the center because of the large difference in field strength. The Fourier transform allows localization by decoding the phase shifts across gradients.

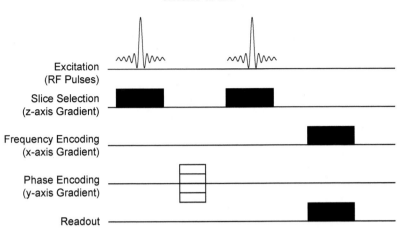

FIGURE 12.10 Simplified representation of a spin-echo magnetic resonance (MR) sequence. The slice-selection gradient (z-axis) is applied during the radiofrequency (RF) excitation pulse. Then, the phase encoding gradient is applied at different levels for each pulse. A second RF pulse is applied for refocusing. Finally, the x-axis gradient is applied for frequency encoding during readout.

on the longitudinal magnetization (M_z) before the RF pulse and the transverse magnetization (M_{xy}) after the RF pulse. The amount of energy emitted as the magnetization exhibits precession is quantified. In other words, only M_{xy} is measured during readout. Yet the value of M_{xy} depends on the M_z when the RF pulse is applied. For small values of M_z (before the magnetization has returned to its original alignment), there is less magnetization to be tipped by the RF pulse and thus a smaller M_{xy}. The imaging strategies exploit the variation in T_1 and T_2 time constants of different materials, as well as the fact that T_2 is always identical to or shorter than T_1. The *longitudinal decay*, T_1, denotes the return of the spin back to its main axis, M_0. The *transversal decay*, T_2, results from the dephasing of the $x-y$-component of the magnetization. T_2 is much shorter than T_1. To simplify, two different types of images are typically generated: T_1- and T_2-weighted images (a third type of image, proton density, is not discussed here and is not typically used). The sequence of gradients (z-direction for selecting slice, x- and y-directions for sampling of Fourier space) is applied at the *repetition time* (TR). The RF pulses used to excite the system generate spin echoes at the *echo time* (TE). The choice of TR and TE values defines the type of image contrast achieved. As discussed earlier, the MR signal decays with time constant T_2 as a result of phase desynchronization. If TE is short, then signals with different T_2 values do not have time to differentiate, and the signal intensity is determined by the time constant T_1. In this case, if TR is too long, the difference in T_1 between two materials will not be detectable, since the magnetization of both materials has returned fully to the original direction. In other words, for T_1-weighted images, the TR is small such that there are pronounced differences caused by different values of T_1. These differences are then captured by a short TE to avoid signal differentiation based on T_2. In contrast, T_2-weighted images use slow TR to complete the longitudinal relaxation such that differences in the T_1 time constant of this process do not contribute to the signal. Also TE is longer to achieve signal differentiation of the tissue as a function of the T_2 time constant (Fig. 12.11). T_1 and T_2 values for different materials are provided in Table 12.1.

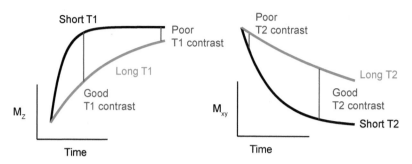

FIGURE 12.11   Images with T_1 or T_2 contrast are collected by adjusting the values of TR and TE. *Vertical blue lines* represent TR (*left*) and TE (*right*).

TABLE 12.1 T_1 and T_2 Values for Different Materials

	T_1 (ms)	T_2 (ms)
Distilled water	3000	3000
Gray matter	810	100
White matter	680	90
Cerebrospinal fluid	2650	280

Of note, materials with long T_1 values appear dark. This is because the magnetization M_z has not fully recovered when the next RF pulse is applied (TR), and the resulting M_{xy} is of lower amplitude. However, materials with long T_2 values appear bright, since less signal decay has occurred at the time of acquisition (TE).

ANALYSIS OF STRUCTURAL MR IMAGING DATA

Image Preprocessing

In a structural MR scan, there is a significant amount of signal from nonbrain structures (eyes, skin, fat, muscle, bone) that needs to be removed after signal acquisition. The removal process is based on *segmentation*, where different regions of the image are assigned to a specific entity as a function of both the known neuroanatomy and visual appearance of the regions in the scan images. Specifically, isolating the brain is referred to as *brain/nonbrain segmentation*. As a next step, the tissue is segmented into gray matter, white matter, cerebrospinal fluid, and any pathological structures. The most basic approach for this segmentation step is to build signal-intensity histograms of all voxels (three-dimensional pixels) and to detect the peaks of a multimodal distribution that reflect the different tissues. In reality, however, the peaks in this histogram can be hard to identify because of technical limitations (such as inhomogeneity of the RF magnetic field used for excitation). Additional information, including the local neighborhood of a given voxel (which is more likely to be of the same than of a different tissue, given the large, continuous shape of the structures of interest), can be taken into account to reduce noise in the segmentation process. The quality of the segmentation process can be further enhanced by combining different image contrasts (eg, T_1- and T_2-weighted images). Once the tissue structures are segmented, it is often required to align images taken with different imaging modalities (eg, structural and functional) from the same participant. This process of *intermodal image registration* is based on the minimization of a cost function that penalizes misalignment as a function of the *registration parameters* (image stretching, translation, and rotation). In its simplest form, *head motion correction* is also performed with image registration.

Quantification of Structural Features

After preprocessing, the goal is to extract features of interest from the neuroanatomical data. One such feature of interest is *cortical thickness*. Estimating cortical thickness by hand is extremely laborious. The field of computer vision has developed algorithms that can be used to automatically detect the white matter/gray matter boundary and the pial surface. The *gyrification index* is the ratio of the entire gyral contour length to the outer, exposed surface (Fig. 12.12). It quantifies the amount of cortical surface "hidden" from the visible surface caused by cortical folding. Large numbers denote brains with extensive cortical folding; small numbers denote brains with limited folding. There is also a localized metric that provides the gyrencephalic index for a local patch of cortex. This metric can be computed for a large number of points on the entire cortical surface to assess regional differences in gyrification. Gyrification is an important aspect of development (in gyrencephalic animals), and it has been suggested that aberrations in gyrification are present in psychiatric illnesses with developmental origin such as schizophrenia.

Comparing structural MR scans between participants presents a new set of challenges. First, the scans of all participants must be brought into the same stereotactic space by spatial normalization. In this image registration process, a spatial transformation is performed to minimize the difference between the individual image and the template image. This process removes differences in overall brain size but does not remove individual (size) variations in

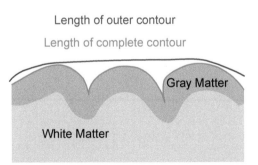

FIGURE 12.12   The gyrification index is computed as the ratio of the length of the complete contour (*red*) to the length of the outer contour (*blue*). Gyrification of the lissencephalic mouse brain is ~1, whereas the human brain is gyrencephalic and has a gyrification index of ~2.5.

local brain structure. This step enables accurate comparison of local features in the absence of large brain size differences. Once the scans are registered to the template, they are preprocessed (smoothed) and segmented. This allows voxel-based volumetry (measurement of volume) of structures of interest as well as comparison between different individuals and groups of individuals.

Quantification of White Matter Tracts: Diffusion Tensor Imaging*

MRI can also be used to visualize white matter tracts by DTI. First, we will examine the underlying physics to better understand exactly what DTI measures [2]. Then we will discuss how fiber tracts are visualized and quantified. *Brownian motion* is a physical process that describes random, temperature-induced molecular motion over time. Brownian motion is associated with *diffusion*, the movement of particles caused by a concentration gradient. DTI measures the degree and directionality of water diffusion in tissue. These diffusion measures indirectly mark the underlying microscopic structure of the tissue. Such particle flux is the direct consequence of a difference between particle concentrations, denoted as $n(r,t)$, a scalar (single number) function of location r and time point t. The stronger the gradient (change as a function of location) of n is, the larger the resulting flux is, which is proportional to the concentration gradient with diffusion constant D. The concentration gradient is a vector composed of the spatial derivatives of the concentration n along all three dimensions. The flux (denoted as vector J) points in the direction opposite from the concentration gradient since particles move from areas of high concentration to areas of low concentration:

$$J = \begin{pmatrix} J_x \\ J_y \\ J_z \end{pmatrix} = -D \begin{pmatrix} \dfrac{\partial n}{\partial x} \\ \dfrac{\partial n}{\partial y} \\ \dfrac{\partial n}{\partial z} \end{pmatrix} = -D\nabla n \tag{12.2}$$

The total number of particles in the system cannot change, so any change in n as a function of time (ie, temporal derivative) must be explained by the total spatial change in the flux J (there is another minus sign because of the direction of the flux relative to the concentration gradient):

$$\frac{\partial n}{\partial t} = -\left(\frac{\partial J_x}{\partial x} + \frac{\partial J_y}{\partial y} + \frac{\partial J_y}{\partial x}\right) = -\nabla \cdot J \tag{12.3}$$

When combining these two equations, the result is Fick's second law:

$$\frac{\partial n}{\partial t} = D\nabla^2 n = D\left(\frac{\partial^2 n}{\partial x^2} + \frac{\partial^2 n}{\partial y^2} + \frac{\partial^2 n}{\partial z^2}\right) \tag{12.4}$$

This law can also be used in the absence of a macroscopic concentration gradient to understand self-diffusion, the process that results from random local concentration differences caused by thermal motion. In this case, the macroscopic concentration gradient is replaced with the conditional probability that a molecule at position r will be at position r' after time t has elapsed:

$$P(r'|r, t) \tag{12.5}$$

We define in analogy with Fick's second law (without detailed justification):

$$\frac{\partial}{\partial t}P(r'|r, t) = D\nabla^2 P(r'|r, t) \tag{12.6}$$

This equation is only true if the diffusion property is the same in all directions. Pure liquids are examples of such *isotropic* media. In the brain, however, the motion of water by diffusion is heavily restricted by the microstructure, in particular the fat-containing myelinated axons. The change in the conditional probability on the left side in Eq. (12.6) (particle flux) can depend on the probability gradient (right side) along all three dimensions. Therefore we need three diffusion coefficients for each direction of the flow to denote its dependence on the gradients in all three dimensions. As a result, we arrive at a 3-by-3 matrix that is called the *diffusion tensor*. In theory, all nine elements could be nonzero. Any matrix D that is not a diagonal matrix with the same coefficient for all three dimensions defines an *anisotropic medium*.

$$\frac{\partial}{\partial t}P(r'|r, t) = \nabla[D\nabla P(r'|r, t)]$$

Diffusion can be measured by MRI using specific pulse sequences that make the signal susceptible to water displacement by diffusion. Thus DTI is based on measurement of the diffusion tensor, which allows the assessment of how unrestricted (isotropic) or restricted (anisotropic) the diffusion is. In the case of restricted diffusion, we can indirectly determine the microstructure that prevented isotropic diffusion. By assessing the location of water molecules at consecutive time points, the diffusion motion can be estimated and the diffusion tensor determined. Enough time must elapse between the two time snapshots. The water molecules in the brain need to have sufficiently moved to ensure that interaction with the constraining microstructure (such as axonal fibers) has occurred. This MR measurement is called the *apparent diffusion coefficient* (ADC). The tensor D can be transformed into a tensor D' with three distinct, nonzero diagonal values (technically termed eigenvalues, we skip the linear algebra of this transformation). This transformation projects the data into a new coordinate system that is aligned with the main dimension of interest, that is, the direction of the axonal bundle (the direction along which diffusion is the largest):

$$D' = \begin{pmatrix} \lambda_1 & 0 & 0 \\ 0 & \lambda_2 & 0 \\ 0 & 0 & \lambda_3 \end{pmatrix} \tag{12.7}$$

The value λ_1 describes the diffusion along the fiber tract and determines the parallel or axial diffusion (ADC *parallel*):

$$\text{ADC}_{\parallel} = \lambda_1 \tag{12.8}$$

The values λ_2 and λ_3 describe the diffusion in the plane orthogonal to the axial direction and are called *radial diffusion*, often quantified by ADC *orthogonal*, the average of the two numbers:

$$\text{ADC}_\perp = 0.5(\lambda_2 + \lambda_3) \tag{12.9}$$

The ratio of the parallel ADC to the perpendicular ADC provides an estimate of the anisotropy. A related metric of anisotropy is called the *fractional anisotropy* (FA), a normalized measure of how diffusion differs from *mean diffusion* (MD). MD is defined as the average of the three eigenvalues of the diffusion tensor:

$$\text{MD} = (\lambda_1 + \lambda_2 + \lambda_3)/3. \tag{12.10}$$

FA is then defined as:

$$\text{FA} = \sqrt{1.5}\,\frac{\sqrt{(\lambda_1 - \text{MD})^2 + (\lambda_2 - \text{MD})^2 + (\lambda_3 - \text{MD})^2}}{\sqrt{\lambda_1^2 + \lambda_2^2 + \lambda_3^2}} \tag{12.11}$$

FA is a normalized quantity that can assume values between zero (isotropic diffusion) and one. Although myelin is probably an important contributor to anisotropic diffusion (and thus FA), anisotropic diffusion of water in the nervous system is more likely a generic property of intact, healthy fibers that occur independently of whether myelination is present. Rather, the axonal membrane in general provides the dominant microstructure that restricts the spontaneous motion of water molecules (ie, anisotropic diffusion).

Measurement of the diffusion tensor not only allows the voxel-by-voxel description of the degree of anisotropy (eg, FA), it also provides the data to reconstruct estimates of white matter tracts in the brain. This approach is called *tractography*. To introduce the principles of tractography, we need to return to the diffusion tensor and introduce two other characterizations besides the anisotropy (FA) defined previously. The first of these characterizations is the overall magnitude of diffusion, a quantity determined as the sum of the three eigenvalues (referred to as *trace* in the language of linear algebra). The second quantity is the direction of the main axis (captured by λ_1) and the two orthogonal axes λ_2 and λ_3 (eigenvectors, by-products of the transformation of matrix D to be aligned with the major axis of diffusion). The basic assumption of tractography is that in any given voxel the direction of the majority of white matter fibers is parallel to the direction of the eigenvector associated with the largest diffusion eigenvalue (λ_1). Visual representation of the dominant eigenvector is achieved by color coding. Every voxel is drawn with a mixture of red, green, and blue corresponding to the three axes of the dominant eigenvector. The intensity of the red is proportional to the x-component of the dominant eigenvector (corresponding to the direction right to left in the DTI images). In the same way, green encodes the anterior–posterior axis, and blue encodes the inferior–superior axis. The overall intensity of every voxel can be scaled by FA to visualize not only the main direction of the fiber, but also the level of anisotropy. Based on this information, tractography algorithms are used to trace individual pathways. In deterministic tractography, a single voxel or multiple "seed" voxels are chosen as starting points. From these starting points, the tracts are reconstructed until a stopping criterion is reached. Simply, deterministic trajectories are reconstructed by using the direction and magnitude of the primary eigenvector as a guide through stepwise integration. The trajectory r is started at location $r(t_0)$, where t_0 parametrizes the trajectory such that increasing values of this parameter track the trajectory. To arrive at the next point of the trajectory, $r(t_1)$, the scaled version of primary eigenvector e_1 is added to the trajectory:

$$r(t_1) = r(t_0) + \alpha e_1(r(t_0)) \tag{12.12}$$

The value e_1 is the eigenvector for the local voxel from where the trajectory is to be continued [at location $r(t_0)$] and α denotes the scaling factor. This piecewise linear reconstruction of a track describes the basic principle of *streamline methods*. These methods, which only utilize the first eigenvector, work well for highly anisotropic tissue where there is a clear, dominant eigenvalue. An alternative method that directly uses the full diffusion tensor is called the *tensor deflection*. In this approach, the incoming direction x of the tract is transformed by left multiplication with the diffusion tensor:

$$x(t_2) = Dx(t_1). \tag{12.13}$$

For example, in the case of isotropic diffusion, the outgoing vector $x(t_2)$ points in the same direction as the incoming vector $x(t_1)$:

$$x(t_2) = D'x(t_1) = \begin{pmatrix} d & 0 & 0 \\ 0 & d & 0 \\ 0 & 0 & d \end{pmatrix} \begin{pmatrix} x_1 \\ x_2 \\ x_3 \end{pmatrix} = \begin{pmatrix} dx_1 \\ dx_2 \\ dx_3 \end{pmatrix} = d \begin{pmatrix} x_1 \\ x_2 \\ x_3 \end{pmatrix} \tag{12.14}$$

In contrast, in the case of an anisotropic voxel, the trajectory reflects the dominant direction of the tensor. For example, if diffusion in the x_2-direction dominates, we find:

$$x(t_2) = D'x(t_1) = \begin{pmatrix} 0 & 0 & 0 \\ 0 & d & 0 \\ 0 & 0 & 0 \end{pmatrix} \begin{pmatrix} x_1 \\ x_2 \\ x_3 \end{pmatrix} = \begin{pmatrix} 0 \\ dx_2 \\ 0 \end{pmatrix} = d \begin{pmatrix} 0 \\ x_2 \\ 0 \end{pmatrix} \tag{12.15}$$

such that the outgoing vector now points in the direction of x_2.

There is a range of more refined and complex algorithms for deterministic tractography. Although it is clear (and easy to determine) that different algorithms provide different results, the main challenge is deciding which method to use. This is difficult since *ground truth data* (eg, postmortem histology) for independent verification is typically not available.

Start and stop criteria need to be defined independently of the exact choice of algorithm for tracing. The starting points are referred to as *seed locations*. Typical strategies involve selecting seeds from all over the brain (*whole-brain seeding*) or limiting the seeds to a specific region of interest to investigate more hypothesis-driven questions. Stopping criteria define the end of a tract. They are needed to prevent the resulting tract from growing infinitely long. Two examples of stopping criteria are the encountering of low FA (to prevent the trajectory being continued into gray matter), and restricting the maximal curvature. Once a tracing algorithm has determined the candidate pathways, postprocessing is used to identify the pathways of interest (eg, by requiring the pathway to pass through specific brain structures). The primary challenge currently being addressed by scientists is the resolution of voxels where different pathways cross. In this scenario, the dominant eigenvector will not point along either of the fibers. To account for the substantial amount of uncertainty in DTI tractography results, probabilistic tractography methods have been developed to provide numerical estimates of the uncertainty of any track identified. These methods are based on the determination of probability density functions of fiber orientation, which can then be used in a Monte Carlo approach where the same track is reconstructed many times (eg, 1000) and a new set of directions is determined for each voxel in each trajectory by its probability density function. Averaging all the candidate trajectories provides the proposed trajectory and its uncertainty.

SUMMARY AND OUTLOOK

In this chapter, we have discussed the basic principles of how MRI works and how it can be applied to the imaging and quantification of both brain structure and brain connectivity. First, we discussed nuclear magnetic resonance and how, in a static magnetic field, protons generate a magnetization vector that can be perturbed by RF magnetic pulses. The response to such perturbations is another RF signal that depends on tissue properties and can be measured. To generate an MR scan, these signals need to be localized in three dimensions, which is achieved by a temporal sequence of additional gradients of the static magnetic field. MR images can be analyzed to extract information about structural features such as gray matter volume and gyrification. MRI can also be used for DTI, an imaging strategy that allows determination of the major white matter bundles and their structure and direction. In the next chapter, we will explore how MRI can be used to measure signals that correlate with neuronal activity.

References

[1] Plewes DB, Kucharczyk W. Physics of MRI: a primer. J Magn Reson Imaging 2012;35(5):1038–54.
[2] Jones DK. Diffusion MRI: theory, methods, and applications. Oxford University Press; 2010.

13

Imaging Functional Networks With MRI

Functional magnetic resonance imaging (fMRI) denotes a set of strategies to noninvasively measure correlates of neuronal activity in the brain. This chapter builds on the previous chapter that introduced the basics of magnetic resonance imaging (chapter: Imaging Structural Network With MRI). First, we will discuss the coupling between neurons and the vascular system that is the source of the signals measured with fMRI. Then we will review the main fMRI acquisition and data analysis strategies, including their advantages and limitations. The toolbox "Graph Theory" introduces concepts used in the discussion of fMRI data analysis in this chapter.

IMAGING CORRELATES OF NEURONAL ACTIVITY

Vascular Coupling of Neuronal Tissue

FMRI measures the change in blood supply in response to neuronal activity, called the *vascular hemodynamic response* [1]. Neuronal activity is associated with an increase in *cerebral blood flow* (CBF), *cerebral blood volume* (CBV), and *cerebral metabolic rate of oxygen* (CMRO$_2$). CBF and CBV represent the vascular response and are correlated; specifically, CBF scales with CBV and velocity. Increase in CBF enhances venous oxygenation level. In contrast, an increase in the CMRO$_2$ (a higher rate of oxygen consumption) leads to a decrease in venous oxygenation level. To maintain venous oxygenation levels, CBF needs to increase together with the increase in CMRO$_2$ during increased neuronal activity. As it turns out, the increase in CBF exceeds the increase in CMRO$_2$ such that in response to neuronal activity venous oxygenation levels actually increase (overcompensation).

BOLD Signal

Oxygen in the blood is carried by hemoglobin. Hemoglobin without the oxygen molecule is called *deoxyhemoglobin* (dHb). Importantly, dHb acts as an endogenous contrast agent that can be measured by MRI. Therefore an increase in neuronal activity is accompanied by a decrease in dHb because of the oversupply of blood in response to an increase in activity. This approach is the basis of the so-called venous *blood oxygenation level-dependent* (BOLD) MRI contrast. Specifically, BOLD contrast is based on the difference in magnetic properties of oxygenated and deoxygenated hemoglobin. Oxygenated hemoglobin is *diamagnetic* (it creates a magnetic field that opposes an applied magnetic field), whereas dHb is *paramagnetic* (it creates a field that is aligned with an applied magnetic field). Paramagnetic dHb alters the static magnetic field in tissue (applied by the MR scanner) and leads to changes in T2-weighted images.

Source of the BOLD Signal

Identification of the causes of fluctuations in the BOLD signal in the brain is crucial for correct interpretation of the measured signals. The most simplistic view is that the BOLD signal corresponds to the hemodynamic response to neuronal spiking activity. There is substantial evidence, however, that the BOLD signal more closely correlates with the local field potential (LFP [2]). The LFP is likely dominated by synaptic activity (see chapter: LFP and EEG). Excitatory, glutamatergic synaptic transmission is the most energy-demanding process in the brain (glutamate production, release, and postsynaptic actions). The dissociation between the coupling of the BOLD signal to the LFP or spiking activity is inherently difficult since synaptic activity and neuronal firing are closely linked as a function of the underlying circuit dynamics.

Comparison of the BOLD Signal to Alternative Imaging Strategies

Alternatively, CBF can be measured with a technique called *arterial spin labeling* (ASL [3]). The basic idea behind ASL is to use the arterial perfusion as an *endogenous tracer* (meaning that no tracer foreign to the body needs to be injected). Both BOLD and ASL are fMRI methods. BOLD is used much more frequently than ASL. The selection of methods requires us to consider the following contrasting properties of the two fMRI techniques. The main advantages of BOLD are the higher signal-to-noise ratio and the higher temporal resolution. Therefore fMRI of task-related events often employs BOLD. As an additional advantage, BOLD is easier to implement than ASL. However, ASL provides better spatial localization, since the BOLD signal is the result of the interplay of CBF, CBV, and $CMRO_2$ that mostly originates from venous structures that do not necessarily align with the neural correlates of interest. The BOLD signal exhibits large fluctuations at very low frequencies (<0.01 Hz), and therefore the typically reported percent change in BOLD signal is hard to interpret for events that are more than 100 s apart since baseline changes during that time period as a result of the slow fluctuations. ASL is free from this limitation, because it is based on the subtraction of two images from consecutive time points. Of note, ASL is based on the same idea as the use of radioactively labeled water in *positron emission tomography* (PET). In PET an exogenous, radioactive tracer is used that diffuses into the brain and emits charged particles named *positrons* that can be detected with PET scanners. It differs from ASL in that it uses radioactively labeled tracers instead of magnetically labeled (endogenous) water.

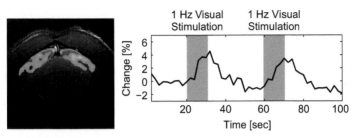

FIGURE 13.1   Functional magnetic resonance imaging (fMRI) scan of an anesthetized ferret during visual stimulation (1 Hz light flicker) shows activation localized to visual cortices (*left*) and delayed onset and slow time scale of fMRI signal (*right*). Periods with visual stimulation (10 s duration) are shaded in gray. Note the low temporal sampling rate of the time series.

Challenges and Limitations of fMRI

The temporal resolution of fMRI is low because of the delayed (later in onset) and slow (lower rise and fall rates) nature of the hemodynamic response compared to neuronal dynamics. Typically, the peak BOLD signal is detected 6–9 s after onset of neuronal (task-related) activity and can take up to 20 s to return to baseline after offset of the neuronal response (Fig. 13.1).

Overall, the relative amplitude of an fMRI response signal is typically in the low single digit percent range. Thus noise is a major issue in these measurements. There are a number of different noise sources. First, the body generates electromagnetic noise caused by the movement of ions. Second, measurement electronics are susceptible to noise. Third, other physiological signals such as heartbeats and breathing modulate the BOLD fMRI signal. Fourth, head movement leads to significant slow drift of the signal that can cause issues for movements as small as 0.1 mm.

RESTING STATE fMRI

Aside from the study of task-evoked activity with fMRI, an entire subfield of fMRI has emerged that focuses on measuring and understanding brain activity "at rest" (*resting state fMRI*, in short RS-fMRI [4]). The definition of the "resting state" as a behavioral state is derived from the instructions given to research participants, which typically are to "relax, stay awake, look straight ahead, and do not think of anything specific." Despite these instructions, the behavioral state of every individual participant can still vary; a number of additional confounds at least in theory could affect measured brain dynamics. Most prominently, visual processing (sometimes a fixation cross is presented), auditory processing (scanner noise), and remnants of previous tasks performed by the participant contribute to contaminations of the RS-fMRI signal. Despite these limitations, RS-fMRI has provided seminal insights into the large-scale organization of brain activity. Perhaps most importantly, the demonstration of spatiotemporally organized brain activity patterns in the absence of behavioral engagement has highlighted the importance of the endogenous dynamics of the brain. Similar dynamics have been found in RS-fMRI in various animal model species.

Networks of the Resting State

RS-fMRI scans typically last a few minutes and the resulting time series of all voxels are analyzed. The basis of the signals are spontaneous fluctuations in the BOLD signal at very low frequencies (<0.1 Hz). The basic analysis approach is based on finding groups of voxels that exhibit correlated BOLD signal time series. Sets of brain areas that undergo such correlated slow fluctuations in the BOLD signal are termed *resting state networks* (RSN). There are several RSNs, most prominently the so-called *default mode network* (DMN [5]). The DMN is composed of the medial prefrontal cortex, the posterior cingulate/retrosplenial cortex, the inferior parietal lobule, the lateral temporal cortex, and the hippocampal formation. Also multiple other networks can be distinguished in RS-fMRI; typically, these other networks correspond to task-based systems (eg, visual). A common conceptual model is that the DMN (also referred to as *task-negative* or *intrinsic* network) is in balance with the opposing task-based systems (*task-positive* or *extrinsic* network). Interestingly, there are a larger number of task-positive networks and only one task-negative network, the DMN. Task-positive networks that relate to sensory processing and motor output include the somatosensory network (primary and higher-order motor and somatosensory areas), the visual network (mostly occipital cortex), the auditory network (Heschl's gyrus, superior temporal gyrus, posterior insula), and the language network (Broca's area, Wernicke's area, and others). Other task-positive networks include brain areas involved in higher-order cognitive function, and are detectable not only in RS-fMRI but also in task-evoked fMRI. Two networks are implicated in attention: the *dorsal attention network* (intraparietal sulcus, frontal eye field) is linked to top-down control of attention. The *ventral attention network* (temporoparietal junction and ventral frontal cortex) is associated with bottom-up attention of salient sensory clues. Two additional networks are implicated in executive function: the *frontoparietal network* (lateral prefrontal cortex and inferior parietal module) and the *cingulo-opercular network* (medial superior frontal cortex, anterior insular, anterior prefrontal cortex).

Data Preprocessing

Analysis of RS-fMRI data requires preprocessing to eliminate unwanted temporal fluctuation of the signal that is not driven by the physiological signal of interest (such as head motion). Typically, the time series are band-pass filtered to remove slow drift and "high-frequency" noise (frequency range of interest spans from 0.009 to 0.08 Hz). Preprocessing steps can, in theory, distort the correlation metrics used to identify the RSNs. Two preprocessing steps have particularly been debated: *whole-brain regression* (also called *global signal regression*) and correction for head motion. Whole-brain regression is achieved by fitting the following equation:

$$s_i(t) = \alpha_i g(t) + x_i(t) \tag{13.1}$$

where $s_i(t)$ is the time course of a given voxel i, α_i is a regression coefficient, and $x_i(t)$ is the signal after this preprocessing step. The global signal average $g(t)$ is the average of the signal across all N voxels of a given scan:

$$g(t) = \frac{1}{N} \sum_{j=1}^{N} s_j(t) \tag{13.2}$$

Eq. (13.1) is solved for α_i to determine to what extent the measured signal $s_i(t)$ is a (linear) function of the global signal average $g(t)$. The residual $x_i(t)$ represents the postprocessed signal. The concern with this commonly used approach is that if before this step all correlations between the raw voxel signals are positive, the global averaging necessarily introduces negative correlations [6]. This makes the interpretation of findings that show anticorrelated networks difficult, as they may be an artifact of this preprocessing. Further clean-up steps include removal of the averaged ventricle and white matter signal by regression following the approach illustrated earlier. In addition, not fully corrected head motion can also lead to artifactual (ie, spurious) correlations. This is of particular concern for data from populations who are challenged by instructions to hold the head still (eg, children with an autism spectrum disorder).

Region of Interest Analysis

After completion of the preprocessing, several distinct algorithms can be used to determine the RSNs. The region of interest (ROI) method is based on the a priori selection of ROIs that correspond to the brain areas of interest; the average BOLD signal of all voxels within each ROI is determined and correlated between all pairs of ROIs. The resulting association matrix describes the functional connectivity between the ROIs. Alternatively, machine learning strategies can be used to cluster the time series of voxels.

Independent Component Analysis

Instead of a hypothesis-driven ROI approach, fMRI data can also be analyzed in a data-driven way by using *independent component analysis* (ICA [7]). This analysis does not make any assumptions about specific brain areas, but instead explores the statistical structure of an fMRI scan. The underlying assumption is that the recorded activity patterns are the weighted sum of different *component maps*. These component maps are spatial maps that are constant throughout the scan and represent the functional networks that are independent of each other. For example, the component map of the DMN would have nonzero values for the voxels that correspond to the brain areas included in the DMN, such as the medial prefrontal cortex. Data from an fMRI scan contain the time series of the signal amplitude of every voxel. We assume that at any given moment in time (time sample), the different components are activated to different degrees, quantified by weighting coefficients that are a function of time. For example, at time $t = 0$, the brain areas of the DMN light up. Thus the corresponding component map heavily contributes to explaining the signal at $t = 0$. At a later time point, the dorsal attention network may be activated and its corresponding component map explains the overall signal. Therefore the DMN component network has a large weighting coefficient at $t = 0$ and a small weighting coefficient for $t = 1$. Accordingly, the weighting coefficient for the dorsal attention network is small for $t = 0$ but large for $t = 1$. In other words, at every time step, the overall fMRI signal can be explained as a linear, weighted sum of the individual component maps. The weighting coefficients are different for each time step since the overall signal fluctuates with time. Each component map is also referred to as a source. ICA determines the component maps by requiring them to be mutually independent, a statistical

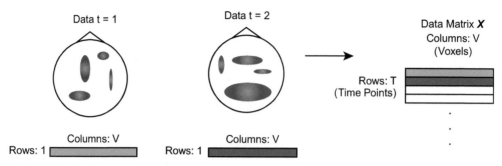

FIGURE 13.2   The data matrix X contains all the raw data from a functional magnetic resonance imaging (fMRI) scan (here shown as topographic maps of the head, *top view*). Each row corresponds to a time point that contains the activation of all voxels. Matrix X therefore has T rows (number of samples) and V columns (number of voxels).

requirement that is stronger than lack of correlation. Next we will discuss the ICA in more detail; first, we develop a matrix-based framework for the process explained earlier.

We begin by organizing the (raw) dataset into a data matrix X, where each row contains the activation of all voxels at a given time point. We denote the number of voxels as V; thus matrix X has V columns. We denote the number of time points at which acquisition was performed as T, which corresponds to the number of rows of matrix X. Every row corresponds to the image acquired at a given time step. In summary, X has T rows (number of acquisition time points or *samples*) and V columns (number of voxels) (Fig. 13.2).

Next, we want to decompose the data matrix X into a weighted sum of the components for each time step. So we introduce a matrix C that contains the components. Note that we will use ICA to determine C. Mathematically speaking, a component map is a row vector with length V. For example, if $V = 10,000$ voxels, each component map or source is a row vector with 10,000 elements (ie, a 1 by 10,000 matrix). We organize all of the component maps into the matrix C where every row corresponds to a component map. ICA requires that the number of component maps equals the number of samples, T, which is determined by the duration of the scan and the sampling frequency (Fig. 13.3).

The component matrix C is the desired result from ICA. To determine C, we need to understand how it relates to the acquired fMRI data of a given scan. The data matrix X relates

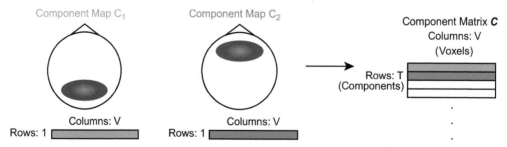

FIGURE 13.3   The component matrix C consists of rows that correspond to the individual components. These component maps are constant for the entire duration of the scan and correspond to the spatial networks that the resting state fMRI (RS-fMRI) approach aims to determine.

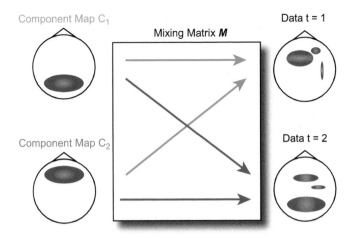

FIGURE 13.4   Mixing. The spatial activation profile measured at any time point is the result of a linear superposition of the different component maps. Mixing matrix *M* defines the relative contribution of every component at every acquisition time point. The light and dark blue arrows correspond to the weighting coefficients for $t = 1$ and $t = 2$, respectively.

to component map matrix *C* through an additional matrix that contains the relative contribution of each component (weighting coefficient) for every time step. This matrix defines how the different components are "mixed" together to generate the acquired data (Fig. 13.4).

This mixing matrix *M* contains a row for each time point of the acquisition. Every column denotes the relative contribution of a component, that is, the weighting coefficient. Thus the dimensions of this matrix are $T \times T$. In summary, the three matrices *X*, *M*, and *C* relate to each other as follows:

$$X = MC \tag{13.3}$$

In other words, the data matrix can be written as the matrix multiplication of the mixing matrix and the component matrix (Fig. 13.5). ICA is now used to determine the mixing matrix *M* and the component matrix *C*.

Using an iterative algorithm based on information theory, ICA provides a component matrix *C* where the individual rows (ie, components) have minimal mutual information and are

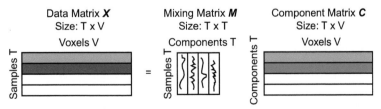

FIGURE 13.5   The data matrix *X* is the matrix product of the mixing matrix *M* and the component matrix *C*. ICA is used to calculate the matrices *M* and *C*.

thus maximally statistically independent. The inner workings of ICA are beyond the scope of this book, but it is important to understand what it means for the component maps to be independent. A weaker requirement would be to make the spatial maps uncorrelated. To illustrate these concepts, we consider the simplified case of only two maps. We further assume that there are 10,000 voxels. We interpret the 10,000 values of each map as 10,000 measurements of a random variable (technically, instantiation of a random process). Since there are two maps, that is, two random variables, we can plot these 10,000 pairs of numbers as a scatter plot. If they are correlated, there would be a significant correlation coefficient and we could fit a straight line. Since correlation is a linear measure, the data from the two component maps can be related by a nonlinear function not captured by the correlation measure. Independence is therefore the more stringent requirement. Mathematically, independence requires that the joint probability density function equals the product of the individual probability density function. For the component maps, such independence is typically achieved by sparse activation, meaning that only few voxels have nonzero values in any given component map. The spatial overlap between the component maps is thus low, but does not have to be zero for the maps to be independent. As a final remark, the analysis outlined here would generate T components, which for a typical RS-fMRI dataset would mean many hundreds of components. Therefore the dimensionality of the data is typically reduced as a further preprocessing step before the application of ICA.

SUMMARY AND OUTLOOK

First, we looked at how MRI can be used to measure vascular signals that correlate with neuronal activity. In particular, the BOLD signal measures an overcompensation in oxygenation supply following increased oxygen consumption caused by neuronal activity. fMRI has the unique strength of noninvasively visualizing neuronal activity. Its main limitation is poor temporal resolution. Second, we focused on one specific use of fMRI that has created an entire subfield of network neuroscience—the study of the resting state. We then linked brain activity during the resting state to fluctuations in the activation of distinct, statistically independent brain networks. We introduced the approach of using ICA to identify these brain networks or component maps. These functional networks can be both task negative, such as the DMN, and task positive, such as the dorsal and ventral attention network. In subsequent chapters, we will encounter the use of RS-fMRI to identify pathological changes in the macroscopic organization of brain activity in patients with neurological or psychiatric illnesses.

References

[1] Buxton RB, et al. Modeling the hemodynamic response to brain activation. NeuroImage 2004;23(Suppl. 1): S220–33.
[2] Logothetis NK. The neural basis of the blood-oxygen-level-dependent functional magnetic resonance imaging signal. Philos Trans R Soc Lond B Biol Sci 2002;357(1424):1003–37.
[3] Borogovac A, Asllani I. Arterial Spin Labeling (ASL) fMRI: advantages, theoretical constrains, and experimental challenges in neurosciences. Int J Biomed Imaging 2012;2012:818456.

[4] Lee MH, Smyser CD, Shimony JS. Resting-state fMRI: a review of methods and clinical applications. AJNR Am J Neuroradiol 2013;34(10):1866−72.

[5] Fox MD, Raichle ME. Spontaneous fluctuations in brain activity observed with functional magnetic resonance imaging. Nat Rev Neurosci 2007;8(9):700−11.

[6] Murphy K, et al. The impact of global signal regression on resting state correlations: are anti-correlated networks introduced? NeuroImage 2009;44(3):893−905.

[7] McKeown MJ, et al. Analysis of fMRI data by blind separation into independent spatial components. 1997. DTIC Document.

14

Deep Brain Stimulation

It has been known for hundreds of years that applying electricity to the nervous system changes its electrical signaling. Nevertheless, it took a long time for invasive brain stimulation to become an established therapy. Essential tremor and Parkinson's disease, both neurological disorders with prominent motor symptoms, were the first conditions for which invasive deep brain stimulation (DBS) became a mainstream therapeutic option. DBS received its name from the fact that electrode leads are implanted to target (subcortical) brain structures. In this chapter, we will discuss what we know about how DBS works, and we will review emerging clinical applications beyond Parkinson's disease (see chapter: Parkinson's Disease), including the treatment of psychiatric disorders. As we will see, unraveling the effects of DBS on interconnected circuits requires the integration of cellular, network, and behavioral studies.

DBS originated in surgical procedures that lesioned the areas now targeted with DBS. It was found that intraoperative stimulation of the target areas had a similar clinical effect to performing a permanent lesion of the area displaying pathological activity. Thus it was assumed that the mechanism of action of DBS was suppression of pathological activity by high-frequency stimulation. As a result, the development of DBS paradigms for other disorders focused on attempting to turn off pathological neural activity in target areas (in a reversible way, in contrast to surgical resection). We will revisit this original hypothesis and find that the mechanism of DBS is much more complex than initially thought.

DEEP BRAIN STIMULATION PROCEDURE

Treating a patient with DBS is a complex process. First, the patient needs to be medically cleared to ensure that the benefits of the procedure outweigh the risks. DBS requires invasive brain surgery that involves risks of bleeding and infection. Once a patient is deemed eligible for DBS, the target location for electrode placement must then be established. This requires detailed structural imaging of the brain by magnetic resonance imaging (MRI, see chapter: Imaging Structural Network With MRI) and of the bone by computed tomography scan. Merging these two datasets provides the required reference points for stereotactic targeting. Once presurgical planning is completed, the patient undergoes surgery for the implantation of the stimulation electrodes. Typically, before insertion of stimulation electrode(s), a smaller recording electrode is used to verify that the target area exhibits pathological network activity. In the case of Parkinson's disease, this activity takes the form of rhythmic bursting at the tremor. For the treatment of other disorders where the pathological network dynamics remain unclear, such recordings play a less significant role in confirming correct electrode placement. Stimulation electrodes are typically implanted bilaterally. In a short follow-up procedure, the stimulation leads are routed under the skin to the chest wall, where the stimulation device, which contains the battery and the control electronics, is implanted. The stimulator battery has a limited lifetime and needs to be replaced in a short outpatient procedure every few years. DBS devices provide a large number of possible stimulation paradigms defined by the choice of active electrode contacts and stimulation parameters such as amplitude and frequency. The stimulator can be programmed with new stimulation paradigms through a wireless device that communicates with it. Once the stimulator is implanted, a lengthy period of parameter adjustment is often required to find ideal settings for maximal symptom relief, since there is only very limited understanding of how different stimulation paradigms affect the network dynamics and thus the symptoms. Given this lack of a rational strategy to choose the right stimulation parameters, the clinician and the patient discuss the development of symptom severity and the clinician then adjusts the stimulation settings in routine outpatient visits that are typically scheduled every few weeks.

CLINICAL APPLICATIONS

To understand the benefits of DBS, it is also important to recognize its limitations. DBS is highly effective for the treatment of motor symptoms such as the tremor, rigidity, stiffness, slowed movement, and walking problems associated with Parkinson's disease. Importantly, DBS typically does not provide relief from cognitive impairment, but in some cases may have a negative impact on cognition. Furthermore, DBS itself does not slow the underlying neurodegenerative causes of the illnesses, but instead provides significant symptom relief. DBS is Food and Drug Administration (FDA) approved for the treatment of essential tremor (1997), Parkinson's disease (2002), dystonia (2003), and obsessive compulsive disorder (2009). DBS is considered a neurosurgical treatment. However, no major, irreversible changes to the brain are induced during surgery (with the exception of limited neuronal damage along the track of the inserted stimulation electrode). Stimulators and stimulation electrodes can be surgically explanted, so the procedure can be reversed. Double-blind studies, in which the stimulator is turned on and off without the treating physician and the patient knowing, can readily be performed (as opposed to conventional surgical manipulations). The stimulator can be turned on and off with prior consent by the patient to be enrolled in a study with a sham arm where no stimulation is applied for the duration of the study (open-label treatment is made available to patients after study completion).

In addition to the FDA-approved applications, DBS is studied for the treatment of other disorders such as chronic pain, major depressive disorder, Tourette syndrome, and epilepsy. These applications are mostly experimental and are not yet ready for clinical use. Several setbacks in the form of negative results from double-blind, placebo-controlled studies have illustrated the difficulties of translating treatment strategies that work in the research laboratory into the real world for treatment of a diverse set of severe neurological and psychiatric illnesses. Nevertheless, DBS is being actively pursued as a more advanced form of functional stereotactic neurosurgery that targets specific dysfunctional circuits. The transition from neurology to psychiatric applications reflects the growing insight that the two fields are less distinct than previously assumed.

Parkinson's Disease

For the treatment of Parkinson's disease, the original target was the ventral intermediate nucleus of the thalamus (VIM). The VIM is part of the ventrolateral nucleus (VL), which represents an important part of the motor thalamus. Stimulation of VIM in patients with Parkinson's disease relieves tremor but does not treat other motor symptoms associated with Parkinson's disease. Today, for the treatment of Parkinson's disease, electrodes are placed either in the subthalamic nucleus (STN) or in the internal segment of the globus pallidus (GPi). The resulting improvements in motor symptoms are comparable across the two stimulation targets. However, cognitive side effects appear to differ between the two targets. While DBS does not lead to improvements beyond what dopaminergic medication—which is the mainstay pharmacological treatment—can achieve, it eliminates the symptom deterioration in between medication doses and permits the reduction of medications and their associated side effects. DBS treatment for Parkinson's disease is more extensively discussed in chapter: Parkinson's Disease.

Essential Tremor

Essential tremor is a neurological condition characterized by involuntary shaking movement with unknown cause. DBS is approved for the treatment of essential tremor. The stimulation electrodes are positioned in the VIM, which together with the nucleus ventralis oralis comprises the VL of the thalamus. In terms of motor function, the VL is a part of two major loops and receives input from two distinct circuits (Fig. 14.1). The first circuit is the loop through the pallidum where input from the cortex is processed in a dopamine-dependent way and then relayed as inhibitory signals to the VL. The second circuit processes input from the cerebellum and provides excitatory input to the VL. There are opposing gradients of these two classes of afferent fibers such that a given location within the VL receives a weighted mix of both types of input. The afferents exhibit somatotopic organization such that fibers that correspond to the same body part terminate in the same location. Thus the anatomy suggests that the VL acts as an integration site, where input from the two pathways is processed and integrated. Because of dopaminergic modulation, the pallidal pathway provides motivational context, while the cerebellar pathway provides proprioceptive context. This convergence of the two pathways in the VL maps onto the divergence of the projections to different areas of the motor cortex. Areas in the VL that receive predominantly cerebellar input project to primary motor, premotor, and presupplementary motor cortex. Areas that receive mostly pallidal input preferentially project to primary motor and premotor cortex.

Dystonia

Dystonia is a neurological illness characterized by involuntary muscle contractions that cause abnormal postures and twisting movements (Fig. 14.2, [1]). Dystonia can affect different parts of the body, and the illness can range from mild to severe. Dystonia is the result of pathological signaling in the basal ganglia. The cause of most forms of dystonia is

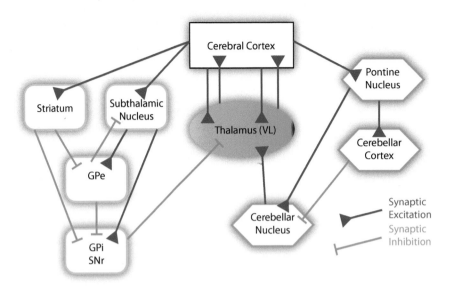

FIGURE 14.1   **Motor loops through the thalamus.** Deep brain stimulation (DBS) of the thalamus is performed in patients with essential tremor and in patients with Parkinson's disease in cases of tremor- dominating symptom manifestation and concerns about neuropsychiatric side effects of stimulating the subthalamic nucleus (STN). The ventrolateral nucleus (VL) exhibits *tremor cells* that fire in discharge patterns with frequencies similar to the tremor frequency. *GPe*, External globus pallidus; *GPi*, internal globus pallidus; *SNr*, substantia nigra pars reticulata. *Rounded yellow squares*, Brain areas that constitute the pallidal pathway. *Green hexagons*, Brain areas that constitute the cerebellar pathway.

FIGURE 14.2   Postures in patients with dystonia are caused by simultaneous contraction of agonist and antagonist muscle groups. Examples are pathological hyperextension of the neck (*left*) and twisting of the hand and fingers (*right*).

unknown; however, there are several genetic and environmental (eg, birth injury) risk factors. *Primary dystonia* refers to dystonia that has no apparent cause and is not associated with any other neurological symptoms aside from tremor and *myoclonus* (sudden, involuntary jerking of muscles). *Secondary dystonia* refers to dystonia caused by other diseases or brain injury. FDA-approved treatment options include the injection of botulinum toxin into the affected muscle groups, since the toxin blocks the release of acetylcholine at the neuromuscular junction, which is required for triggering muscle contractions. Also medications that (1) block the cholinergic system (anticholinergic agents), (2) enhance GABAergic signaling (eg, benzodiaz-epines), or (3) modulate dopaminergic tone are prescribed *off-label*, meaning that these med-ications are only FDA approved for other clinical indications.

DBS has emerged as a promising alternative given its success with treating Parkinson's disease and essential tremor by targeting pathological activity in the basal ganglia. DBS for dystonia has been shown to be effective in randomized, sham-controlled studies and is approved under a humanitarian device exemption for chronic, medically intractable dystonia in the United States. DBS is particularly effective for primary dystonia, and the start of DBS early in the disease course is associated with better response. Both children and adults with dystonia benefit from DBS. Using the pallidum as a target is motivated by the fact that pal-lidotomy (surgical lesioning of the pallidum) is an effective strategy to reduce dystonia symp-toms. Most often the GPi is targeted. The higher the firing rate of neurons in the GPi, the less severe the dystonia, suggesting that dystonia results from decreased inhibition of the motor thalamus. In contrast to DBS for Parkinson's disease, DBS of the GPi for dystonia takes several months for symptom improvement to be observed. Therefore DBS-induced plasticity may play an important role in reconfiguration of the impaired circuits. Further support for the role of plasticity comes from reports of patients who were able to discontinue DBS without clinical worsening after a prolonged period of DBS treatment. Leveraging this plas-ticity to design more effective stimulation paradigms is of high priority. Today's DBS para-digms for dystonia require high amplitude stimulation, which leads to accelerated battery depletion and the need for frequent (surgical) battery replacement.

Depression

Treatment of depression (see chapter: Major Depressive Disorder) with DBS is less straightforward than the treatment of movement disorders for two reasons. First, it requires the reconceptualization of psychiatric illnesses in a neurobiological framework as circuit or network disorders. Second, it presents the challenge of finding a specific region of the brain as a target that can be modulated with DBS. The subcallosal cingulate (area SCC 25) was one area that appeared to be a promising target. Functional imaging has shown increased activity in SCC 25 with depressed mood, for example, in response to activation of sad memories. Suc-cessful treatment of depression with selective serotonin reuptake inhibitors, repetitive trans-cranial magnetic stimulation, and even placebo all reduced activity in this area. DBS of that area led to quite remarkable improvement in initial (open-label) studies, although not all implanted patients benefited equally [2]. However, a larger, double-blind, sham-controlled study of area SCC 25 as a target was halted prematurely because of lack of efficacy of the stimulation (not yet published). Furthermore, another randomized, sham-controlled study of a different target (ventral capsule/ventral striatum) also failed [3]. This second target had emerged from the observation that patients implanted with DBS electrodes in those brain regions for the treatment of obsessive compulsive disorder had also shown improvement in depressive symptoms. These results were a setback for the field. Together, the results empha-sized the importance of (1) a better understanding of the causes of outcome heterogeneity, (2) further improvements in targeting strategy and stimulation delivery, (3) elucidating how high-frequency stimulation interacts with the surrounding tissue, and (4) continued research of the neurobiological circuits of depression.

MECHANISMS OF DEEP BRAIN STIMULATION

The initial assumption was that DBS prevents neuronal firing (often referred to as inhibition, not to be confused with synaptic inhibition) since stimulation was effective for the location at which lesion surgery was typically performed. Subsequently, this initially proposed mechanism of action has evolved into a broader discussion about whether DBS is "inhibitory," "excitatory," or something else. No consensus has emerged about how DBS causes clinically beneficial changes to brain activity. A number of different hypotheses postulate specific mechanisms [4,5]. Originally, it was assumed that high-frequency stimulation sufficiently depolarizes neurons to achieve *depolarization block*, a state in which voltage-gated sodium channels are inactivated to a point where action potential can no longer be generated (see chapter: Dynamics of the Action Potential). Alternatively, it has been proposed that the stimulation reduces neuronal activity by preferentially activating inhibitory interneurons. Similarly, stimulation has been hypothesized to trigger glutamate release from astrocytes, causing activation of inhibitory interneurons. Also, in theory, stimulation could preferentially modulate the activity of nearby axons (*fibers of passage*) and not of the activity of the somata within the target structure. Lastly, high-frequency stimulation may interfere with pathological oscillations such as those associated with Parkinson's disease and essential tremor. Adding to the complexity of the discussion is the fact that different brain areas (at least in theory) may respond differently to the same stimulation paradigm, and that the effect of the stimulation critically depends on exact stimulation parameters such as stimulation frequency.

The most basic questions concern the nature of the neuronal elements activated by DBS. For example, does DBS preferentially activate axons, dendrites, or somata? One classical approach to characterize the neuronal excitability in response to electric stimulation is based on the principal that amplitude and duration of stimulation relate to each other in terms of their effects. To achieve the same stimulation outcome, lower-amplitude stimulation needs to last longer. The lowest stimulation amplitude that has an effect (for asymptotically long stimulation) is called the *rheobase*. The steeper the decay of the required amplitude for increasing pulse duration, the more excitable a given neuronal element is (Fig. 14.3). This decay is measured as a time constant termed *chronaxie*, which is defined as the required stimulation pulse length for activation with twice the rheobase current. For myelinated fibers, the chronaxie is orders of magnitude smaller than for dendrites and somata. As a result, DBS can be expected to primarily modulate electric activity in axons.

Studies addressing the neurophysiological mechanisms of DBS have been performed both in slice preparation (in vitro) and in intact animals (in vivo). Slice preparation allows the study of a network in isolation, and offers unique experimental access for recording, stimulation, and pharmacological manipulations. For unknown reasons, the findings from in vitro studies are quite heterogeneous. For example, some studies found that synaptic transmission plays an important role in mediating the effects of DBS, where other studies found no effect of blocking synaptic transmission by antagonist application. A common thread in these studies is that high-frequency stimulation in the DBS range (above 100 Hz) does not necessarily block neuronal firing but instead may override endogenous activity dynamics by replacing neuronal firing with stimulation-locked firing. However, these animal studies use rodents,

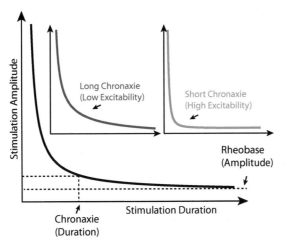

FIGURE 14.3 Excitability defined as the required stimulation amplitude and duration to elicit a neuronal response. Long (*blue*) and short (*red*) chronaxie denote low and high excitability, respectively.

and there are substantial differences in the circuitry between rodents and primates. For example, the rodent motor thalamus does not contain local inhibitory interneurons, whereas the equivalent structures in the primate do. In contrast, the in vivo model provides the opportunity to probe the effect of stimulation in the primate Parkinson's disease model (see chapter: Parkinson's Disease). High-frequency DBS of the STN increases the firing rate in both the GPi and external globus pallidus (GPe). These effects are likely mediated by increased output of the STN to the GPi and antidromic activation of projection fibers from GPe. However, firing rate decreases have been observed in STN in response to brief epochs of DBS stimulation in humans. Overall, the one firm conclusion that arises from all electrophysiological studies is that it is likely that the effect of DBS depends on multiple variables, including brain area, stimulation amplitude, duration, and frequency. DBS of the STN is likely to increase activity in the GPi. Thus the effect of DBS is not "inhibitory" in the sense that it silences the target area. Rather, DBS disrupts pathological signal in multiple, complex ways as a function of the stimulation parameters.

In addition to the previously discussed electrophysiological strategies, functional imaging provides another important tool to study the circuit-level effect of DBS. However, several important issues need to be considered. First, turning the stimulator off can cause an increase in tremor, which results in increased sensory feedback not directly caused by the stimulation. Second, functional MRI studies are challenging because of concerns about the effect of the magnetic fields of the scanner on the stimulation hardware. However, a number of studies have addressed these limitations and provide support for the findings from electrophysiological studies that associate DBS of the STN with an increase in activity in the circuit.

One school of thought about how DBS works focuses on neuronal oscillations (see also unit: Cortical Oscillations). In the so-called oscillation model of the basal ganglia, rhythmic activity is divided into *antikinetic* (hindering movement, frequencies in the theta and beta frequency bands) and *prokinetic* (enabling movement, gamma frequency band oscillations). Indeed, both human patients and animal models with motor symptoms exhibit pathologically increased beta oscillations in both the basal ganglia and in the motor cortex. Medication that increases dopaminergic tone (eg, the dopamine precursor L-Dopa) reduces beta oscillations and increases gamma oscillations. Discontinuation of dopaminergic medication increases the beta oscillation recorded from the DBS stimulation electrodes in the GPi and the STN. The belief is that DBS reduces pathological beta oscillations and synchronization between the STN and the motor cortex. Indeed, there is evidence that clinically effective DBS reduces the power and coherence of beta oscillations. Early results suggest that targeting individual oscillation frequencies may provide similar relief to conventional high-frequency DBS with the benefit of reducing power consumption and increasing battery life. Increasing battery life is of high clinical relevance since every battery replacement requires an outpatient invasive procedure. In further support of DBS targeting beta oscillations, closed-loop control DBS (application of DBS whenever beta oscillation power crossed a threshold, Fig. 14.4) was shown to be more effective than continuous (open-loop) DBS or randomly applied DBS [6]. This suggests that individualized targeting of oscillatory activity is a promising, emerging direction.

Electrocorticography (ECoG, see chapter: LFP and EEG) over the cortex, carried out by placing grid electrodes using the burr hole for inserting the DBS electrodes, enables a more direct assessment of interaction dynamics between activity in the STN and the primary motor cortex [7]. These recordings point toward pathological *phase-amplitude coupling* (PAC)

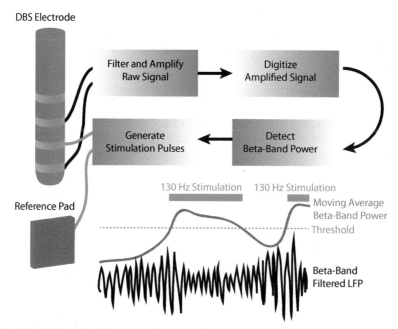

FIGURE 14.4   Closed-loop deep brain stimulation (DBS) that provides stimulation during periods of increased beta-band activity recorded from contacts of the DBS macroelectrode. The recorded local field potential (LFP) is amplified, digitized, and processed by a computer. When the moving average (in *gray*) of the beta-band filtered LFP (*black*) surpasses a threshold (*red, dashed line*), 130-Hz stimulation is applied through the stimulation electrode. Details in Ref. [6].

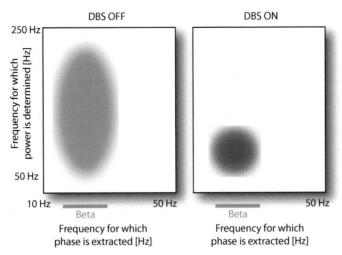

FIGURE 14.5   Phase-amplitude coupling (PAC) determines the power of an oscillation at a higher frequency as a function of the phase of an oscillation at a lower frequency. PAC of electrocorticography (ECoG) signals recorded from motor cortex without (*left*) and with (*right*) deep brain stimulation (DBS) show that in the "DBS off" condition, the phase of beta oscillations is coupled to high-frequency oscillations up to the high gamma range (see chapter: Gamma Oscillations). In the "DBA on" condition, PAC of the beta phase with the gamma oscillation is more limited. Details in Ref. [7].

between the phase of beta-band activity and power of the gamma-band activity in the primary motor cortex. PAC measures to what extent the occurrence of a faster oscillation is gated by the phase of a lower-frequency oscillation. DBS reduced this pathological PAC, and the stimulation-induced reduction correlated with improvements of some patients' *symptom scales* (numerical assessment of symptom severity) (Fig. 14.5). Magnetoencephalography (MEG, see chapter: LFP and EEG) has also been used to investigate the changes in cortical network dynamics by DBS. The advantages of MEG in this context are the relatively large number of sensors that provide better spatial resolution, and improved capability to suppress the stimulation artifact.

SUMMARY AND OUTLOOK

In this chapter, we learned about the application of DBS to several different neurological and psychiatric illnesses such as essential tremor, Parkinson's disease, dystonia, and depression. While the application of DBS for movement disorders is relatively easy to explain in terms of the targeted circuit, the case for depression (and other psychiatric illnesses) is much more complex. The failure of several controlled clinical trials underlines the complexity of the effect of stimulation. In terms of mechanisms, DBS has evolved from being considered an alternative to a surgical lesion by "inhibiting" neuronal activity to becoming a tool to disrupt pathological synchronization in large-scale brain circuits.

Numerous innovations are poised to further improve efficacy and expand the number of clinical applications of DBS. Improved neuroimaging strategies can enhance targeting precision. Intraoperative MRI is not straightforward because of induction of unwanted current in the implanted electrode. However, it provides improved real-time targeting during implantation and may enable surgery under general anesthesia because of the achieved targeting precision. Today, whenever possible, anesthesia is partially reversed to verify stimulation electrode location by symptom improvement on the operating table. Diffusion tensor imaging (see chapter: Imaging Structural Network With MRI) can be used to identify fiber bundles that have been implicated as optimal targets for certain investigational applications of DBS. Also the device hardware is undergoing many changes. Basic improvements such as wireless charging can improve patient quality of life, since fewer battery replacement surgeries are needed.

Perhaps the most fundamental advance is stepping away from continuous stimulation that does not take into account (pathological) neuronal activity or how activity is modulated by stimulation. Such open-loop stimulation was adopted for DBS from cardiac pacemaking. Closed-loop, adaptive stimulation automatically adjusts stimulation in real time based on objective neurophysiological measures. Newer devices have this capability, in theory, since neuronal activity can be recorded and processed simultaneously by the device. However, it remains unclear what the stimulation algorithms should look like. As discussed earlier, stimulation controlled by the detection of beta oscillations, which are associated with inhibition of movement, is more effective than standard DBS and also consumes less power.

Together, these innovations are likely to lead to an expansion in clinical applications of DBS to the benefit of patients with severe, medication-refractory disorders of the central nervous system.

References

[1] Breakefield XO, et al. The pathophysiological basis of dystonias. Nat Rev Neurosci 2008;9(3):222—34.
[2] Crowell AL, et al. Characterizing the therapeutic response to deep brain stimulation for treatment-resistant depression: a single center long-term perspective. Front Integr Neurosci 2015;9:41.
[3] Dougherty DD, et al. A randomized sham-controlled trial of deep brain stimulation of the ventral capsule/ventral striatum for chronic treatment-resistant depression. Biol Psychiatry 2015;78(4):240—8.
[4] Kringelbach ML, et al. Translational principles of deep brain stimulation. Nat Rev Neurosci 2007;8(8):623—35.
[5] Alhourani A, et al. Network effects of deep brain stimulation. J Neurophysiol 2015;114(4):2105—17.
[6] Little S, et al. Adaptive deep brain stimulation in advanced Parkinson disease. Ann Neurol 2013;74(3):449—57.
[7] de Hemptinne C, et al. Therapeutic deep brain stimulation reduces cortical phase-amplitude coupling in Parkinson's disease. Nat Neurosci 2015;18(5):779—86.

15

Noninvasive Brain Stimulation

We can characterize and ultimately understand a system by establishing the relationship between applied inputs and the resulting system outputs. This principle also applies to the brain. Perturbing brain activity allows us to test specific hypotheses about the role of neuronal network dynamics in behavior. For example, the stimulation of specific neuronal populations to identify their role in behavior has become a powerful approach in animal experiments with the advent of optogenetics (see chapter: Optical Measurements and Perturbations). Using a similar rationale, noninvasive brain stimulation (NIBS) provides tools for perturbation of brain activity in humans. Although spatial targeting with NIBS is less specific than in the invasive strategies commonly applied in animal models, its use in humans has become a critical tool in network neuroscience. In addition to serving as a technique to establish the importance of specific brain areas and the causal role of activity patterns in behavior, NIBS can be used to correct pathological activity patterns in the brain, effectively treating disorders of the central nervous system. In fact, some forms of NIBS have received approval by the Food and Drug Administration (FDA) as medical treatments for illnesses such as depression. In this chapter, we will focus on two predominant forms of NIBS: transcranial magnetic stimulation (TMS) and transcranial electrical stimulation (tES, also called transcranial current stimulation, TCS).

TRANSCRANIAL MAGNETIC STIMULATION

TMS is an NIBS modality that applies a focal, targeted electromagnetic pulse to the brain (Fig. 15.1). A TMS device consists of the main unit, which includes the control electronics and user interface, and a stimulation coil that is connected to the main unit (Fig. 15.2). The magnetic field is generated by passing a strong current through a tightly wound wire within the coil. The geometry of the coil and the current delivered to the coil determine the magnetic field distribution. The stimulation pulse is very brief (less than a millisecond), and the applied magnetic field induces an electric field of around 100 V/m in the brain. Typical TMS protocols utilize pulses with rise times of about 100 µs and peak field strengths of 1 T or more. We will see later that the electric field induced in the brain with TMS is about 100 times stronger than the field generated by tES. Most importantly, the field produced by TMS is strong enough to trigger action potentials (ie, to achieve *suprathreshold stimulation*). For example, a single TMS pulse over the primary motor cortex

197

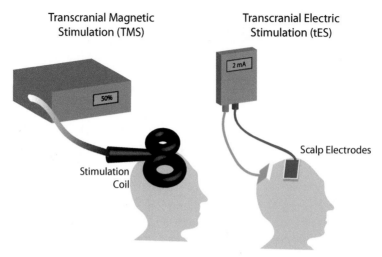

FIGURE 15.1   Noninvasive brain stimulation (NIBS). *Left*: Transcranial magnetic stimulation (TMS). Stimulation is applied with a stimulation coil that is positioned next to the scalp. *Right*: Transcranial electrical stimulation (tES). Stimulation is applied through scalp electrodes.

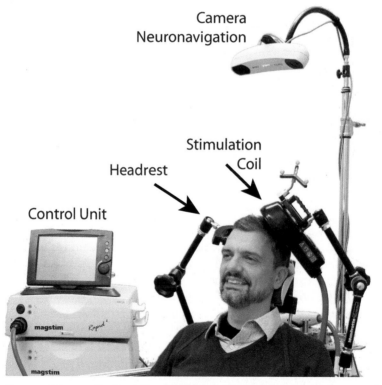

FIGURE 15.2   Transcranial magnetic stimulation (TMS) setup. The participant sits in a custom chair that enables fixation of the head position with a headrest. The stimulation coil is positioned close to the head. Computer vision software combined with a camera is used for neuronavigation to enable spatial targeting based on previously acquired neuroimaging data of the participant. The control unit provides the user interface to apply stimulation.

(M1) causes a twitch in the muscle group corresponding to the targeted region within M1. This muscle contraction is quantified by measuring the electric activity evoked in the muscle, a signal called the *motor-evoked potential* (MEP, Fig. 15.2). In TMS, the stimulation amplitude is often indicated by the device as a percentage of the maximum stimulation possible, a value known as the output intensity. The *motor threshold* is determined by finding the stimulation amplitude that causes a clearly detectable MEP (typically at least 20 μV in amplitude) in at least half the stimulation trials. Typically, the motor threshold is determined to individualize the stimulation amplitude. In *paired-pulse stimulation*, two TMS pulses are applied in rapid succession. Depending on the interval between the pulses, the amplitude of the resulting MEP is either decreased (short interpulse intervals up to 5 ms) or increased (intervals between 6 and 25 ms) (Fig. 15.3). This approach is limited to M1, since most brain areas do not provide such a direct physiological readout of stimulation effects. Nevertheless, stimulation parameters for other areas are often chosen based on insights from stimulating the motor cortex. Electroencephalography (EEG) can also be used to determine responses evoked by TMS.

TMS was originally developed as a research tool to probe network dynamics and brain function [1]. TMS has revolutionized cognitive neuroscience by enabling, for the first time, direct and safe probing of the role of different cortical circuit nodes in generating activity patterns that drive cognition and behavior. For example, several studies have employed single-pulse TMS to disrupt neuronal function at specific time points as a means of demonstrating the relevance of the targeted area at a given moment in time during a task. Single-pulse TMS induces brief *virtual lesions*, which are temporary and reversible. The term virtual lesion stems from the fact that this transient behavioral effect of TMS is similar to what is seen in patients with lesions in the structure of interest. Conceptually, the activity induced by the stimulation pulse is seen as additional neuronal activity that is not task related and therefore causes disruption of function. In a commonly utilized study design, the participant repeatedly solves a task (multiple trials) while precisely timed stimulation is applied at different time points in each trial. Performance impairment at some time points but not others points to the temporal order of neuronal processing and the functional role of the targeted area.

Several TMS paradigms differ in terms of the number and frequency of stimulation pulses (Fig. 15.4). Stimulation protocols are typically separated into "excitatory" and "inhibitory" paradigms (no relation to synaptic excitation and inhibition). These rather broad terms describe whether or not the stimulation enhances or decreases activity. In an attempt to develop TMS paradigms that cause long-lasting changes in network function, repetitive TMS (rTMS) was introduced. rTMS applies TMS pulse trains with up to few thousand pulses per session. It is commonly assumed that the frequency of the pulse train determines if a protocol is excitatory or inhibitory. Low-frequency (1 Hz) rTMS is considered "inhibitory." Continuous theta-burst stimulation (cTBS) consists of bursts of three 50 Hz pulses that are applied every 200 ms (ie, 5 Hz) for a total duration of several tens of seconds. Although it has been determined that electrical stimulation in the theta frequency range (4—8 Hz) is ideal for inducing synaptic plasticity, cTBS is considered "inhibitory." The effect of 10 Hz rTMS is less clear, but is typically assumed to be excitatory. rTMS of the left prefrontal cortex at 10 Hz [2] is approved by the US FDA for the treatment of certain types of depression (see toolbox: Psychiatry). As a treatment for depression, rTMS is typically applied for many sessions spread out over multiple weeks. Currently, rTMS is being evaluated for the treatment of other

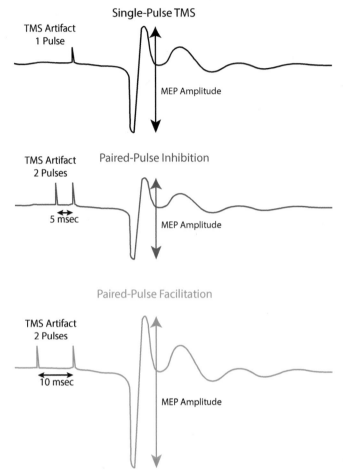

FIGURE 15.3   Single-pulse and paired-pulse transcranial magnetic stimulation (TMS). A single TMS pulse elicits a motor-evoked potential (MEP) in the target muscle (*black*). If two pulses are applied in rapid succession, the duration between the two pulses determines if the MEP amplitude is decreased (*blue*, inhibition, intervals up to 5 ms) or enhanced (*red*, facilitation, interval larger than 5 ms).

psychiatric illnesses, although the results so far have been mixed [3]. rTMS has also been successfully used for cognitive enhancement [4].

TMS enables targeted modulation of brain activity and behavior, making its use ideal in network neuroscience studies in humans. However, since TMS is accompanied by clicking noises and rather intense skin sensations, not all behavioral results (eg, improved cognitive performance) are a consequence of the direct manipulation of brain activity by the TMS pulse alone. Rather, the stimulation and its accompanying noises and sensations can alter overall arousal and thus influence attention. Additionally, the sensory experience of TMS can facilitate the processing of task-related sensory stimuli. This makes the careful design of experimental controls such as alternate stimulation locations and

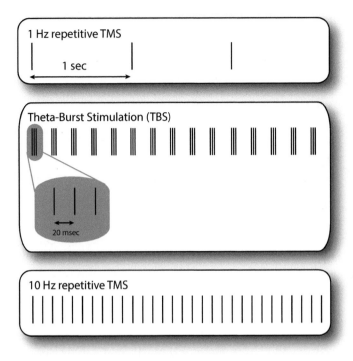

FIGURE 15.4 Transcranial magnetic stimulation (TMS) paradigms with repetitive pulses. *Top*: 1 Hz rTMS is considered inhibitory. *Middle*: Theta-burst stimulation (TBS), which employs triplets of pulses (at 50 Hz, ie, spaced at 20 ms) applied at 5 Hz (theta frequency band), is considered inhibitory. *Bottom*: 10 Hz rTMS is considered excitatory.

sham conditions of paramount importance. Examples of such controls are noise cancellation, which removes the auditory stimulation caused by the noises of the TMS machine, and electrical stimulation of the skin, which is used to mimic the somatosensory experience of *verum stimulation*.

One strength of TMS is its ability to selectively target a specific brain region. Several different targeting methods can be used, ranging from the simple and inaccurate to the very sophisticated. The simplest of these approaches is to use standard EEG coordinates (ie, the 10–20 system, see chapter: LFP and EEG). This approach uses a standardized coordinate system on the scalp for targeting stimulation and not the actual brain structure of interest. However, variability among individuals means that the scalp coordinates across individuals can correspond to different structures. More sophisticated approaches require individual structural MR scans of the participants (see chapter: Imaging Structural Network With MRI), which allow the use of *neuronavigation* to target a brain area based on the individual neuroanatomy. Neuronavigation systems capture the location of the stimulation coil with a 3D computer vision system, and map the position of the coil onto the previously acquired MR scan of the participant. Simulation of the resulting electric field as a function of the coil position is then projected directly onto the scan. Both structural and functional MR scans can be used with such neuronavigation approaches. Monitoring the neurophysiological effects of

TMS can be performed with EEG and magnetic resonance imaging (MRI). In EEG-TMS, the main technical problem that must be solved is the suppression of strong artifacts induced by the TMS pulse. Amplifier *saturation*, a state in which the amplifier no longer responds to incoming input, is the primary challenge of using EEG-TMS since recovery from saturation takes time. Saturation makes it likely that an electrophysiological signal of interest immediately after a TMS pulse will be missed. Successful solutions to this artifact problem include protection circuits built into the EEG amplifier (which protect the amplifier from saturating in response to the TMS-induced field) and blanking strategies (where the amplifiers are briefly switched to ground instead of the recording electrodes during the stimulation). TMS can also be combined with functional MRI (fMRI). In the easier approach, fMRI scans are performed briefly before and/or after stimulation. In the more difficult approach, TMS is performed inside the scanner. This approach requires careful engineering to avoid artifacts and, more importantly, injury from movement of the TMS coil in the scanner caused by the magnetic field of the scanner.

The effect of stimulation depends on a multitude of factors that can easily introduce heterogeneity between trials and participants. This heterogeneity can be sufficient to mask the phenomenon of interest. At first glance, applying TMS is seemingly similar to the most fundamental way of probing systems: application of a very brief perturbation that causes an "impulse response," which in the case of a linear time-invariant system provides a full characterization of the system. However, this conceptual view does not capture the effects of TMS, as well as other forms of brain stimulation, because the brain is almost never quiet. Instead, it is constantly engaged in generating sophisticated, large-scale activity patterns that respond to stimulation differently. As a rule, the effect of stimulation is *state dependent*, that is, a function of the activity present at the moment of stimulation. For example, alpha oscillations (see chapter: Alpha Oscillations) appear to influence perception by periodically modulating excitability of the visual system [5a]. Similarly, applying TMS pulses to the visual system at different phases of alpha oscillation causes different physiological responses [5b]. Also a number of factors known to modulate plasticity have been identified as factors able to alter response to TMS. For example, aging is associated with reduced plasticity and reduced response to plasticity-inducing TMS paradigms. Gender and fluctuations in sex hormones in women also influence plasticity and the response to TMS [6].

TRANSCRANIAL ELECTRICAL STIMULATION

tES is another form of NIBS modality (Fig. 15.1). In its most basic form, a weak electric current (typically 2 mA or less) is passed through the head via two relatively large (eg, 5×7 cm) sponge or rubber-carbon electrodes [7a]. There are multiple types of tES, distinguished by their stimulation waveform, that is, the time series of the stimulation current (Fig. 15.5).

The most basic and frequently studied form of tES is *transcranial direct current stimulation* (tDCS), which employs a constant current (or direct current, in the language of electrical engineering). In contrast, *transcranial alternating current stimulation* (tACS) utilizes sine-wave stimulation waveforms at different frequencies. A third type of tES is transcranial random noise stimulation (tRNS), which uses band-pass filtered noise waveforms. It is likely that

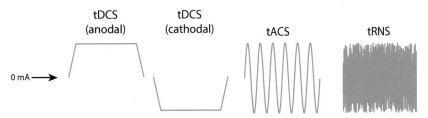

FIGURE 15.5 tES stimulation waveforms (*from left to right*). Transcranial direct current stimulation (tDCS) uses a constant current with a "ramp in" and "ramp out" to reduce skin sensation. Anodal stimulation denotes stimulation with a positive current, whereas cathodal stimulation denotes stimulation with a negative current. Transcranial alternating current stimulation (tACS) uses sine waves to target cortical oscillations of interest. Transcranial random noise stimulation (tRNS) uses a broadband noise stimulation waveform.

the different waveforms engage neurons and circuits in unique ways; thus the mechanisms of action are expected to differ. Next, we will discuss each of the aforementioned forms of tES.

Transcranial Direct Current Stimulation

tDCS uses a constant current for stimulation. By convention, the two stimulation electrodes are referred to as the *anode* and *cathode*. The current flows from the anode (typically marked in *red*) to the cathode (*blue*). The current traveling between the two electrodes is a function of the tissue between the electrodes, making it a function of both electrode location and anatomy. The current preferentially follows pathways through tissues of low electrical resistance (see toolbox: Physics of Electric Fields) such that a significant fraction of the current never reaches the brain but is shunted through the scalp and cerebrospinal fluid. The resulting electric field in the cortex is estimated to be about 0.3 V/m for each 1 mA of stimulation. Neurons with pronounced somatodendritic axes act like antennae, and change their membrane voltage in response to an applied electric field parallel to their somatodendritic axes. This change in membrane voltage is not uniform across the cell and is called *polarization*. When imagining a pyramidal cell that is parallel to the direction of the electric current (and thus the electric field), neurons under the anode exhibit a depolarized soma and hyperpolarized apical dendrites (see toolbox: Neurons). In contrast, neurons under the cathode exhibit the opposite polarization; the soma is hyperpolarized and the apical dendrites depolarized [7b]. This change in membrane voltage has been studied extensively by intracellular recordings of the somatic membrane voltage of neurons in vitro in the presence of homogeneous electric fields with different amplitudes. Overall, the amplitude depends on the cell morphology (and the electric state of the neuron), but it is typically less than 0.5 mV per 1 V/m of applied electric field. Therefore the change in membrane voltage is clearly not sufficient to elicit action potentials in quiet neurons, because the required depolarization from the resting membrane voltage (typically −70 to −60 mV) to spiking threshold is usually about 20 mV. This means that the mechanism of tDCS does not involve direct activation of quiet neurons. Since neurons are virtually never at their resting membrane voltage but are much more likely engaged in complex network activity patterns, their membrane voltage exhibits fluctuations that include values very close to or above the threshold for

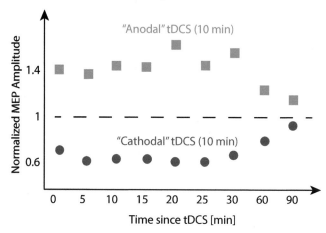

FIGURE 15.6   Modulation of motor-evoked potentials (MEPs) by 10 min of anodal (*red squares*) and cathodal (*blue circles*) tDCS. Anodal stimulation increased the MEP amplitude whereas cathodal stimulation had the opposite effect. The effects lasted for about an hour. *Adapted from Nitsche MA, Doemkes S, Karaköse T, Antal A, Liebetanz D, Lang N, Tergau F, Paulus W. Shaping the effects of transcranial direct current stimulation of the human motor cortex. J Neurophysiol April 2007;97(4):3109—17. Epub January 24, 2007.*

action potential firing. In this case, even a very weak perturbation of the membrane voltage can alter the firing rate caused by the highly nonlinear mechanism of spike initiation (see chapter: Dynamics of the Action Potential).

The first experiment that propelled tDCS into the limelight was based on neurophysiological validation of a change in motor cortex excitability [8]. In this experiment, 10 min of tDCS was applied to the motor cortex, followed by probing of network excitability with TMS and measurement of the MEP (Fig. 15.6). Positive current over M1 (anodal stimulation) increased the excitability, whereas a negative current decreased excitability (cathodal stimulation). These experiments triggered an unabated wave of enthusiasm for tES. However, such a bidirectional modulation of excitability in the motor cortex does not imply that other cortical areas will respond the same way. Most tDCS studies are based on this oversimplified approach of "anodal" and "cathodal" stimulation to enhance and decrease neuronal activity, respectively. A number of pharmacological studies have been performed to tease apart the mechanisms through which tDCS alters network function. As yet no clear explanation has emerged, but it has become evident that the effect of tDCS can be pronouncedly modulated by pharmacological agents such as benzodiazepines (see chapter: Synaptic Transmission), which are routinely used in neurology and psychiatry. Regarding the mechanism of action, there is a gap in understanding between the well-known (nearly) instantaneous effect on membrane voltage and the persistent effects after many minutes of stimulation (sometimes also called "offline" or "outlasting" effects). Synaptic plasticity is an obvious candidate for these outlasting changes. Evidence from animal experiments and human studies suggests an important role of the protein *brain-derived neurotrophic factor* (BDNF) in these effects. Since BDNF is an essential protein for synaptic plasticity, the persistent changes are thought to be mediated by synaptic plasticity (see chapter: Synaptic Plasticity). Indeed, in vitro studies have shown that the combination of weak electric fields and stimulation of afferent pathways caused increased long-term potentiation compared to stimulation alone [9].

EEG and fMRI can be combined with tDCS, and significant changes in those markers of brain function have been found. The literature on behavioral effect is growing rapidly, but results vary quite substantially from one study to the next. Many factors can in theory contribute to such heterogeneity. For example, the significance of the state of the network during stimulation has become increasingly clear. This is not surprising given the proposed mechanism of action, which involves interaction of the applied perturbations with endogenous brain dynamics. This state dependence translates from the single neuron through the network level to the behavioral level. For example, receiving stimulation during a working memory training paradigm is likely to be very different from receiving stimulation before performing a working memory test. Spatial targeting is another interesting aspect of tES. With conventional tES electrodes, large parts of the cortex are stimulated, and so the spatial resolution of the stimulation is quite poor (Fig. 15.7).

In contrast, choosing smaller stimulation electrodes (Fig. 15.8) allows more precise targeting of specific locations [11]. This approach is typically combined with a detailed (mathematical) head model used to calculate the electric field distribution in the entire head to determine optimal electrode placement. It is important to take electrode size into account when dosing across tES studies is compared; from a regulatory viewpoint, the *dose* is defined as current density, which is computed as the stimulation amplitude divided by the area (size) of the electrode. A more precise approach would be to compare the resulting electric field distributions in the brain instead of the current density at the stimulation electrode.

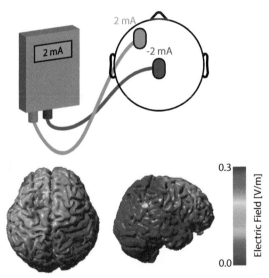

FIGURE 15.7 Detailed simulation of electric field distribution for transcranial electrical stimulation (tES) paradigm with electrodes on F3 and Cz. Stimulation causes an electric field that covers a substantial fraction of the cortical surface and lacks the spatial specificity of transcranial magnetic stimulation (TMS). Details in Ref. [10].

Single Device Configuration

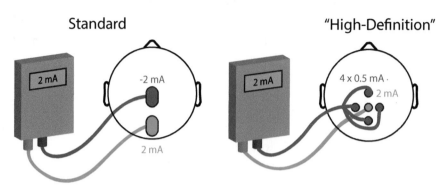

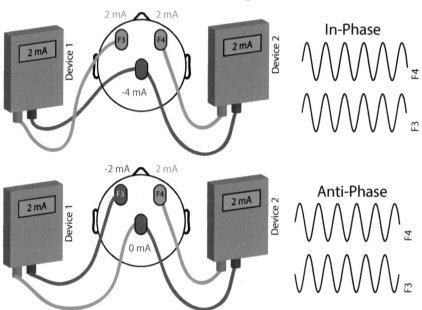

FIGURE 15.8 Transcranial electrical stimulation (tES) configurations. *Top*: Standard configuration with two electrodes (*left*). "High-definition" configuration with increased spatial focality (*right*). *Bottom*: Transcranial alternating current stimulation (tACS) configurations to modulate synchronization of frontal sites F3 and F4. In-phase stimulation applies two sine waves with zero phase lag to the two frontal sites. Antiphase stimulation applies two sine waves with 180 degree phase offset.

Transcranial Alternating Current Stimulation

tACS uses essentially the same hardware as tDCS and is based on the same concept of weak perturbations to the membrane voltage of neurons. But the targeting strategy of tACS is different [12,13], because the main aim is not to modulate overall activity levels (as in tDCS) but rather to engage specific aspects of the temporal structure of cortical network activity, namely, cortical oscillations (see unit: Cortical Oscillations). The stimulation waveform used is a sine wave with a defined amplitude and frequency. By convention, the stimulation amplitude is specified as a peak-to-peak amplitude and assumes values of up to 2 mA, similar to the amplitudes used in tDCS studies. The choice of frequency is perhaps the most important parameter in designing a tACS study. The stimulation frequency is chosen to match the targeted oscillation frequency and its associated brain state. Computer simulations have shown that matching the stimulation frequency to the endogenous oscillation frequency of the network is crucial for targeted enhancement of that oscillation [14]. In the model, if the frequencies match reasonably well, the network can be entrained such that the external periodic "force" exerted by the stimulation aligns neuronal activity, which causes *phase locking* with the stimulation. The larger the stimulation amplitude, the less critical the exact frequency matching of the stimulation and endogenous frequencies becomes. This general principle applies to a large number of oscillating systems and their responses to periodic stimulation. The principle is called *Arnold tongue*, in reference to the inverted triangular shape of the parameter combinations of stimulation, amplitude, and frequency that permit successful entrainment (Fig. 15.9 [15]). To what extent this entrainment applies to tACS remains unclear, but there is evidence from a number of other experimental approaches (mostly neurophysiological measures in acute brain slices) that such tuning of the stimulation is a core aspect of the mechanism and thus the rational design of tACS. tACS has become the tool of choice to demonstrate the functional role of a number of different (thalamo-)cortical

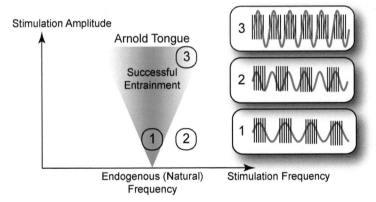

FIGURE 15.9   Arnold tongue. Oscillation entrainment requires the matching of the stimulation frequency to the endogenous frequency. The larger the stimulation amplitude, the larger the range of frequencies at which a periodic perturbation can entrain an oscillator (*shaded in red*). The Arnold tongue delineates the area of parameter values that allow entrainment. (1) Successful entrainment for very weak stimulation at the endogenous frequency. (2) Failed entrainment for the same stimulation amplitude but mismatched stimulation frequency. (3) Successful entrainment to a frequency different from the endogenous frequency by stimulation with stronger amplitude.

activity patterns. Behavioral effects have been demonstrated in a range of behaviors [16], from basic sensory perception of threshold sensory stimuli [17] to higher-order cognitive functions such as fluid intelligence [18] and creativity [19a].

Several practical aspects of tACS are worth considering. First, the terminology "anode" and "cathode" makes little sense since the direction of current flow switches during each cycle of stimulation. Also considering one electrode as the "active" electrode (over the target area) and the other as the "passive" return electrode is fundamentally wrong since both electrodes provide equal stimulation (just with a 180 degree phase offset). One promising application of tACS is the modulation of coherence, that is, oscillatory interaction between regions (see chapter: Functional Connectivity). In this approach (Fig. 15.8, bottom), two stimulators are used that either act in-phase or antiphase (180 degree offset of stimulation waveform).

Transcranial Random Noise Stimulation

tRNS is a less-studied form of tES compared to tDCS and tACS. The stimulation waveform is a noise signal (generated by picking a random value for each time sample) that is filtered to contain frequency components in the range of cortical oscillations and beyond (up to frequencies above 1000 Hz). The mechanism of action is not understood, although phenomena such as *stochastic resonance*, a general principle by which adding noise to a subthreshold signal pushes the signal over the threshold and thereby enhances it, have been proposed. Of note, individual cortical pyramidal cells respond with strong temporal fidelity to intracellular noise current injections [19b], so it would be reasonable to hypothesize that tRNS modulates spike timing by a similar mechanism.

"Brain Hacking"

Given the seemingly simple technical aspects of transcranial electric stimulation (in contrast to TMS), there is a growing community of do-it-yourself brain stimulation enthusiasts. Commercially available devices marketed directly to consumers (often with unsubstantiated claims about effectiveness for a long list of applications) further fuel interest in tES. This do-it-yourself trend is concerning [20] because the long-term effects of tES on brain function are not sufficiently well understood. Particularly, nothing is known about the effect of repeated applications of stimulation at frequencies higher than in the typical designs of research studies. Notably, not all laboratory studies have resulted in improved brain function, and the enthusiasm about tES may in part be because of publication bias, where positive findings are published more often than negative findings. Further concerns arise from device safety questions. The often-portrayed belief that tDCS is just a 9 V battery hooked to the scalp is technically incorrect. Accidents with devices that do not adhere to safe design and implementation practices for biomedical devices are a real possibility.

SUMMARY AND OUTLOOK

In this chapter, we discussed TMS and tES, the two most prevalent forms of NIBS. TMS applies brief stimulation pulses that trigger action potentials in the brain. Relatively small volumes of brain tissue can be targeted by TMS since the magnetic field is mostly unaffected by the tissue. The most basic application of TMS is the administration of a single stimulation pulse to the motor cortex with simultaneous measurement of the MEP in the target muscle group. Different sequences of TMS pulses are being used in research. Typically, low-frequency pulse trains decrease neuronal activity, whereas as higher-frequency pulses increase neuronal activity. TMS has given cognitive neuroscience a major boost, because it allows for the causal manipulation of the circuit activity that underlies sophisticated cognitive abilities in humans. Furthermore, repetitive TMS at 10 Hz is successfully used in the clinic to treat major depressive disorder. tES is a newer stimulation modality that has gathered a large following of enthusiasts. tES is administrated by a low-amplitude electric current to the scalp. The stimulation current spreads through the head as a function of the bioelectric tissues. As a result, the electric fields caused by tES are spatially very diffuse and of very weak amplitude (less than 1 V/m). tDCS appears to recruit plasticity and enable changes in circuit excitability. In contrast, tACS employs sine-wave stimulation waveforms that can be designed to target specific endogenous oscillations (see also unit: Cortical Oscillations). In both tDCS and tACS, the stimulation amplitude is too weak to cause action potentials in quiet neurons. Rather, the stimulation (synergistically) interacts with the endogenous network activity. Optimization of stimulation waveforms will likely improve the efficacy of tES paradigms.

References

[1] Hallett M. Transcranial magnetic stimulation and the human brain. Nature 2000;406(6792):147—50.

[2] George MS, et al. Daily repetitive transcranial magnetic stimulation (rTMS) improves mood in depression. Neuroreport 1995;6(14):1853—6.

[3] George MS, Lisanby SH, Sackeim HA. Transcranial magnetic stimulation: applications in neuropsychiatry. Arch Gen Psychiatry 1999;56(4):300—11.

[4] Luber B, Lisanby SH. Enhancement of human cognitive performance using transcranial magnetic stimulation (TMS). NeuroImage 2014;85:961—70.

[5] [a] Romei V, et al. Causal implication by rhythmic transcranial magnetic stimulation of alpha frequency in feature-based local vs. global attention. Eur J Neurosci 2012;35(6):968—74.
[b] Dugué L, Marque P, VanRullen R. The phase of ongoing oscillations mediates the causal relation between brain excitation and visual perception. J Neurosci 2011;31:11889—93.

[6] Smith MJ, et al. Menstrual cycle effects on cortical excitability. Neurology 1999;53(9):2069—72.

[7] [a] Nitsche MA, Paulus W. Transcranial direct current stimulation—update 2011. Restor Neurol Neurosci 2011;29(6):463—92.
[b] Bikson M, Inoue M, Akiyama H, Deans JK, Fox JE, Miyakawa H, Jefferys JGR. Effects of uniform extracellular DC electric fields on excitability in rat hippocampal slices in vitro. J Physiol 2004;557:175—90. http://dx.doi.org/10.1113/jphysiol.2003.055772.

[8] Nitsche MA, Paulus W. Excitability changes induced in the human motor cortex by weak transcranial direct current stimulation. J Physiol 2000;527(Pt 3):633—9.

[9] Fritsch B, et al. Direct current stimulation promotes BDNF-dependent synaptic plasticity: potential implications for motor learning. Neuron 2010;66(2):198—204.

[10] Sellers KK, et al. Transcranial direct current stimulation of frontal cortex decreases performance on the WAIS-IV intelligence test. Behav Brain Res 2015;290:32—44.

[11] Bikson M, Rahman A, Datta A. Computational models of transcranial direct current stimulation. Clin EEG Neurosci 2012;43(3):176—83.

[12] Frohlich F. Experiments and models of cortical oscillations as a target for noninvasive brain stimulation. Prog Brain Res 2015;222:41—73.

[13] Herrmann CS, et al. Transcranial alternating current stimulation: a review of the underlying mechanisms and modulation of cognitive processes. Front Hum Neurosci 2013;7:279.

[14] Ali MM, Sellers KK, Frohlich F. Transcranial alternating current stimulation modulates large-scale cortical network activity by network resonance. J Neurosci 2013;33(27):11262–75.

[15] Pikovsky A, Rosenblum M, Kurths J. Synchronization: a universal concept in nonlinear sciences. The Cambridge nonlinear science series. Cambridge: Cambridge University Press; 2001. xix, p. 411.

[16] Frohlich F, Sellers KK, Cordle AL. Targeting the neurophysiology of cognitive systems with transcranial alternating current stimulation. Expert Rev Neurother 2015;15(2):145–67.

[17] Helfrich RF, et al. Entrainment of brain oscillations by transcranial alternating current stimulation. Curr Biol 2014;24(3):333–9.

[18] Santarnecchi E, et al. Frequency-dependent enhancement of fluid intelligence induced by transcranial oscillatory potentials. Curr Biol 2013;23(15):1449–53.

[19] [a] Lustenberger C, et al. Functional role of frontal alpha oscillations in creativity. Cortex 2015;67:74–82.
[b] Mainen ZF, Sejnowski TJ. Reliability of spike timing in neocortical neurons. Science June 9, 1995;268(5216):1503–6.

[20] Bikson M, Bestmann S, Edwards D. Neuroscience: transcranial devices are not playthings. Nature 2013;501(7466):167.

16

Network Interactions

In the previous chapters in this unit, we learned about different methods to record and perturb neuronal activity. These experimental methods allow us to simultaneously record neuronal activity from several locations. Such recordings are the basis for the study of network interactions. Simultaneous recordings can reveal how different sites in the brain interact with each other. For example, if one brain area drives the activity in another area, we expect that the recorded signals from the two locations are correlated. Thus, by analyzing how similar two signals are, we can unravel what is called the *functional connectivity* between the two recording sites. One limitation of functional connectivity is that it does not reveal the direction of the interaction between two locations. To establish directionality, different methods referred to as measures of *effective connectivity* need to be used. They quantify the interaction between two simultaneously recorded activity traces, and provide the direction of the interaction. Notably, the presence of functional or effective connectivity does not require the existence of underlying, direct structural connection between the two locations. In this chapter, we will introduce the main metrics of functional and effective connectivity. The toolboxes "Graph Theory" and "Time and Frequency" are recommended in preparation for this chapter. This chapter contains some sections that require more mathematical background than the remainder of the book and are marked by an asterisk.

FUNCTIONAL CONNECTIVITY

The original insights into how the central nervous system may work were derived from careful anatomical studies delineating neurons and their connections. However, the mere presence of structural connections does not necessarily imply that the connected neurons significantly interact in terms of their electric activity. Knowing if and how different neurons or networks of neurons interact with each other is fundamental to understanding how the brain processes inputs and generates outputs [1]. Such neuronal interaction is referred to as *functional connectivity*. More technically, functional connectivity is defined as the quantification of the lack of independence of at least two neuronal signals. Thus functional connectivity assesses the similarity or correlation (and therefore the interaction) of simultaneously recorded neuronal signals. Here we will expand on this basic concept and review different analysis strategies and their applications for the study of functional connectivity in the nervous system.

The goal of functional connectivity measures is to determine a graph (see toolbox: Graph Theory) that describes the interactions among the activities measured in multiple locations. Remember that a graph consists of nodes that connect with each other by edges. Once a graph is established, graph theory is used to describe the structure of the graph, for example, if there are nodes that play a particularly important role because of their widespread connectivity within the network (*hubs*). Here we focus on the process of determining the nodes and edges for such a functional connectivity graph. Building a graph of functional connectivity is a two-step process. First, the nodes are assigned based on the method used to record brain activity. If a method such as calcium imaging (see chapter: Optic Measurements and Perturbations) is used, the individual cells represent the nodes. However, for other methods that lack such single-cell resolution [eg, functional magnetic resonance imaging (fMRI), electroencephalogram (EEG)], some type of anatomic segregation of the brain into distinct nodes (ie, *parcellation*) is required. The parcellation process is particularly challenging for methods such as EEG, where the signal itself is a mixture of signals from different, spatially distributed sources. In such a case, it makes sense first to preprocess the recorded signals to identify the underlying sources (see chapter: LFP and EEG), which will serve as the nodes of the functional network(s) of interest. Once the nodes are assigned and the signal associated with the nodes is determined, the second and last step of establishing functional connectivity among these nodes is performed. Later, we will discuss this process in more detail and review the most commonly used methods to determine the functional connectivity.

Determining functional connectivity requires the simultaneous measurement of neuronal activity in two or more locations of the brain. The concepts discussed here can be applied to both electrophysiological and imaging modalities, and can therefore determine functional connectivity at different spatial and temporal scales. Fundamentally, the basic concept is that if two areas are functionally connected, their activity should fluctuate together or correlate. Therefore, in its most basic form, the time series of activity simultaneously measured from two locations are used to determine the correlation coefficient. The correlation coefficient is calculated assuming all samples from each signal are drawn from a corresponding probability distribution. Therefore no information about the temporal order is maintained in this analysis. Positive correlation coefficients indicate that the two locations are functionally connected such that activity in one area increases activity in the other area. Zero or very small correlation coefficients indicate the lack of a measurable functional connectivity. Lastly, negative correlation coefficients indicate a functional connection where the activity is antagonistic, such that activity in one area reduces activity in the other area. Several important caveats need to be considered when applying this most basic assessment of functional connectivity. Correlation-based metrics do not provide any information about the direction of the interaction. A positive correlation could be caused by a one-way directional link in either of the two directions (A → B or B → A), or by the simultaneous presence of both of these links (A ↔ B). Also a correlation can arise in the absence of any direct connection between A and B but rather as the result of a shared common input from a third (possibly not recorded from) area, or a *transitive* relationship of the form A → C → B (Fig. 16.1). In this last case, there is no functional connection present, despite the correlation of the two signals.

In the case of more than two time series (as in fMRI, eg, see chapter: Imaging Functional Networks With MRI), the pairwise correlations between all time series (or voxels in the

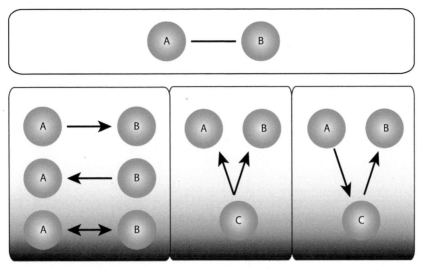

FIGURE 16.1   *Top*: Functional connectivity measures, such as cross-correlations and coherence, do not provide information about the direction of the interaction. Effective connectivity identifies both unidirectional connections and bidirectional connections (*bottom left*). Shared input (*bottom middle*) or an indirect connection via a third node (*bottom right*) can lead to overestimations of functional connectivity.

case of fMRI) are computed and visualized as a square matrix with the recording locations as rows and columns (*N* by *N* matrix for *N* recording locations or time series). By definition, the diagonal of this matrix assumes values of one, since the correlation of the signal with itself is one. Also the matrix is symmetric since the correlation is a nondirectional, symmetric measure. These color-coded matrices (typically the hotter the color the stronger the connection) provide an easy-to-process map of network interactions. The question though is to what extent the determined correlations are real. Subsequent statistical analysis is required to identify the functional connections that are not merely caused by chance.

Cross-Correlogram

If action potentials are recorded from several units at the same time (see chapter: Unit Activity), the interaction between the units can be studied. The most common approach to determine functional connectivity is to calculate the *cross-correlogram* between the poststimulus time histograms (PSTHs) of pairs of neurons. In this approach, one spike train (sequence of action potentials) is taken as a reference and the other as the target (Fig. 16.2). For every spike in the reference train, a PSTH of the target train is constructed (a short window centered on the reference spike). The cross-correlogram is then determined by averaging together the PSTHs for each reference spike (Fig. 16.3). If the neuron with the reference spike train excited the neuron in the target spike train, the cross-correlogram will exhibit a peak shifted from zero with a delay of a monosynaptic connection (a few milliseconds). Similarly, if the reference neuron inhibits the target neuron, there will be a trough. This process is repeated for all

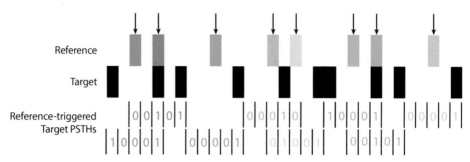

FIGURE 16.2   Relating a target spike train (*spikes in black*) with a simultaneously recorded reference spike train (*rainbow colors*) by building **reference-triggered** target poststimulus time histograms (PSTHs).

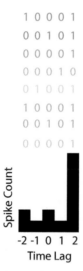

FIGURE 16.3   Computing the raw cross-correlogram by averaging the reference-triggered target poststimulus time histograms (PSTHs). This example demonstrates an excitatory connection from the reference to the target unit.

trials of a given experiment, and the trial average of the cross-correlogram for individual trials represents the final cross-correlogram. In the case of common input (eg, sensory stimulus) driving the two neurons, the firing rates of both neurons will increase, which can lead to a peak in the cross-correlogram in the absence of a connection. To control for this, the target spike trains are shuffled and the cross-correlograms for mismatched pairs of reference and target spike trains are computed. This way, the modulation of the firing rate by the stimulus is preserved, but the relationship between spikes in the reference and target spike trains is removed, because now spike trains from different trials are correlated. The resulting cross-correlogram is called the *shift predictor*, and this is subtracted from the raw, original cross-correlogram to determine the *corrected cross-correlogram*.

Cross-correlograms remain a ubiquitous approach to determine functional connectivity between single units. Next, we will discuss how to assess frequency-specific functional connectivity for LFP and EEG data.

Coherence, Phase-Locking Value, and Intertrial Phase Coherence*

The previously discussed correlation-based approach detects functional connections based on fluctuations in amplitude or power at different network sites. In contrast, *spectral coherence* measures frequency-specific similarities between two signals. Spectral coherence is based on the power spectrum, the determination of signal power as a function of frequency (see toolbox: Time and Frequency, for a more comprehensive and gentler introduction to the topic). The power spectrum (or more correctly, the power spectral density) is defined as the magnitude squared of the Fourier transform of the signal of interest. This measure is a real number and denotes the signal power as a function of the frequency. It can also be written as the product of $X(f)$ and $X^*(f)$, where $X(f)$ is the Fourier transform of signal $x(t)$, and $X^*(f)$ is the complex conjugate of $X(f)$. Remember that the complex conjugate is the same complex number but the imaginary part has its sign flipped. If we assume that the Fourier transform $X(f)$ of signal $x(t)$ is given by:

$$X(f) = a + ib \tag{16.1}$$

where a is the real part and b is the imaginary part, we can determine the power spectral density as follows:

$$X(f)X^*(f) = (a + ib)(a - ib) = a^2 + b^2 = |X(f)|^2 \tag{16.2}$$

For coherence, we determine whether there is a consistent relationship between two signals as a function of frequency. We start in the time domain and revisit the idea of correlating two signals $x(t)$ and $y(t)$. The correlation coefficient r is determined for n pairs of values $\{x(t_i), y(t_i)\}$:

$$r = \frac{\frac{1}{n-1} \sum_{i=1}^{n} (x(t_i) - \bar{x})(y(t_i) - \bar{y})}{\sigma(x)\sigma(y)} \tag{16.3}$$

where $\bar{x}$ and $\bar{y}$ denote the means and $\sigma(x)$ and $\sigma(y)$ the standard deviations of signals x and y, respectively. This correlation coefficient assesses how the two signals correlate for a zero time offset. If we assume a mean of zero and a standard deviation of 1, the correlation coefficient is reduced to:

$$r = \frac{1}{n-1} \sum_{i=1}^{n} x(t_i)y(t_i) \tag{16.4}$$

If we assume that signals $x(t)$ and $y(t)$ are continuous, we can replace the summation with the integral:

$$r_{xy} = \int_{-\infty}^{\infty} x(t)y(t)\, dt \tag{16.5}$$

We can generalize this definition to define the correlation coefficient for any time offset τ between the two time series:

$$r_{xy}(\tau) = \int_{-\infty}^{\infty} x(t)y(t + \tau)\, dt \tag{16.6}$$

This equation defines r_{xy}, which is called the *correlation function*. If we compute the correlation function for the signal $x(t)$ with itself, this is called the *autocorrelation*:

$$r_{xx}(\tau) = \int_{-\infty}^{\infty} x(t)x(t+\tau)\, dt \tag{16.7}$$

We now transform this equation into the frequency domain (derivation not shown). We find that the Fourier transform of the autocorrelation function is nothing else than the power spectrum of the signal. Similarly, we can compute the cross-spectrum as the Fourier transform of the cross-correlation function, which equals the product of the Fourier-transformed signal $X(f)$ and the complex conjugate of signal $Y(f)$. We use these quantities to define *magnitude-squared coherence* (MSC):

$$C(f) = \frac{|S_{xy}(f)|^2}{S_{xx}(f)S_{yy}(f)} \tag{16.8}$$

where we define:

$$\begin{aligned} S_{xx}(f) &= X(f)X^*(f) \\ S_{yy}(f) &= Y(f)Y^*(f) \end{aligned} \tag{16.9}$$

And, finally, in complete analogy:

$$S_{xy}(f) = X(f)Y^*(f) \tag{16.10}$$

We now assume again:

$$\begin{aligned} X(f) &= a + bi \\ Y(f) &= c + di \end{aligned} \tag{16.11}$$

And plug these into the formula for MSC:

$$C(f) = \frac{|(a+bi)(c-di)|^2}{(a+bi)(a-bi) + (c+di)(c-di)} = \frac{a^2c^2 + b^2d^2 + a^2d^2 + b^2c^2}{(a^2+b^2)(c^2+d^2)} = 1 \tag{16.12}$$

Therefore the MSC has a value of one when determined for a single time epoch. In reality, MSC is computed either for many time windows or for multiple trials. In this case, it is important to remember that coherence is a complex number; thus it not only has an amplitude but also a phase. Coherence is typically computed as an average over multiple time windows or across trials. In that case, the phase matters, since averaging is performed on the original complex number. If coherence for all trials points in the same direction, that is, the phase difference for a given frequency, is stable, the MSC will be high. Mathematically, assuming N trials, we write the coherence for a given frequency f_0 as:

$$\langle C(f_0) \rangle = \frac{|\langle S_{xy}(f_0) \rangle|}{\sqrt{\langle S_{xx}(f_0) \rangle}\sqrt{\langle S_{yy}(f_0) \rangle}} = \frac{\left| \sum_{n=1}^{N} A_n e^{\varphi i} B_n e^{-\psi i} \right|}{\sqrt{\sum_{n=1}^{N} A_n^2} \sqrt{\sum_{n=1}^{N} B_n^2}} \tag{16.13}$$

where we used:

$$X_n(f_0) = A_n e^{\varphi_n i}, \quad Y_n(f_0) = B_n e^{\psi_n i} \tag{16.14}$$

In this equation, the angle brackets indicate trial averages. If we assume that the amplitude was stable and identical for all trials and for both signals:

$$A_n = B_n = D \tag{16.15}$$

We find that the trial-averaged coherence function is:

$$\langle C(f_0) \rangle = = \frac{\left| \sum_{n=1}^{N} D e^{\varphi_n i} D e^{-\psi_n i} \right|}{\sqrt{\sum_{n=1}^{N} D^2} \sqrt{\sum_{n=1}^{N} D^2}} = \frac{D^2 \left| \sum_{n=1}^{N} e^{(\varphi_n - \psi_n)i} \right|}{\sqrt{ND^2}\sqrt{ND^2}} = \frac{1}{N} \left| \sum_{n=1}^{N} e^{(\varphi_n - \psi_n)i} \right| \tag{16.16}$$

Therefore the trial-averaged coherence is the average over the phase differences between the two signals, which is determined by summing the unity vectors with angles:

$$\phi_n = \varphi_n - \psi_n \tag{16.17}$$

If the phase difference between signals is exactly the same for all trials, then the vector will be of maximal length, N, and the coherence will be 1. If, however, the phase difference varies between trials, the coherence will be smaller.

Importantly, MSC is caused by amplitude correlations and a stable phase relationship between the two signals. Therefore MSC cannot be interpreted as a measure of *phase synchronization*, that is, how stable the phase difference between two signals is. We saw this earlier, where we had to assume that the amplitude of the two signals is the same and does not change over time (ie, across trials) to find that coherence measures phase synchronization.

Other metrics have been developed to exclusively determine the phase relationship between two signals. These metrics are ideally not affected by amplitude fluctuations of the two signals and provide a metric of phase synchronization as a function of time [2]. In particular, we will briefly introduce the *phase-locking value* (PLV) and *intertrial phase coherence* (ITPC). PLV determines whether there is a stable phase difference between two signals, in other words if they are phase synchronized. To compute the PLV, first the phase is extracted from both signals, followed by setting the amplitude to unity for both signals. Then the difference in phase is determined by subtraction of the two phase values, sample by sample. It is then assessed whether the phase difference is stable over time. By computing the average of the complex number with the sample-by-sample phase difference as phase, the PLV value for N trials can be calculated as:

$$\text{PLV}(t) = \frac{1}{N} \left| \sum_{n=1}^{N} e^{(\varphi_n(t) - \psi_n(t))i} \right|$$

In the case of perfect phase synchronization, the phase difference is constant and therefore the PLV assumes a value of 1. If there is no phase relationship, the phase-difference values are randomly distributed on the unit circle and the average converges to zero. Note that this approach is very similar to the special case of coherence already described. The main difference is that we require the phase as a function of time for the two signals. This phase information can be determined from a wavelet decomposition or Hilbert transform of the signal.

ITPC follows the same approach. But instead of assessing how stable the phase relationship between two different signals is, ITPC assesses how stable the phase relationship across multiple trials of the same signal is. Technically this is not a functional connectivity metric, but we introduce it here because of its conceptual similarity to PLV.

EFFECTIVE CONNECTIVITY

Most of biology is based on computing the correlation between variable pairs to understand their relationship. The most frequently used methods to determine network interactions based on correlations were introduced previously. The limitation of such functional connectivity metrics is that they establish that two signals relate to each other, but they do not provide insight into who influences whom (ie, the direction of the interaction between two sites). Directionality of the interaction is of great interest in the context of network neuroscience. Understanding network interactions requires us to be able to determine to what extent brain activity in one area causes activity changes in another brain area. Measures of effective connectivity attempt to overcome this limitation. We will start by developing a working definition of causality, the conceptual basis of effective connectivity. Then, we will consider three main strategies to establish effective connectivity—*Granger causality* (GC), *dynamic causal modeling* (DCM), and *information theory* approaches—and their applications to network neuroscience.

Granger Causality

The definition of *causality* is of continued debate. We will not dwell on the theoretical arguments about what exactly constitutes causality, but rather we will discuss the straightforward, most prevalent conceptual definition of causality. It was pioneered by Norbert Wiener and was further developed and applied by Clive Granger, after whom the method is named [3].

Before we define GC, we need to establish how to predict the future value $x(t)$ of a time series if we know all previous values $x(t-1)$, $x(t-2)$, and so on. In the simplest case, we predict the next value of $x(t)$ by multiplying the last known value $x(t-1)$ with the coefficient a_1:

$$x(t) = a_1 x(t-1) \tag{16.18}$$

To make more accurate predictions, we can take into account more values from the past of $x(t)$, for example, we can go back five samples and thereby build a linear equation with five coefficients a_1 to a_5:

$$x(t) = a_1 x(t-1) + a_2 x(t-2) + a_3 x(t-3) + a_4 x(t-4) + a_5 x(t-5) \tag{16.19}$$

We can write this compactly as the summation of the five terms:

$$x(t) = \sum_{i=1}^{5} a_i x(t - i) \tag{16.20}$$

Most of the time, our prediction will not be completely accurate. In other words, the true value $x(t)$ will differ from the predicted value, determined by the weighted sum of the past values of $x(t)$. The difference between the two is called the model error, denoted as $\varepsilon_x(t)$.

$$x(t) - \sum_{i=1}^{5} a_i x(t - i) = \varepsilon_x(t) \tag{16.21}$$

We can rewrite for the more general case of taking into account N samples for the prediction. N is called the model order. This approach of predicting $x(t)$ by a linear sum of the weighted past values of $x(t)$ is called *autoregressive modeling*:

$$x(t) = \sum_{i=1}^{N} a_i x(t - i) + \varepsilon_x(t) \tag{16.22}$$

Since we are interested in the interaction of two time series $x(t)$ and $y(t)$, we now again predict $x(t)$ by taking into account not only the past values of process $x(t)$, but also the past values of the second process $y(t)$. We again include a noise term $\varepsilon_{xy}(t)$:

$$\begin{aligned}
x(t) &= a_1 x(t - 1) + a_2 x(t - 2) + \cdots + a_n x(t - n) + b_1 y(t - 1) + b_2 y(t - 2) \\
&\quad + \cdots + b_n y(t - n) + \varepsilon_{xy}(t) \\
&= \sum_{i=1}^{N} a_i x(t - i) + \sum_{i=1}^{N} b_i y(t - i) + \varepsilon_{xy}(t)
\end{aligned} \tag{16.23}$$

We now have two predictions of $x(t)$. The first prediction only uses the past of $x(t)$, whereas the second also uses the past of $y(t)$. We can now define a causal interaction between $x(t)$ and $y(t)$. If by including $y(t)$ into the model we reduce the model error, then we can say that $y(t)$ has a causal influence on $x(t)$. To compare the two model errors, we measure the magnitude of the errors $\varepsilon_x(t)$ and $\varepsilon_{xy}(t)$ by computing their variance across all time samples.

Several variants of this basic measure are used in neuroscience. First, GC can also be performed in the frequency domain to determine the *spectral Granger causality*, which provides information on causality as a function of frequency. Second, GC cannot distinguish between direct interactions of two connected nodes and indirect interactions of two nodes via an intermediate node. The *partial Granger causality* addresses this issue. The underlying approach is the same, and two different models are used for assessment of indirect interactions. Specifically, we fit a linear model to nodes x and z as before (note that we renamed the second node to ensure that our notation is consistent with the literature). We compare the error of this model with the error of an expanded model that also directly takes into account the intermediate node y. If including y does not decrease the model error, then the connection from X to Z is the only one. If the model error decreases by including Y, then we know that some information transfer from X to Z goes via node Y. More formally, we start with the regression equations for the two cases:

$$x(t) = \sum_{i=1}^{N} a_i x(t-i) + \sum_{i=1}^{N} c_i z(t-i) + \varepsilon_{xz}(t) \qquad (16.24)$$

And compare it to the case where Y is also explicitly modeled:

$$x(t) = \sum_{i=1}^{N} a_i x(t-i) + \sum_{i=1}^{N} b_i y(t-i) + \sum_{i=1}^{N} c_i z(t-i) + \varepsilon_{xyz}(t) \qquad (16.25)$$

The error (residuals) of the model, ε_{xy} and ε_{xyz}, are again compared.

Dynamic Causal Modeling

One of the limitations of GC is that it is not based on an underlying biological model of how the signals are generated. Instead, it assumes that the next sample can be predicted as a linear sum of the previous values. This is a phenomenological model that does not represent any neurobiological processes. DCM addresses this limitation by including a so-called generative model that aims to explain how the recorded data are generated [4]. This is particularly important for fMRI studies where the change in blood flow/oxygenation lags the underlying neuronal changes in activity by several seconds, and therefore drawing conclusions about the temporal order/directed influence of neuronal signals from vascular signals is difficult. The model has two components. The first is a set of differential equations that define the change in rate of activity of each area. The rate change is a function of (1) the weighted activity in the other areas (effective connectivity), (2) the weighted sum of the products of an input signal and the activity of every area, and (3) the weighted input itself. These three terms enable us to model not only the effect of the other state variables on a given state variable (connection), but also how afferent input alters connections (second term), and, finally, how input directly modulates the activity state variable. The second component is the transformation of this (simplified) neuronal signal into a vascular response. This transformation is achieved by a detailed model with neuronal activity $x(t)$ as an input and several coupled differential equations that model blood flow, blood volume, and deoxyhemoglobin. The model of this neurovascular coupling is typically region specific and so avoids misinterpretation of signal lags that may not be caused by directed flow of information but rather are an artifact of different delays of the hemodynamic response in different areas. Therefore DCM enables the dissection of different connectivity patterns and the role of afferent input.

In practice, experimental data are fit to different models and the quality of the fit is then compared. As with all data-fitting procedures, models with a larger number of parameters (eg, connections) in general fit the data better, but lose generality and turn out not to be helpful (overfitting). This tradeoff is taken into account in DCM, and models are assessed with a metric that includes both accuracy and model complexity. The procedure to estimate the model parameters is referred to as *model inversion*.

It is worth discussing how GC and DCM differ. GC is simpler, since it does not consider the underlying neurobiology or the nature of how the signals are generated. GC determines the direction of connectivity exclusively based on the time series. GC is based on a discrete-time linear model that is fitted to best predict the future based on past values of the considered time series. DCM, on the other hand, is more specific and builds on a continuous time set of differential equations that describe the neuronal state dynamics. DCM acknowledges that the signals measured with fMRI are the consequence of neuronal activity that itself is not directly measurable (so-called hidden state variables).

Mutual Information and Transfer Entropy*

Following our approach of borrowing methods from very different fields to study and understand brain activity, we turn to information theory [5]—the theoretical basis of all modern telecommunication and computer networks—to establish yet another family of methods to measure effective connectivity. Information theory provides a framework to assess how information flows between "senders" and "receivers" (in our case, two interconnected brain structures). First, we will develop an intuitive understanding of what is meant by information. Then, we will develop the main quantities used in information theory. Finally, we will apply these concepts to develop *transfer entropy*, a measure of effective connectivity.

Intuitively, if you are provided with *information* ("facts"), you know more than you did before. To develop a mathematical definition of information, we must first understand the concept of *entropy*. Entropy is a fundamental property that has proven useful in many different disciplines such as physics and telecommunication. Here we simply define entropy as a quantification of the "randomness" of a variable or process. A deterministic process (no randomness) has by definition entropy of zero. A random variable that follows a uniform distribution where all outcomes are equally likely (eg, the toss of a perfectly balanced coin) has the highest entropy. Furthermore, we request that entropy always assumes nonnegative values. How do we compute entropy? Intuitively, we will want to include the individual probabilities of all outcomes in our entropy measure. Thus we could simply sum all of the probabilities. However, this will not work, because all of the probabilities of all possible outcomes by definition always add to one, independent of the values of the individual probabilities. The logarithm enables us to get an entropy of zero for a deterministic process that has only a single outcome with a probability of 1 since $\log(1) = 0$. Thus we could add the logarithms of the individual probabilities. A quick remark about the notation used in this section: our use of "log" will correspond to the logarithm to the base 2, which will allow us to interpret the calculated quantities as "bits" of information. However, any event with zero probability would cause a serious problem since $\log(0)$ is negative infinity. An elegant solution is to weigh the individual logarithms with the actual probabilities and sum them. We need to add a minus sign to make entropy a positive quantity since all probabilities are values between 0 and 1 and therefore their logarithms are negative. We can now define entropy $H(X)$ of random variable X as:

$$H(X) = -(p_1 \log p_1 + p_2 \log p_2 + \ldots) = -\sum_{i=1}^{N} p_i \log p_i \tag{16.26}$$

Next, consider an example. A perfectly balanced coin has equal likelihood to land heads or tails up, such that Pr(heads) $= .5$ and Pr(tails) $= .5$. Entropy of a single coin is thus given by:

$$H(\text{Coin}) = -\text{Pr(heads) logPr(heads)} - \text{Pr(tails) logPr(tails)}. \tag{16.27}$$

Plugging in the numbers, we find that $H(\text{Coin})$ is:

$$H(\text{Coin}) = 2(-.5\log .5) = -1\log .5 = -\log\left(\frac{1}{2}\right) = -\log(1) - -\log(2) = 0 + \log 2 = 1. \tag{16.28}$$

using the relationship:

$$\log\left(\frac{a}{b}\right) = \log a - \log b \tag{16.29}$$

Our coin toss has entropy of 1 bit.

The definition of entropy can be generalized to two random variables, a measure called *joint entropy*. The same principles we introduced for entropy of a single variable also apply to joint entropy. Given two random variables x and y with a joint probability density function $p(x,y)$, we again take the logarithm of the probability for all possible outcomes (x,y) and build the weighted sum:

$$H(X,Y) = -\sum_x \sum_y p(x,y)\log p(x,y)$$

Finally, we define *conditional entropy* $H(Y|X)$. Before that, let us briefly review *conditional probabilities*. If two random variables X and Y are not independent, then knowing the value of one will give us some clue about the value of the other one. The probability distribution of Y given that we know the value of X is written as $p(y|x)$ and is called conditional probability. If X and Y are independent, then:

$$p(y|x) = p(y) \tag{16.30}$$

Formally, the conditional probability is defined as the joint probability of X and Y, $p(x,y)$ divided by $p(x)$:

$$p(y|x) = \frac{p(x,y)}{p(x)} \tag{16.31}$$

Conditional entropy is defined as the sum of the entropy of the conditional probability density function $p(y|x)$ weighted by the probability $p(x)$ for all outcomes of X:

$$H(Y|X) = \sum_x p(x)H(Y|X = x) \tag{16.32}$$

The joint entropy of two random variables $H(X,Y)$ can be shown to be equal to the sum of the entropy of X, $H(X)$, and the conditional entropy of Y given X, $H(Y|X)$.

$$H(X,Y) = H(X) + H(Y|X)$$

Next, we introduce *relative entropy*, which is a "distance measure" of two probability distributions. Relative entropy is also called the *Kullback—Leibler distance*, and is computed as follows for two distributions $p(x)$ and $q(x)$:

$$D(p\|q) = \sum_x p(x)\log\left(\frac{p(x)}{q(x)}\right) \tag{16.33}$$

The key properties of relative entropy D are: (1) D is never negative, (2) D is zero if and only if $p(x) = q(x)$, and (3) D is not symmetric such that $D(p\|q)$ is not necessarily the same as $D(q\|p)$. Equipped with this tool to measure the "distance" (technically not a distance because of the lack of symmetry), we now finally define mutual information of two random variables $I(X;Y)$ as the relative entropy between the joint probability distribution $p(x,y)$ and the product of the marginal distributions $p(x)$ and $p(y)$:

$$I(X;Y) = \sum_x \sum_y p(x,y)\log\frac{p(x,y)}{p(x)p(y)} \tag{16.34}$$

With some rewriting of this equation that defines mutual information and by using $p(x,y) = p(x|y)p(y)$, we find:

$$I(X;Y) = H(X) - H(X|Y) = H(Y) - H(Y|X) \tag{16.35}$$

This equation tells us two things: first, the information that X contains about Y is the same as the information Y contains about X, and, second, intuitively we can conceptualize mutual information as the reduction in uncertainty (ie, entropy) about X when we know Y. If variables X and Y are independent, then the reduction will be zero, since the uncertainty about X knowing Y, $H(X|Y)$ will be the same as the uncertainty about X, $H(X)$. Therefore mutual information does not provide a measure to determine directed connectivity, since it is symmetric:

$$I(X;Y) = I(Y;X) \tag{16.36}$$

Transfer entropy resolves this concern. Transfer entropy compares the entropy of the distribution of $x(t)$ if we only know the past of $x(t)$, and the entropy of $x(t)$ if we know the past of both $x(t)$ and $y(t)$. The larger the difference, the larger the transfer entropy from Y to X, denoted as $T(Y \rightarrow X)$. Similarly, transfer entropy $T(X \rightarrow Y)$ measures the effect of X on Y. The derivation and mathematical definition of transfer entropy exceeds the scope of this book. The main advantage of transfer entropy is that no specific structure of the interaction between X and Y is assumed. This differs from GC, which assumes an underlying linear model.

SUMMARY AND OUTLOOK

In this chapter, we discussed methods to analyze simultaneously measured neuronal activity. Both functional and effective connectivity metrics quantify the interaction of neuronal activity recorded at different sites. Functional connectivity metrics such as cross-correlation, coherence, and phase-locking value determine the extent of the interaction of a pair of recordings, but do not provide any information on the direction of the interaction. In contrast, effective connectivity metrics such as GC, DCM, and transfer entropy assess the direction of the interaction between pairs of recordings. DCM stands out because it uses a (simple) neuronal model with dynamics. We will encounter several of these methods in subsequent chapters, when we discuss pathological changes to network structure and function.

References

[1] Bullmore E, Sporns O. Complex brain networks: graph theoretical analysis of structural and functional systems. Nat Rev Neurosci 2009;10(3):186–98.

[2] Lachaux JP, et al. Measuring phase synchrony in brain signals. Hum Brain Mapp 1999;8(4):194–208.

[3] Granger CWJ. Testing for causality. J Econ Dyn Control 1980;2:329–52.

[4] Friston K. Causal modelling and brain connectivity in functional magnetic resonance imaging. PLoS Biol 2009;7(2):e33.

[5] Cover TM, Thomas JA. In: Hoboken NJ, editor. Elements of information theory. 2nd ed. Wiley-Interscience; 2006. xxiii, 748 p.

UNIT III

CORTICAL OSCILLATIONS

Rhythmic structure is a ubiquitous feature of the physical processes that define our environment—all the way from the large scales of galactic and planetary motion to the small scale of subatomic particle motion. So it can be argued that living beings—adapted for survival through evolutionary pressures—also inherently exhibit rhythmic processes to optimally engage with the environment. For example, the sleep–wake cycle in humans is (at least partially) hardwired, and probably represents an adaption to the day–night cycle imposed by planetary motion. Since behavior is a consequence of brain activity, it is reasonable to expect that the brain also exhibits rhythmic processes in terms of electric activity patterns of neurons and networks of neurons. Indeed, the brain exhibits rhythmic activity patterns (referred to as oscillations) at different spatial and temporal scales. In this unit, we focus on cortical oscillations, which are rhythmic activity patterns measured in the hippocampus and the neocortex.

The time scale of cortical oscillations ranges from seconds to milliseconds; thus it is interesting to speculate about what processes these oscillations are tuned to. Since the brain is optimized to efficiently sample the environment and take action to respond to it, one key factor that defines brain rhythms could be the frequency of the interaction with the surrounding environment. For example, low-frequency cortical oscillations are related to the breathing cycle, possibly since these oscillations are optimized to process olfactory information that arrives at discrete, approximately equally spaced time points, that is, at each consecutive sniff [1]. This then leaves us with the question of why the olfactory system has been optimized to periodically sample the environment. Likely, periodic sampling of an inherently analog process (odor concentration as a function of time) saves processing resources, and the sampling rate is matched to the temporal structure of the information content provided by olfactory cues in the environment. On the one hand, the sampling rate of such active sensing needs to exceed the rate at which the environment changes so that no important information is lost because of undersampling. On the other hand, oversampling would represent costly redundancy. Active, periodic sampling can also be imagined for other senses, particularly those that include a strong active component such as vision, where primates perform saccades to sample the environment. Similar arguments can be made for faster oscillation frequencies in a way that they define periodic sampling of incoming input. For example, alpha oscillations (8–12 Hz) appear to cyclically modulate sensory processing so that, for example, perception of a threshold-level sensory stimulus depends on the alpha phase [2]. Therefore within each alpha cycle (approximately 100 ms), visual input is preferentially sampled and then consecutively suppressed. The resulting time scale could again be dictated by the amount of temporal resolution of the visual system needed to process sensory input. As we will see in this unit, cortical oscillations serve a much broader functional role than that timing of sensory sampling, but they may have originated for that reason.

When considering the functional role of cortical oscillations beyond sampling of the environment, it is helpful to remember that they mostly reflect periodically modulated membrane voltages of neurons that interact by synaptic transmission. Therefore oscillations in essence reflect an alternating pattern of depolarization and hyperpolarization that enables or disables effective translation of incoming synaptic input into postsynaptic action potential firing. So if a subset of neurons is undergoing such cyclical depolarization together, the corresponding communication links between them are privileged in the sense that all depolarized neurons can easily generate action potentials [3]. Also multiple communication channels can exist in the same frequency band if different groups of neurons are preferentially active at different phases of the overall oscillation (Fig. 1).

In addition, multiple frequencies often occur simultaneously. Typically, the lower frequency "gates" the occurrence of faster oscillations (Fig. 2).

For example, the theta oscillation (4–8 Hz in humans) often gates the gamma oscillation (>30 Hz [4]). In other words, the amplitude of the gamma oscillation is influenced by the phase of the theta oscillation. Such so-called *phase amplitude coupling* leads to a hierarchical organization of cortical rhythms. Although the reason for such organization of multiple oscillations remains an open question, one hypothesis derives from the observation that there is typically an inverse association between oscillation frequency and spatial distance between the communication partners. Low-frequency oscillations are a rather global phenomenon, such as the slow oscillation (<1 Hz) during slow-wave sleep that can encompass the entire cortex. In strong contrast, the gamma oscillation (>30 Hz) appears to be (mostly) a local rhythm within a limited cortical territory or circuit. This inverse relationship makes intuitive sense given that propagation delays are a significant obstacle in oscillation genesis and synchronization. The shorter the distances, the smaller the propagation delays of action potentials along the axon, and therefore the easier it is to synchronize at faster frequencies. As a result, low-frequency oscillations that orchestrate and organize activity across cortical areas occur. At the more local scale, individual circuits perform their communication and computation on given cycles of the global, slower oscillation.

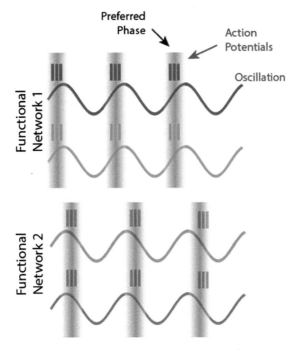

FIGURE 1 Multiple functional networks can be formed by selective synchronization of neurons to different phases of the shared oscillation (here an example of two functional networks is shown). The *gray shading* indicates the phase of the oscillation during which neurons from a given functional network preferentially fire action potentials (*symbolized as vertical lines*).

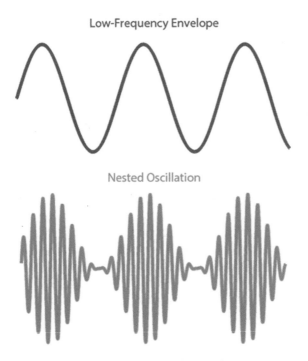

FIGURE 2 Modulation of the oscillation power of a faster oscillation by a slower oscillation. The amplitude of the faster oscillation is a function of the phase of the slower oscillations. *Top*: Low-frequency envelope that "gates" faster oscillation. *Bottom*: Nested fast and slow oscillation.

How do oscillations mechanistically arise in the nervous system? Oscillations at different frequencies are implemented in different "biological hardware." For example, the best-understood oscillation is the gamma oscillation, which arises through local interactions between fast-spiking inhibitory interneurons and excitatory pyramidal cells [5]. In essence, the inhibitory interneurons prevent the pyramidal cells from firing action potentials for the duration of the inhibitory postsynaptic potential. As the pyramidal cells are released from the incoming synaptic inhibition, they fire action potentials that in turn recruit again the inhibitory interneurons that then start the next oscillatory cycle. From a more theoretical viewpoint, oscillations emerge from the interaction of positive and negative feedback. During each oscillation cycle, activity is generated through positive feedback in a way that activity begets more activity. At the same time (with some delay), a set of negative feedback processes are recruited that reduce neuronal activity in an activity-dependent way. This dynamic interplay of positive feedback followed by negative feedback enables robust responses to changes in the input to a circuit through amplification (positive feedback), but also ensures that overall activity levels do not grow boundlessly (negative feedback). Such fundamental balance between positive and negative feedback is an omnipresent dynamic mechanism that easily leads to oscillations.

In this unit, we will discuss the different rhythmic activity patterns in the cortex. Originally, the different oscillations were defined based on their frequencies, using Greek letters (from low to high frequencies: delta, theta, alpha, beta, gamma). We spend a chapter on each "classical" frequency band and discuss the occurrence of specific cortical oscillations, their possible functional roles, and the underlying cellular and synaptic mechanisms. We will discuss evidence and insights from computer simulations, in vitro and in vivo animal experiments, and human studies. The study of brain rhythms has rapidly evolved and has become an important aspect of modern systems neuroscience. Previously, oscillations in neuronal networks have been considered a side product of neuronal activity with little functional importance. More recently, however, targeted manipulation with tools such as noninvasive brain stimulation in humans and optogenetic stimulation in animals has enabled unprecedented insight into the possible functional roles of cortical oscillations. Additionally, increasing evidence implicates pathological changes to cortical oscillations as network-level causes of neurological and psychiatric illnesses such as schizophrenia (see next unit: Network Disorders). At the conceptual level, a consensus is emerging in which oscillations provide a hierarchical system of nested oscillations at different frequencies that define excitability and enable timed and selective communication between populations of neurons.

Discussing oscillations by classical frequency bands as we do in this unit of the book is not without problems. First, the nomenclature of oscillations is not very well developed, and is full of conflicting definitions. Particularly damaging is the lack of agreement about whether to classify oscillations by their mechanism, functional role, or frequency structure (today's most prevalent approach, which is also mostly adopted for this unit). Second, the most interesting and functionally relevant oscillatory processes may be the coupling between different simultaneously occurring oscillations, since fast oscillations are typically grouped or patterned by lower-frequency oscillations. The chapter-by-chapter presentation by frequency band should not distract from that fact.

References

[1] Kay LM, et al. Olfactory oscillations: the what, how and what for. Trends Neurosci 2009;32(4):207–14.
[2] Busch NA, Dubois J, VanRullen R. The phase of ongoing EEG oscillations predicts visual perception. J Neurosci 2009;29(24):7869–76.
[3] Fries P. A mechanism for cognitive dynamics: neuronal communication through neuronal coherence. Trends Cognitive Sci 2005;9(10):474–80.
[4] Canolty RT, et al. High gamma power is phase-locked to theta oscillations in human neocortex. Science 2006;313(5793):1626–8.
[5] Tiesinga P, Sejnowski TJ. Cortical enlightenment: are attentional gamma oscillations driven by ING or PING? Neuron 2009;63(6):727–32.

Low-Frequency Oscillations

Low-frequency rhythms (<4 Hz) are a fundamental activity mode of the cortex. Several different behavioral states are associated with such low-frequency oscillations, most prominently deep sleep and anesthesia. Recent work has suggested that the waking animal can also exhibit low-frequency oscillations during resting in the absence of salient sensory input. Importantly, there are several, probably distinct, oscillation patterns with frequencies below 4 Hz, most prominently the classical *delta oscillation* (1–4 Hz), the *slow oscillation* (<1 Hz), and even slower, the so-called *infra-slow oscillations*, which we do not further consider in this chapter. We will start by discussing the key features of sleep and anesthesia in terms of low-frequency oscillatory activity in both human and experimental animal studies. Then, we will briefly discuss the presence of low-frequency oscillations in the awake state, in patients with disorders of the CNS, and during early development.

ANESTHESIA

Low-frequency oscillations are the dominant activity pattern in cortical recordings from the anesthetized animal. *Anesthesia* is a state defined by loss of consciousness, absence of behavioral responses, and cardiovascular and metabolic suppression. The motivation for the study of anesthesia and its underlying mechanisms is manifold. First, understanding network dynamics during anesthesia is of translational importance given the routine administration of anesthetics in clinical settings. Maintaining appropriate depth of anesthesia is one of the key tasks of anesthesiologists. Light anesthesia is associated with awareness during procedures caused by insufficient loss of consciousness, and very deep anesthesia can lead to side effects such as transient cognitive impairment and longer recovery times. Monitoring brain function during anesthesia provides an important tool for assessing depth of anesthesia. As our understanding of neuronal network activity as a function of depth of anesthesia increases, these insights can be used for the design of more advanced electroencephalogram (EEG)-based monitoring solutions. Second, from a basic science viewpoint, the fact that anesthesia induces loss of consciousness is interesting in its own right, because it provides a window into the neural processes that underlie consciousness. Third, systems neuroscience has a long history of using anesthetized animals, particularly as an experimental model of sleep.

231

Slow Oscillation of Anesthesia

During anesthesia, cortical network dynamics are patterned into brief, alternating episodes of activity and quiescence. The frequency of the transition between these two activity modes defines the dominant frequency of EEG and local field potential (LFP) signals during deep anesthesia. These bouts of activity and quiescence are referred to as *Up* and *Down* states, respectively [1,1a]. During Up states, cortical neurons are depolarized and fire action potentials, whereas during the Down state, neurons are hyperpolarized and do not fire action potentials. As a result, a histogram of the membrane voltage measured from a neuron during low-frequency activity exhibits a characteristic bimodal distribution. The peak of more depolarized voltages corresponds to the Up state and the peak of more hyperpolarized voltages corresponds to the Down state. These slow rhythmic fluctuations of the membrane voltage are synchronized across neurons. In general, low-frequency cortical oscillations such as the delta oscillation (<4 Hz) are synchronized across long distances (albeit with some phase offset) and therefore can be considered a rather global activity pattern. Of note, however, these oscillatory patterns differ in their properties across cortical networks, with the most prominent Up and Down states (ie, the clearest separation between the two states) typically recorded from frontal cortical areas. Accordingly, anesthesia monitoring systems based on EEG compute spectral properties from electrodes attached to the forehead. Although slow rhythmic activity is a shared feature of general anesthesia, typically a combination of the anesthetic ketamine and the sedative xylazine is used in animal studies for the study of low-frequency oscillations during anesthesia. Another particularly interesting anesthetic agent is urethane (only used in animal experiments because of toxicity concerns in humans). Under urethane anesthesia, animals also exhibit low-frequency cortical oscillations. But epochs of Up and Down states are interleaved with epochs of desynchronized cortical activity that more resemble the awake state, despite the animal being fully anesthetized (Fig. 17.1). This desynchronized, awake-like state is also referred to as *cortical activation* (see chapter: Neuromodulators). Thus urethane allows for the study of state dynamics and state dependence of sensory inputs in the anesthetized

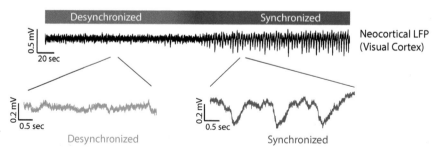

FIGURE 17.1 *Top*: Neocortical local field potential (LFP) recorded in the visual cortex of an animal anesthetized with urethane. The state of the cortex switched from "desynchronized" (*red*) to "synchronized" (*blue*). *Bottom*: Enlarged activity traces. Activity during the desynchronized (activated) state lacks high-amplitude, low-frequency fluctuations (*left, red*). Activity during the synchronized state is dominated by low-frequency oscillations mediated by periodic transitions between Up and Down states (*right, blue*).

animal with the advantage of more stable and easier preparation compared to the study of the same dynamics in the awake and naturally sleeping animal.

SLEEP

Sleep Stages

During sleep, the brain cycles through a well-defined sequence of distinct states referred to as stages. Each sleep stage exhibits a set of distinct physiological features captured in the EEG, electromyogram (EMG), and electrooculogram (EOG), which measures electric activity of the dipole formed by the front and the back of the eye (Fig. 17.2).

Briefly, upon eye closing (still awake), alpha oscillations (8–12 Hz) increase (see chapter: Alpha Oscillations). At sleep onset, *stage N1* sleep occurs. This sleep stage is defined by the disappearance of alpha oscillations and a "slowing" of the EEG, that is, a decrease in frequency. State N1 is followed by *stage N2*. This sleep stage is characterized by the occasional occurrence of *K-complexes* that likely correspond to individual, isolated Down states (with a subsequent transition to an Up state). In *stage N3* sleep (slow-wave sleep, SWS), the EEG is dominated by slow periodic activity, which is called *slow-wave activity* (SWA, 0.75–4.5 Hz). Stages N1–N3 are termed nonrapid eye movement (NREM) sleep (Fig. 17.3). *Rapid eye movement (REM) sleep* follows stage NREM sleep. During REM sleep, the EEG resembles the awake EEG (low-amplitude, high-frequency activity), but muscle tone greatly decreases (reduced EMG activity) and the eyes make pronounced ballistic movements (called *rapid eye movements*) that are captured by the EOG. The occurrence of an epoch of REM sleep concludes one cycle and a new cycle follows (stage N1 → stage N2 → stage N3 → REM sleep). These cycles are repeated several times throughout the night and are visualized in the form of a hypnogram (Fig. 17.1). Historically, sleep stages were divided into "Waking," stages 1–4, and REM sleep [1b]. The current classification employs the letter "N" to refer to NREM sleep stages [2]. Note that the previous stages 3 and 4 are combined into N3.

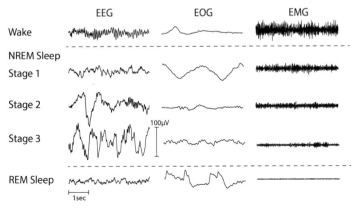

FIGURE 17.2 Sleep stages. *EEG*, Electroencephalogram (measures brain activity); *EOG*, electrooculogram (measures eye movements); *EMG*, electromyogram (measures muscle activity).

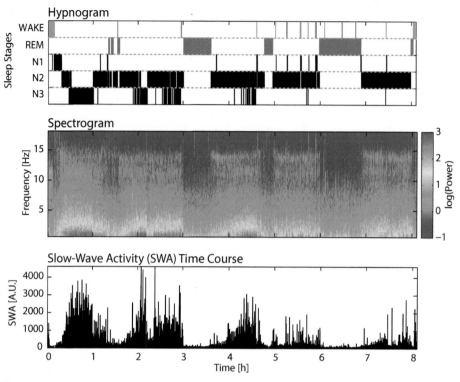

FIGURE 17.3   Sleep is characterized by cycling through stereotyped activity states. *Top*: Hypnogram illustrates the cycling through sleep stages as a function of time during the night (here shown for a total of 8 h). N1—N3 denote nonrapid eye movement (NREM) stages 1—3. Scoring was performed on 20-s epochs according to the AASM scoring rules [2]. The hypnogram shows the cyclic alternations of NREM and REM sleep. Middle: Spectrogram for frequencies between 0.75 and 20 Hz. The logarithm of the oscillation power is color coded (physical units: $\mu V^2/Hz$). *Bottom*: Specific frequency bands dominate as a function of sleep stage. For instance, slow-wave activity (SWA) (0.75—4.5 Hz) and spindle activity (11—16 Hz) are prevailing during NREM sleep, and are clearly lower during REM sleep. Time course of SWA (0.75—4.5 Hz) illustrates that SWA decreases in the course of a night of sleep.

Oscillations During NREM Sleep

NREM sleep is associated with three distinct rhythmic activity patterns. Two of them are the focus of this chapter: slow oscillations (<1 Hz) and delta oscillations (1—4 Hz). The third (transient) rhythmic activity pattern, sleep spindles (7—15 Hz), is discussed in more detail in the chapter "Alpha Oscillations." The terminology around low-frequency oscillations during sleep can appear to be quite confusing, especially as the names of certain oscillations differ between research communities that employ different neurophysiological recording methods. Here, we consider the slow oscillation (terminology from animal models) to mediate the slow waves detected in scalp EEG (terminology from human EEG studies). Specifically, a slow wave detected in the EEG corresponds to an Up and a Down state of the slow oscillation. Slow waves are traveling waves, and the most frequent direction of propagation is from anterior to posterior. The specific pattern of origin of slow waves across the cortical surface is a stable trait that manifests itself as near-identical for repeat recordings from the same participant [3]. Overall, however, mounting evidence suggests that slow waves and spindles are less a global but more a local phenomenon [3a]. Slow waves and spindles are selectively modulated as a function of the activation patterns during the day before. In further support of a local regulation of these sleep rhythms, recording methods with higher spatial resolution than EEG, such as electrocorticography in epilepsy patients, have demonstrated that distant cortical areas are often independently in an Up or Down state, and that spindles are more localized than originally assumed. Interestingly, this local structure of oscillatory sleep network activity patterns becomes more prominent as the night progresses. Therefore sleep oscillations are global activity patterns that are gradually replaced by more localized patterns as the night progresses.

Delta oscillations and their differences to the slow oscillation are a matter of debate. In the anesthetized animal, periods of delta oscillations appear at the slow-oscillation frequency, suggesting that the slow oscillation "groups" organizes the occurrence of delta oscillations. While the slow oscillation is often considered to be of cortical origin (see later), the delta oscillation is typically considered to be a thalamocortical (TC) rhythm (Fig. 17.4). Briefly, this

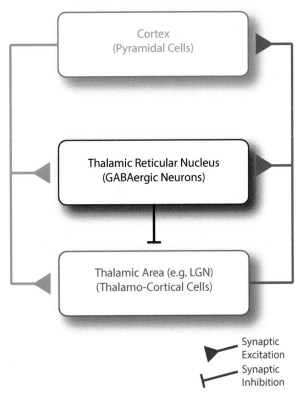

FIGURE 17.4   Thalamocortical (TC) circuit. Pyramidal (PY) cells in the cortex and TC cells in the thalamus excite each other and GABAergic neurons in the reticular nucleus of the thalamus [reticular (RE) cells]. TC cells receive inhibition from the RE cells. *LGN*, Lateral geniculate nucleus.

circuit includes three main players. The two excitatory types of neurons excite each other: TC neurons that project to the cortex and cortical pyramidal (PY) cells project to the thalamus. Also inhibitory neurons in the reticular nucleus of the thalamus (TRN) receive excitation from TC and PY cells and provide inhibition to the TC cells.

In the case of the delta oscillation, the interaction of two intrinsic ion currents (low-threshold calcium current I_T and hyperpolarization-activated depolarizing current I_h) endows TC neurons with resonance properties in the delta frequency range when sufficiently hyperpolarized. In this model, cortical Up states drive inhibitory neurons in the TRN that then cause the required hyperpolarization of TC cells for the occurrence of delta oscillations. Notably, this grouping of delta and slow oscillations may not be present during natural sleep. Nevertheless, the two rhythms can still be different since they have different dynamics during a night of sleep; in contrast to the slow oscillation, delta oscillations decrease with time spent sleeping during the night.

Mechanisms of the Slow Oscillation

Most of what we know about the cellular and network mechanisms of low-frequency oscillations is derived from the anesthetized animal, in vitro slice preparation, and computer simulations. It is likely but not certain that the mechanisms found in these model systems also apply to slow cortical oscillations during SWS and waking. From a dynamic systems viewpoint, networks that generate Up and Down states exhibit a pronounced bistability. Transitioning between these two states requires transient amplification and stabilization of activity during the Up state. The following Down state is the result of a collapse of activity that is temporally and spatially orchestrated and is eventually followed by a recovery leading to the next Up state. At the network level, the main question is to what extent such periodic transitions are a uniquely cortical feature or to what extent interconnected brain areas, in particular the thalamus, also contribute. This has become a controversial question, with evidence for both a prominent cortical role but also a thalamic contribution [4]. A main argument in favor of a cortical oscillation is the presence of slow rhythmic activation of cortical slices in vitro that share important properties with the in vivo slow oscillation, in particular the localization of Up state initiation to neocortical layer V [5]. Similarly, cortical networks isolated from subcortical brain structures by a surgical undercut also display oscillatory activity, with smaller slabs exhibiting more infrequent occurrence of Up state-like activity [6]. Larger, undercut gyri exhibited in vivo-like slow oscillations, but no spindles, because of the lack of TC interactions; such undercut preparations have the advantage that the tissue remains relatively intact in its natural environment. Arguments in favor of a role of the thalamus in the genesis of such pronounced oscillations center on the fact that neurons in thalamic slices can also exhibit slow periodic activity in the absence of patterned input from the cortex [4]. However, pharmacological activation of metabotropic glutamate receptors (particularly mGluR1), which are located postsynaptically on the corticothalamic synapses, is required. Activation of mGluR1 decreases the leak current, which sufficiently depolarizes the TC neurons to exhibit slow rhythmic activity as the result of the interaction of the low-threshold calcium current I_T and the hyperpolarization-activated depolarizing current I_h. At the onset of Up states, activation of I_T causes a low-threshold calcium potential during which TC neurons fire high-frequency bursts of action potentials. In conclusion, thalamic circuits may be conditional oscillators. Therefore slow rhythmic activity in the intact brain may arise from the interaction of the thalamus and the cortex, which can both intrinsically generate slow rhythmic activity.

At the more microscopic level, the main questions about the mechanisms of slow oscillatory activity have been (1) the mechanism by which a silent network becomes active again (transition from Down to Up state), (2) the mechanism(s) that boost and then suppress network activity during the Up state, (3) the mechanism that keeps the network silent during the Down state, and finally (4) the mechanism by which the networks switch from low-frequency oscillations to the desynchronized state. Computational models of networks of neurons with relatively detailed descriptions of intrinsic and synaptic physiology have taken center stage in answering these questions.

First, activation of a quiet network requires the emergence of action potentials that then trigger further action potentials, a typical positive feedback loop. Multiple sources of this initial activity at the onset of the Up state have been proposed. The presence of cells that do not cease firing action potentials during the Down state could create the momentum required for triggering the next Up state. Indeed, at least in the in vitro preparation, layer V pyramidal cells appear to fire action potentials at a low rate during the Down state [7]. Since it has remained unclear if such Down-state activity occurs in vivo, another mechanism based on spontaneous synaptic events has been proposed. Presynaptic terminals release neurotransmitter molecules in the absence of presynaptic action potentials (see chapter: Synaptic Transmission). In vitro, these *miniature synaptic potentials* (*minis*) are isolated by application of the sodium channel blocker TTX, which suppresses action potential firing by blocking voltage-gated sodium channels. If such minis temporally coincide in cells that are quiet but relatively close to threshold, temporal summation could trigger action potential firing. Computational models of this mechanism predict that with growing network size the oscillations become more regular and their frequency increases, a finding that is in agreement with the previously mentioned undercut experiments [8]. Despite a lack of direct experimental evidence, this mechanism is appealing, since no initial action potential firing during the Down state is required. The frequency of minis is increased in recently potentiated synapses, which would therefore imply that during slow oscillations neurons with recently potentiated synapses are preferentially activated. This directly relates to the theory of depotentiation of synapses during sleep (see earlier).

Second, synaptic and ionic processes on the time scale of the duration of an Up state (typically a few hundred milliseconds) enable transient stabilization followed by a collapse of network activity. Excitatory synaptic interactions clearly play an important role in the buildup of activity, together with the persistent sodium channel and high-voltage-activated calcium channels that help to stabilize the depolarized membrane voltage. Furthermore, synaptic excitation is rapidly balanced by synaptic inhibition, meaning that the total excitatory and inhibitory currents scale together. Sustained neuronal firing during the Up state activates hyperpolarizing outward currents such as the calcium-activated potassium current (I_{KCa} mediated by BK channels) and possibly a sodium-activated potassium current (I_{KNa}) as well. Both channels are activated by the increase of intracellular ion concentrations driven by neuronal depolarization and action potential firing. Therefore these two ion currents represent a negative feedback mechanism.

Third, the sustained hyperpolarization defining the Down state is not the result of sustained GABAergic inhibition, since inhibitory interneurons are also quiescent during the Down state. Instead, recovery from the negative feedback processes that terminate the Up state mediates the Down state.

Fourth, activation of cholinergic or noradrenergic projections abolishes the slow rhythmic activity patterns in the cortex and is closely related to behavioral arousal, waking, and attention (see chapter: Neuromodulators). These neuromodulators cause a decrease in potassium leak conductance. As a result, neurons depolarize and exhibit increased input resistance, which increases the change in membrane voltage caused by synaptic currents.

In the end, slow rhythmic activity needs to be understood in the framework of the cortico-TC circuit. Despite the usual focus on ascending information through the canonical feedforward TC circuit, in fact the descending ("feedback") projections from the cortex to the thalamus widely outnumber the feedforward connections. The descending feedback pathway targets both reticular neurons (in the TRN) and TC neurons. The effect of cortical activity, however, vastly differs in reticular (RE) and TC neurons. RE neurons respond with a spindle-frequency oscillation, whereas in TC cells, the inhibition from the RE cells dominates the direct excitation. The resulting hyperpolarization of the TC neurons enables the occurrence of a low-threshold calcium spike typical for the thalamic bursting mode. The transition to the activated (ie, awake) state in the TC system is supported by the differential effect of cholinergic neuromodulation. TC neurons depolarize and increase their input resistance in response to increased cholinergic tone; this provides the basis for the switch to an activated, depolarized state. In contrast, acetylcholine hyperpolarizes RE cells.

Functional Role of Low-Frequency Oscillations During NREM Sleep

NREM sleep is critical for the transformation of new memories formed during waking to be consolidated into more stable, long-lasting memory traces. Sleep likely enhances both memories of facts that require conscious recall (*declarative memory*) and skills that result from repeated practice that do not require conscious recall (*procedural memory*). Oscillations during sleep have been directly linked to learning and memory. For example, learning selectively increases the amplitude of the slow oscillation in the following night in a brain area-specific way [9]. Sleep spindles exhibit similar area-specific enhancement for both declarative and procedural learning paradigms. As we have seen, however, slow oscillations and sleep spindles are intimately related, and clear experimental separation is difficult to achieve. The strongest support for a functional role of sleep oscillations in memory consolidation comes from experiments where the oscillations were exogenously enhanced by noninvasive brain stimulation. In particular, transcranial current stimulation (see chapter: Noninvasive Brain Stimulation) during SWS with a waveform that exhibited periodic patterning at 0.75 Hz significantly enhanced both slow oscillations and sleep spindles [10]. Such boosting of slow rhythmic activity increased memory consolidation, specifically for the memorization of word pairs, which is a declarative memory task. Conceptually, similar results were found applying phase-locked auditory stimuli during SWS [11]. Such sensory stimulation as a tool to manipulate cortical oscillations for probing their functional role is particularly attractive, because it allows for simultaneous high-quality EEG (no stimulation artifacts).

Pharmacological agents are also used to perturb sleep network dynamics. However, a drawback of such experimental manipulations is that they may induce many other changes to the underlying molecular and cellular substrate.

Two main (but not mutually exclusive) hypotheses have emerged with regard to the functional role of low-frequency activity during SWS. The first hypothesis proposes that slow oscillations play a causal role in active memory consolidation by selectively repeating activity patterns that occurred during waking and therefore redistributing and strengthening the *memory trace* across the neocortical–hippocampal system (*active system consolidation hypothesis*). One very important finding in support of the consolidation hypothesis was the discovery of reactivation in SWS of neuronal firing patterns that had occurred during the previous day in the awake state [12]. Rats exposed to novel environments or spatial tasks exhibited sequences of neuronal firing that were *replayed* during sleep. Furthermore, in the hippocampus, *sharp-wave ripple* events are linked to spindles in the TC system. Sharp waves are fast depolarizing events accompanied by local high-frequency oscillations in CA3 of the hippocampus (see chapter: Microcircuits of the Hippocampus). Together, slow waves, spindles, and sharp-wave ripples are organized such that during an Up state, sharp-wave ripples and spindles co-occur, and so may provide communication channels between the hippocampus and the neocortex for transfer and consolidation of memories into long-term storage [13]. The second hypothesis focuses on synaptic downscaling, in the form of synaptic homeostasis as the main mechanism for memory consolidation during sleep (*synaptic homeostasis hypothesis*). In this framework [14], synapses become reduced in strength during sleep, and therefore the weakly potentiated synapses fall below threshold, whereas the more strongly potentiated synapses "survive" and remain strengthened (albeit at reduced amplitude) after a night of sleep. In this model, synapses relevant for a memory are spared, while other, irrelevant synapses selectively decrease in strength ("downselection"). Such a process would enhance the signal-to-noise ratio. This hypothesis is supported by evidence gathered with molecular and electrophysiological methods that shows that synapses potentiate during waking and depotentiate during sleep.

AWAKE STATE

Traditionally, the occurrence of slow rhythmic activity has been exclusively associated with sleep and pathological states such as epilepsy or structural brain damage. However, animal studies have demonstrated the presence of slow rhythmic activity in the cortex of awake animals that are resting. Details of the experimental context appear to modulate the presence of slow rhythmic activity such that reward anticipation, attention, and possibly stress-induced arousal suppress the occurrence of slow rhythmic activity. As a result, the presence of slow rhythmic activity in the cortex of awake animals has led to a model in which there is a graded presence of slow rhythmic activity [15]. One end of the spectrum is occupied by a state of high vigilance, with maximal engagement of the environment and virtually no slow rhythmic activity. The other end of the spectrum corresponds to the classical scenario of slow oscillations during SWS. Animals appear to dynamically regulate their state on this continuum and behavioral output such as whisking and locomotion correlate with where the animal is on this continuum. Cortical state is strongly associated with overall neuromodulatory tone. Most prominently, acetylcholine defines neuronal state (see chapter: Neuromodulators). Other neuromodulators such as serotonin may also play a role. Additionally, the separation between awake and asleep may be less complete and global as commonly assumed. Sleep-deprived rats, for example, appear to exhibit individual, localized Down states in the cortex during wakefulness [16].

Quite surprisingly, delta frequency-band oscillations may also play a role in active cognitive processing of sensory input. In attention tasks that have an inherent rhythmic structure, delta oscillation activity appears to amplify response in a phase-dependent manner by entrainment of the delta oscillation to the rhythmic structure of the sensory input [17]. These results are in apparent conflict with the notion that attention and vigilance decrease slow rhythmic activity (see earlier). This difference can possibly be explained by the nature of tasks. The entrainment of the cortex to the periodic structure of the stimulus makes sense to selectively boost processing of the incoming activity. In contrast, in tasks where the stimulus occurs at an unpredictable time in the future, periodic modulation of excitability by low-frequency network activity may be detrimental and therefore suppressed. Such results lead to more questions than answers, but they indicate that low-frequency oscillation may need to be considered in a broader and more sophisticated context than only sleep, anesthesia, and wakeful quiescence.

DISEASE STATE

Before the advent of modern neuroimaging, EEG was broadly used as a diagnostic tool. In contrast to today, EEG was also employed to identify pathological changes to neuronal structure, such as brain tumors. In the case of neurological disorders, the most prominent change in macroscopic network dynamics is the occurrence of slow rhythmic activity during the awake state (*cortical slowing*). Such focal slow activity was originally discovered in a patient with a brain tumor in one cerebral hemisphere and has become a commonly encountered clinical EEG marker of a localized, structural lesion. Continuous versus intermittent slow activity distinguishes larger from smaller lesions. Interestingly, lesions limited to gray matter tend not to generate cortical slowing. In contrast, a decrease in afferent drive because of white matter or subcortical damage causes cortical slowing. The underlying cellular and network mechanisms remain unclear; computational modeling suggests that maladaptive homeostatic plasticity can lead to enhanced slow rhythmic activity in the cortex after (partial) deafferentation [17a]. Such cortical slowing represents a nonepileptic pathological activity pattern. Slow rhythmic activity patterns are also fundamental to epileptic seizures, but we will defer that discussion to chapter "Epilepsy."

DEVELOPMENT

During the third trimester of pregnancy in humans and during the first few postnatal days in rats, cortical network activity does not show differential patterns as a function of behavior state. Instead, cortical neurons are at the resting potential, with the exception of bursts of action potentials that sporadically occur both during wakefulness and sleep

[18]. This is a major difference from the awake, more mature animal, where neurons are depolarized during wakefulness and exhibit low-frequency oscillations during sleep. During these bursts in the immature cortex, membrane voltage does not exhibit a stable depolarization as during Up states in the adult animal. In the rat, this fundamental switch in activity patterns occurs before eye opening during the second postnatal week. Synaptic inhibition in response to visual stimulation greatly increases during that short time window [19]. Therefore maturation of feedforward inhibition may play a central role in this switch from immature bursting and mature, continuous activity. In humans, this switch occurs in the perinatal period, since immature bursting can be detected by EEG in preterm but not full-term infants. During postnatal development, SWA in NREM sleep exhibits a developmental time course in humans. SWA starts out as most prominent in posterior brain areas, and then gradually shifts in the anterior direction, in agreement with the overall posterior–anterior maturation pattern of gray matter in the cerebral cortex. It remains unknown whether this time course of the topography of low-frequency oscillations plays a causal role in circuit maturation [20].

SUMMARY AND OUTLOOK

In this chapter, we discussed low-frequency (thalamo-)cortical oscillations in different states: anesthesia, sleep, and waking. We also learned about the occurrence of low-frequency oscillations during development and in disorders of the CNS. In terms of functional roles of such oscillations, most importantly, slow rhythmic activity during sleep may play an active and important role in memory consolidation. The mechanisms of low-frequency oscillations are relatively well understood. Both cortical and thalamic circuits can generate such oscillations in isolation. The connectivity of the TC system organizes the activity of these oscillators as a function of overall behavioral state.

References

[1] Sanchez-Vives MV, McCormick DA. Cellular and network mechanisms of rhythmic recurrent activity in neocortex. Nat Neurosci October 2000;3(10):1027–34.

[1a] Steriade M, Nuñez A, Amzica F. A novel slow (< 1 Hz) oscillation of neocortical neurons in vivo: depolarizing and hyperpolarizing components. J Neurosci August 1993;13(8):3252–65.

[1b] Rechtschaffen A, Kales A. A manual of standardized terminology, techniques and scoring system for sleep stages of human subjects. Neurological Information Network; 1968. p. 204.

[2] Silber MH, et al. The visual scoring of sleep in adults. J Clin Sleep Med 2007;3(2):121–31.

[3] Massimini M, et al. The sleep slow oscillation as a traveling wave. J Neurosci 2004;24(31):6862–70.

[3a] Huber R, Ghilardi MF, Massimini M, Tononi G. Local sleep and learning. Nature July 1, 2004;430(6995):78–81.

[4] Crunelli V, Hughes SW. The slow (<1 Hz) rhythm of non-REM sleep: a dialogue between three cardinal oscillators. Nat Neurosci 2010;13(1):9–17.

[5] Sanchez-Vives MV, McCormick DA. Cellular and network mechanisms of rhythmic recurrent activity in neocortex. Nat Neurosci 2000;3(10):1027–34.

[6] Timofeev I, et al. Origin of slow cortical oscillations in deafferented cortical slabs. Cereb Cortex 2000;10(12):1185–99.

[7] Compte A, et al. Cellular and network mechanisms of slow oscillatory activity (<1 Hz) and wave propagations in a cortical network model. J Neurophysiol 2003;89(5):2707–25.

[8] Bazhenov M, et al. Model of thalamocortical slow-wave sleep oscillations and transitions to activated states. J Neurosci 2002;22(19):8691—704.

[9] Huber R, et al. Local sleep and learning. Nature 2004;430(6995):78—81.

[10] Marshall L, et al. Boosting slow oscillations during sleep potentiates memory. Nature 2006;444(7119):610—3.

[11] Ngo HV, et al. Auditory closed-loop stimulation of the sleep slow oscillation enhances memory. Neuron 2013;78(3):545—53.

[12] Wilson MA, McNaughton BL. Reactivation of hippocampal ensemble memories during sleep. Science 1994;265(5172):676—9.

[13] Diekelmann S, Born J. The memory function of sleep. Nat Rev Neurosci 2010;11(2):114—26.

[14] Tononi G, Cirelli C. Sleep and the price of plasticity: from synaptic and cellular homeostasis to memory consolidation and integration. Neuron 2014;81(1):12—34.

[15] Harris KD, Thiele A. Cortical state and attention. Nat Rev Neurosci 2011;12(9):509—23.

[16] Vyazovskiy VV, et al. Local sleep in awake rats. Nature 2011;472(7344):443—7.

[17] Lakatos P, et al. Entrainment of neuronal oscillations as a mechanism of attentional selection. Science 2008;320(5872):110—3.

[17a] Fröhlich F, Bazhenov M, Sejnowski TJ. Pathological effect of homeostatic synaptic scaling on network dynamics in diseases of the cortex. J Neurosci February 13, 2008;28(7):1709—20. http://dx.doi.org/10.1523/JNEUROSCI.4263-07.2008.

[18] Colonnese MT, et al. A conserved switch in sensory processing prepares developing neocortex for vision. Neuron 2010;67(3):480—98.

[19] Colonnese MT. Rapid developmental emergence of stable depolarization during wakefulness by inhibitory balancing of cortical network excitability. J Neurosci 2014;34(16):5477—85.

[20] Ringli M, Huber R. Developmental aspects of sleep slow waves: linking sleep, brain maturation and behavior. Prog Brain Res 2011;193:63—82.

Theta Oscillations

The chapters in this unit are sorted by oscillation frequency. Here, we discuss the oscillation with the next-higher frequency than the slow and delta oscillations discussed in the previous chapter—the theta oscillation. The theta oscillation is very prominent and quite well understood in rodents. In nonhuman primates and humans, however, the theta oscillation possibly plays different roles. This difference already becomes apparent in the definition of the theta frequency band. In rodents, theta oscillations span the 4–15 Hz range. In humans, theta oscillations are more narrowly defined as ranging from 4 to 8 Hz. In this chapter, we will first examine rodent theta oscillations and learn about their roles in complex, hippocampus-dependent behaviors such as navigation and learning. The chapter "Microcircuits of the Hippocampus" provides the required background information on the circuits we discuss here in the context of theta oscillations. We will then discuss what is known about theta oscillations in the more complex brains of primates, where theta oscillations have been found in various neocortical areas as a function of behavioral demands.

THETA OSCILLATIONS IN RODENTS

Mechanisms of Theta Oscillations in the Hippocampus

In the rodent, theta oscillations [1] are most pronounced in CA1 of the hippocampus, specifically the stratum lacunosum-moleculare, where CA1 neurons receive direct input from the entorhinal cortex via the perforant path (see chapter: Microcircuits of the Hippocampus). Theta oscillations can also be detected in recordings from many other rodent brain structures, including other areas of the hippocampal formation (CA3, dentate gyrus, entorhinal cortex, subiculum, etc.). The medial septum, a cholinergic structure in the basal forebrain, plays a key role in the genesis of theta rhythms in the rodent hippocampus. Both lesioning and inactivation of this structure, which is bidirectionally connected with the hippocampus, disrupt theta oscillations. GABAergic inhibitory interneurons in the medial septum express hyperpolarization-activated cyclic nucleotide-gated (HCN) channels that mediate the hyperpolarization-activated depolarizing current I_h. Importantly, these neurons fire at the theta frequency. In the classical model of theta oscillations, two distinct inputs drive oscillation in hippocampal neurons. First, the inhibitory input from the medial septum targets local inhibitory interneurons in the hippocampus that provide somatic inhibition to CA1 pyramidal cells. Second, input from the entorhinal cortex (layer III) provides an excitatory input to the distal dendrites. This model is based on the current source density (CSD) analysis of depth recordings with laminar probes (see chapter: LFP and EEG). The main feature of this classical model is that the medial septum acts as a pacemaker. However, this model cannot explain all experimental evidence. Classically, a combination of surgical lesions and pharmacological manipulations has been used to test the models of theta mechanisms in the hippocampus. Given the complexity of the circuit, pharmacological probing is only of limited help, since it is unclear which targets of the complex circuit are being perturbed. Nevertheless, historically there is a distinction between atropine-sensitive and atropine-insensitive theta oscillations. Atropine blocks muscarinic (cholinergic) receptors (see chapter: Neuromodulators). This distinction was derived from the fact that under anesthesia, theta oscillations are abolished by atropine, whereas in the awake, moving animal, little effect of atropine on theta was found. Surgical elimination of input from the entorhinal cortex removes the CSD signature of excitatory input to the distal dendrites and makes the resulting theta oscillation atropine sensitive. Therefore the entorhinal cortex input is atropine insensitive, and is probably mediated by NMDA synaptic excitation onto the distal dendrites of CA1 pyramidal cells. A third but smaller input can also be detected in the form of a current sink in the stratum radiatum, where the Schaffer collateral input from CA3 targets CA1 pyramidal cells.

Within the local microcircuit, inhibitory interneurons in the CA1 circuit play a fundamental role in the generation of theta oscillations [2]. The exact role of different types of interneurons remains to be fully dissected. One obstacle is the fact that pyramidal cells and basket cells fire action potentials at the same phase in the freely behaving rat, but at opposite phases during anesthesia. As mentioned earlier, interneurons are the sole recipient of inputs from the medial septum and also are the only type of neurons that project back to the medial septum. In addition to these network-level mechanisms, several cell types in this circuit exhibit intrinsic frequency preference that can lead to resonance in the theta frequency range. For example, cholinergic neurons of the medial septum fire bursts of action potentials grouped by low-threshold spikes that occur at the theta frequency. The GABAergic cells in the medial septum exhibit pronounced subthreshold oscillations that are interleaved with bursts of action potentials. Pyramidal cells also exhibit voltage-dependent intrinsic oscillations. Specifically, hippocampal pyramidal cells exhibit subthreshold resonance in vitro—with synaptic transmission blocked, the injection of a weak sine-wave current with increasing frequency (chirp) elicits the strongest modulation of the membrane voltage for stimulation frequencies in the theta band [3]. The main ionic current implicated in subthreshold resonance is the hyperpolarization-activated depolarizing current I_h. However, little is known about how this subthreshold behavior for hyperpolarized membrane voltages in the absence of synaptic input translates to the in vivo condition. For example, hippocampal pyramidal cells did not exhibit theta resonance when optogenetically stimulated in vivo. Instead, periodic stimulation of parvalbumin-positive inhibitory interneurons caused a theta resonance peak in firing of the pyramidal cells. These results indicate that the theta oscillation arises from a complex interaction of cellular and network properties.

Memory and Navigation

The two best-understood functions of the hippocampal formation are to support memory formation and consolidation, and spatial navigation. These two seemingly distinct functions in fact share common properties [4]. As we will see, both functions are closely related to the occurrence of theta oscillations. Spatial navigation is enabled by static maps that encode locations in a reference coordinate system. Such maps arise from the integration of motion and knowledge about previous locations. Map-based representations are referred to as *allocentric*. Path integration for navigation is referred to as an *egocentric* process. Animals may use a dynamic combination of the two reference systems as a function of the momentary circumstances. Absence of external cues (eg, in the dark) favors egocentric navigation that relies on endogenously generated signals such as speed, elapsed time, and head direction. Allocentric navigation strategies are more favorable in cue-rich environments. These two spatial navigation modes have been suggested to correspond to two types of declarative memories. Specifically, declarative memories can also be divided into subclasses that exhibit a similar distinction to the one that distinguished allocentric from egocentric navigation: *semantic* and *episodic memories*. Semantic memories encompass facts about objects and events that do not require individual temporal context. Such memories may be conceptually related to allocentric navigation. Episodic memories, on the other hand, share the first-person qualities of egocentric navigation, because they are concerned with individual experiences.

Given this proposed analogy between navigation and memory [4], we can ask if these behavioral functions share underlying neuronal mechanisms at the level of circuits and dynamics. The neuronal bases of navigation in the hippocampal formation are *place cells* and *grid cells* (Fig. 18.1).

Place cells are neurons in the hippocampus that respond to a specific spatial location of the animal. Grid cells fire at multiple spatial locations that are arranged in a hexagonal

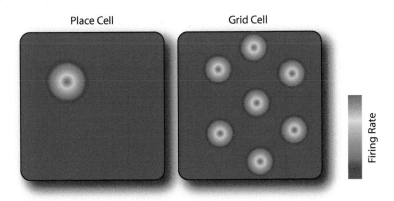

FIGURE 18.1 Firing rate of a neuron as a function of the spatial location of the animal. (Left) Place cells respond to a specific spatial location that is referred to as place field. (Right) Grid cells respond to a set of different locations that are arranged on a hexagonal grid.

pattern. Grid cells are most prominent in the superficial layers of the entorhinal cortex (input to the hippocampus, see chapter: Microcircuits of the Hippocampus). Grid cells are active in all environments, and therefore they provide a universal representation of space across environments. Conversely, place cells change their activity pattern as a function of the environment. Together, the neurons in the hippocampal system have the right properties to build context-specific, map-based representations of the environment. Even small changes in the environment can lead to complete remapping of the encoding of space in the hippocampus. The ability to generate and maintain these spatial maps has been proposed to represent the origin of more general purpose memory functions such as declarative memory. Today, however, no similar framework of what the activity of individual neurons encodes in terms of declarative memories exists. The one similarity that has been noted is that both episodic memory and egocentric navigation strategies fundamentally rely on temporal sequences (of events or locations). This is of interest in the context of this chapter since it appears that the theta oscillation in the rodent provides a general timing or reference signal for the representation of sequences by the firing patterns of clusters of neurons, so-called *cell assemblies*. Therefore the theta oscillation may provide the shared, underlying neuronal mechanism of both spatial navigation and declarative memory.

Cell Assemblies and Theta Oscillations

Cell assemblies are groups of neurons that are active together within a short time window. The duration of the activity of a cell assembly is limited such that the postsynaptic potentials caused by the activity of individual cells within the assembly interact in the downstream neurons. Theta oscillations appear to group cell assemblies, meaning that different cell assemblies are activated on different cycles of the theta oscillation. For example, the place cells in rats exposed to two different environments defined by light cues encoded different locations as a function of the environment. The rat was subsequently moved to a new enclosure in which either of the previous two environments could be mimicked (by activation of corresponding lights). Depending on which environment was simulated, the place cells rapidly switched between their respective representations. Most importantly, these two distinct representations were activated on subsequent cycles of the theta oscillation and were never mixed within a given theta cycle [4a].

Place cells that fire within a theta oscillation cycle encode paths that begin behind the animal and end before the animal. The locations at which a place cell fires action potentials are collectively referred to as *place field*. Firing of place cells occurs at different phases of the theta oscillation as a function of the position of the animal within the place field. When first entering the place field, action potentials occur at late phases of the oscillation cycle. The phase of subsequent action potentials then gradually decreases as the animal crosses through the place field. This phenomenon is called *phase precession* (Fig. 18.2).

Both place cells and grid cells exhibit phase precession. Two classes of models relate theta oscillations to phase precession in grid and place cells. The two models are not mutually exclusive, but nevertheless are the basis of a spirited scientific debate [5].

In the *dual oscillator model*, theta precession arises from the interaction of two oscillations within a single cell: a perisomatic (ie, at the soma) and a dendritic oscillation.

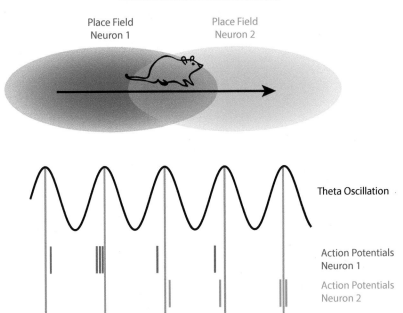

FIGURE 18.2   When a rat traverses a place field (top), the action potentials of the place cell fire earlier and earlier in subsequent theta cycles (bottom). This phase precession is shown for two place cells with nearby place fields (blue and red). *Black sine wave* symbolizes theta oscillation. *Gray vertical bars* are drawn to visually emphasize the shift of the individual action potentials relative to the theta oscillation.

The perisomatic oscillation occurs at the network theta frequency, whereas the dendritic oscillation increases in frequency above baseline (network frequency) when the animal passes through the place field. Therefore phase precession emerges from the interaction of the network theta oscillation and a slightly faster oscillation frequency of an individual neuron. On every cycle, the phase advance of the oscillation in the neuron of interest and the network increases. As a result, action potentials appear on earlier and earlier phases of the network theta oscillation as the animal traverses the place field. In agreement with this model, intracellular recordings from place cells revealed subthreshold membrane fluctuation in the theta frequency that increased when the animal passed through the place field. Such an acceleration of the oscillation within a place cell corresponds to the faster oscillation postulated by the dual oscillator model.

In the *spreading activation model*, place cells receive input from other place cells that encode fields earlier in the route of the animal. When the animal enters the place field, the neurons receive input from farther-back place cells, which in turn have longer transmission delays, so spiking occurs at a later theta phase. Toward exiting the place field, the place cell receives input from place cells closer to the current place field. Thus this input exhibits less of a delay. Neither model can explain all of the main experimental observations. Nevertheless, the study of phase precession by computational models is a successful demonstration of how computational neuroscience can advance the mechanistic study of cellular and network-level phenomena that directly relate to behavior.

THETA OSCILLATIONS IN PRIMATES

Memory and Navigation

Despite the long history of studying theta oscillations in rodents, insights into theta oscillations in humans are mostly based on recent discoveries. One of the first studies that have demonstrated the presence of theta oscillations in humans [6] was performed in epilepsy patients implanted with subdural electrocorticography electrodes (see chapter: LFP and EEG). Patients learned to navigate a virtual spatial maze—a task that was designed to mimic the type of behavioral tasks used in rodent studies. During performance of this task, pronounced but short episodes of theta activity were detected on many cortical electrodes, suggesting that theta oscillations are also associated with spatial learning in humans. Subsequent studies then demonstrated recruitment of theta oscillation during the Sternberg task, which is a working memory task. Participants need to memorize a set of consecutively presented letters and are then asked if a probe letter presented after a delay was part of the previously presented set (Fig. 18.3).

Theta oscillations have also been implicated in top-down control of encoding and retrieval of episodic memory [7]. In humans, theta oscillations increase in frontal areas during encoding of items that are subsequently correctly remembered. Phase-locking (chapter: Network Interactions) between frontal and parietal/temporal areas correlates with successful memory

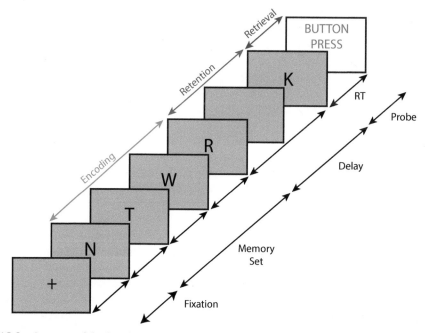

FIGURE 18.3   Structure of the Sternberg working memory task. After initial fixation, a series of letters (memory set) are displayed that need to be encoded in working memory. After a retention period, a probe stimulus is presented and the participant presses a button if the probe stimulus is part of the memory set. *RT*, Reaction time.

encoding. Simultaneous recordings of action potentials and local field potential in the human hippocampus and amygdala showed that pictures that elicited action potentials time locked to the theta oscillation were better remembered in a recognition task (at least 15 min later) than the ones that fail to elicit a phase-locked response [8]. However, no difference in theta power was found. Oscillation power and functional connectivity in the theta band appear to mediate a similar top-down control during memory retrieval processes [9]. It is likely that there may be different theta sources that contribute to different processes that together enable successful performance of memory tasks. Indeed, correlating theta power at different locations with different task variables (task performance, reaction time, memory load) points to such complex, distributed organization of theta oscillations. In nonhuman primates, neurons in the occipital visual area V4 phase lock to the theta oscillation during the retention period of a working memory task [10]. These results suggest that theta oscillations may act as an overall organizing signal that orchestrates not only frontal—temporal interactions but also more basic sensory processing areas.

Cognitive Control

In addition to memory, theta oscillations in humans have been strongly implicated in *cognitive control*, which is defined as the ability to overcome a habitual response pattern to execute a goal-directed action. Theta oscillations measured from frontal midline EEG electrodes (over the medial prefrontal cortex) correlate with situations that call for cognitive control [11]. Thus they are prominent in a range of scenarios, including novelty, conflict, punishment, and error. All of these types of events are associated with a prediction error, which is defined as the mismatch between anticipated and received award. These theta signals are localized to the midcingulate cortex and the presupplemental motor area. Low-frequency oscillations are ideal to enable communication over longer distances since propagation delays are less detrimental to synchronization. Accordingly, the midline frontal theta oscillation exhibits synchronization with distant sites that could reflect large-scale organization of cortical activity. It remains unclear to what extent such organization of activity in the theta band serves as a general signal of prediction error or actually carries specific task-related information.

SUMMARY AND OUTLOOK

In this chapter, we discussed theta oscillations. We learned that the definition of theta oscillations is a challenge, because of species and brain area differences. Theta oscillations provide an important timing signal that relates to spatial navigation and memory in the rodent hippocampus. Theta oscillations are state dependent and dominate the awake, exploring state and rapid eye movement sleep in rodents. We reviewed the cellular and circuit mechanisms that generate theta oscillations in the rodent hippocampus. These oscillations arise from the dynamic circuit interactions between the medial septum and the hippocampus. Intrinsic resonance properties mediated by HCN channels also potentially contribute to the genesis of theta oscillations. There is also ample evidence of theta oscillations in the human neocortex; these rhythms appear to play a major role in orchestrating long-range interaction of multiple cortical areas. Overall, a unifying framework of human theta oscillations that includes their appearance, functional roles, and underlying mechanisms is lacking.

References

[1] Buzsaki G. Theta oscillations in the hippocampus. Neuron 2002;33(3):325—40.

[2] Colgin LL. Mechanisms and functions of theta rhythms. Annu Rev Neurosci 2013;36:295—312.

[3] Hu H, Vervaeke K, Storm JF. Two forms of electrical resonance at theta frequencies, generated by M-current, h-current and persistent Na^+ current in rat hippocampal pyramidal cells. J physiology 2002;545(Pt 3):783—805.

[4] Buzsaki G, Moser EI. Memory, navigation and theta rhythm in the hippocampal-entorhinal system. Nat Neurosci 2013;16(2):130—8.

[4a] Jezek K, Henriksen EJ, Treves A, Moser EI, Moser MB. Theta-paced flickering between place-cell maps in the hippocampus. Nature 2011;478:246—9.

[5] Burgess N, O'Keefe J. Models of place and grid cell firing and theta rhythmicity. Curr Opin Neurobiol 2011;21(5):734—44.

[6] Kahana MJ, Seelig D, Madsen JR. Theta returns. Curr Opin Neurobiol 2001;11(6):739—44.

[7] Raghavachari S, et al. Gating of human theta oscillations by a working memory task. J Neurosci 2001;21(9):3175—83.

[8] Rutishauser U, et al. Human memory strength is predicted by theta-frequency phase-locking of single neurons. Nature 2010;464(7290):903—7.

[9] Anderson KL, et al. Theta oscillations mediate interaction between prefrontal cortex and medial temporal lobe in human memory. Cereb Cortex 2009:bhp223.

[10] Lee H, et al. Phase locking of single neuron activity to theta oscillations during working memory in monkey extrastriate visual cortex. Neuron 2005;45(1):147—56.

[11] Cavanagh JF, Frank MJ. Frontal theta as a mechanism for cognitive control. Trends Cogn Sci 2014;18(8):414—21.

Alpha Oscillations

Alpha oscillations are a rhythmic thalamocortical activity pattern (8–12 Hz in humans) first described by Hans Berger, who discovered electroencephalography. Alpha oscillations are the most obvious rhythmic activity pattern recorded from the cortex of awake human participants, and they can be easily identified by eye without any signal processing. Alpha oscillations were originally assumed to reflect cortical "idling" or "inactivity," since they prominently occur in posterior (visual) areas in the electroencephalogram (EEG) of participants with closed eyes. More recently, alpha oscillations have been recast as a mechanism that establishes long-range interactions between distant cortical areas, and they have been implicated in gating of sensory processing. In this chapter, we first introduce the terminology used for describing alpha oscillations, delineate what is known about the underlying mechanisms of alpha oscillations, discuss the possible functional roles of alpha oscillations, and conclude with an outlook on noninvasive brain stimulation paradigms that can modulate alpha oscillations.

TERMINOLOGY

Alpha oscillations have been recorded in humans since the discovery of the EEG in the 1920s [1]. Alpha oscillations can often easily be detected in the raw signal recorded from scalp electrodes positioned over parietal and occipital areas. Simply closing the eyes elicits a pronounced power increase in the alpha frequency band. Accordingly, recording EEG activity during *eyes open* and *eyes closed* conditions is standard for most clinical usages of EEG (Fig. 19.1).

The terminology used to describe basic modulation of the power of the alpha oscillation can be confusing (Fig. 19.2).

First, any rhythmic signal in the EEG reflects the synchronized activation of a large number of neurons, whose small extracellular electric fields sum to form the macroscopic EEG signal (see chapter: LFP and EEG). Therefore, high-amplitude alpha oscillations are also referred to as *alpha synchronization*. Accordingly, reduced alpha power is called *alpha desynchronization*. If such changes in alpha oscillation are associated with external events (eg, during a cognitive task), then an increase and a decrease in alpha oscillations are called *event-related synchronization* and *event-related desynchronization*, respectively. This nomenclature is less than ideal because we are often interested in how the alpha oscillation at different electrode sites

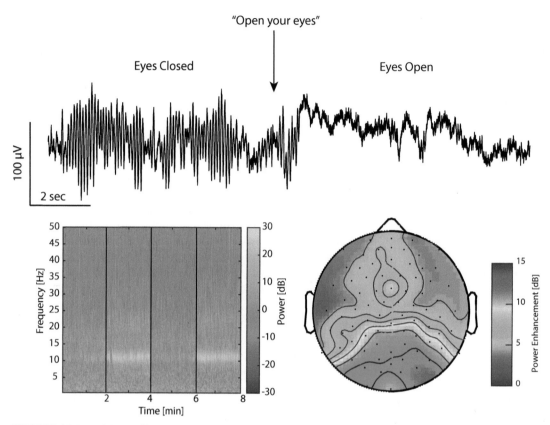

FIGURE 19.1   Alpha oscillations are prominent when eyes are closed and are suppressed when eyes are open. *Top*: Raw electroencephalogram (EEG) trace. *Bottom left*: Spectrogram of alternating epochs of "eyes open" and "eyes closed." *Bottom right*: Topographic map of changes in alpha oscillations when closing eyes.

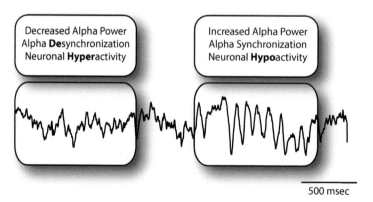

FIGURE 19.2   Sample electroencephalogram (EEG) trace with low alpha oscillation power (*left*) and high alpha oscillation power (*right*).

interacts and synchronizes. In this case, synchronization refers to the temporal alignment of different (spatially distant) EEG signals and not of the underlying neuronal processes recorded from a single EEG electrode. Second, an increase in the power of alpha oscillations typically correlates with a decrease in neuronal firing rates. Thus an increase in alpha oscillation is also called (neuronal) *hypoactivity* or simply decrease in activity. This is another source of confusion, because an increase in EEG signal in the alpha band is referred to as a decrease in activity.

It is important to remember that not all activity within the 8−12 Hz frequency band corresponds to a monolithic alpha oscillation. Other activity signatures that play functionally different roles from the "classical" alpha oscillation occupy the same frequency band. Aside from the different functional roles, these different oscillation patterns are also likely to be driven by different mechanisms. For example, the sleep spindle (11−16 Hz) is a transient activity pattern that overlaps with the alpha frequency band and occurs during nonrapid eye movement sleep. The first and most obvious difference is the sustained occurrence of alpha oscillations in posterior cortical areas in contrast to the transient occurrence of spindles that are of short duration, typically 1−2 s long and most pronounced in frontal and central brain areas.

MECHANISMS OF ALPHA OSCILLATIONS

The study of the cellular and network-level mechanisms of alpha oscillations has been hampered because mice and rats do not exhibit alpha oscillations. In rodents, the alpha frequency is occupied by theta oscillations, which are defined as ranging from 4 to 15 Hz (for humans theta oscillations are confined to 4−8 Hz). In human studies, source localization demonstrates that the sources of the alpha signal recorded by EEG reside in the parietal cortex. However, this does not demonstrate that the driver of the signal resides in the cortex. Similarly to the case of low-frequency oscillations during sleep (see chapter: Low-Frequency Oscillations), the relative role of the thalamus and the cortex in the genesis of alpha oscillations is a topic of ongoing debate. Clearly, the thalamus is somehow involved in the genesis of alpha oscillations, which is supported by findings of changed alpha oscillations in humans with thalamic lesions. Also correlations of activation of the thalamus measured by functional magnetic resonance imaging and EEG recordings of alpha oscillations have been reported. Simultaneous recordings in the thalamus and cortex of the dog show alpha oscillations that were dominant in posterior cortex areas and occurred during eyes closed periods [2]. Importantly, in some periods these recordings showed alpha oscillations in the lateral geniculate nucleus (LGN) of the thalamus without a matched alpha oscillation in the visual cortex, suggesting that the thalamus can generate alpha oscillation in the absence of a contribution by the visual cortex. However, functional connectivity in the alpha frequency band was more pronounced for two cortical sites than for the LGN and the visual cortex. Similar alpha oscillations were also found in cats. But in contrast to the results from the dog studies, alpha oscillations in the cat were associated with strong functional connectivity between the LGN and the visual cortex [3]. A subset of thalamic neurons fire action potentials phase locked to the cortical alpha oscillations. Interestingly, similar behavior can be recorded in vitro from cat thalamic slices [4] when exposed to a metabotropic glutamatergic agonist (see chapter: Synaptic Transmission). This pharmacological manipulation mimics the afferent drive from corticothalamic afferents, because mGluR1 receptors are located postsynaptically on the synapses formed by corticothalamic afferents on thalamic neurons. The neurons that fire in synchrony with the alpha oscillation fire high-threshold bursts of action potentials. In addition, these neurons are connected by gap junctions, which further enhance their synchronization.

FUNCTIONAL ROLES OF ALPHA OSCILLATIONS

Sensory Input Gating

The best-understood functional role of alpha oscillations in cognition is gating of sensory information. Specifically, alpha oscillations modulate sensory perception. Three properties of the alpha oscillations are correlated with modulation of sensory perception: amplitude, phase, and functional connectivity. We will discuss each.

Alpha oscillations fluctuate over time. The amplitude of the alpha oscillation at stimulus onset modulates the sensory detection threshold. The higher the power in the alpha band, the higher the detection threshold. Accordingly, the detection of threshold sensory stimuli (typically weak stimuli of very short duration) is modulated by the *envelope* of the alpha oscillation. The envelope describes the changes in signal power over time (Fig. 19.3). Note

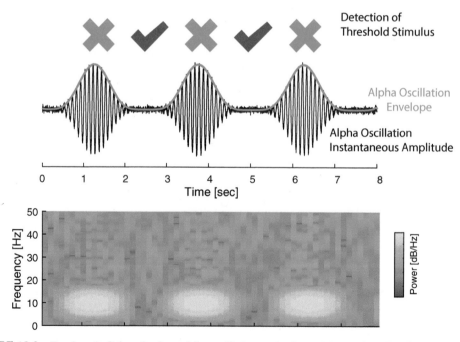

FIGURE 19.3 Envelope (*red*) describes how alpha oscillation power fluctuates over time. *Top*: Time series. *Bottom*: Spectrogram. Periods of high alpha power are associated with misses and periods of low alpha oscillations with hits in a sensory threshold paradigm. Data shown are computer-generated.

that the envelope differs from instantaneous amplitude, which fluctuates at the alpha frequency.

The phase of the alpha oscillation also shapes the response to threshold perceptual stimuli [5]. Within each ~100-ms cycle of the alpha oscillation, two opposing windows of preferred ("online") and nonpreferred ("offline") sensory processing occur. In the online period, the perceptual threshold is lower, that is, participants have a higher probability of detecting a weak sensory stimulus. In the offline period, the perceptual threshold is higher (Fig. 19.4). The presence of such a modulation of the perception of threshold sensory stimulus by the alpha oscillation phase is determined by calculating the *phase bifurcation index*. This index determines the differences in phase distributions for the stimuli that were not detected (misses) and stimuli that were detected (hits).

The strength of phase locking of alpha oscillations across cortical sites (in particular between frontal and parietal sites) also correlates with perceptual performance. Phase locking in this context refers to the phase-locking value (PLV) introduced in the chapter "Network Interactions." Briefly, PLV assesses how stable the phase difference between two signals is over time. Low PLV values of alpha band oscillations are associated with lowered perceptual threshold. Correspondingly, high PLV values are associated with increased perceptual threshold. It is interesting to note that most studies that investigated the role of alpha phase in determining behavioral responses to threshold stimuli used simple sensory stimuli (light flashes, brief tones, etc.), but the same mechanism was also found for more sophisticated discrimination tasks [6] such as distinguishing real words from similarly sounding pseudowords ("banana" vs "banena"). The localization of the EEG electrodes that showed this effect

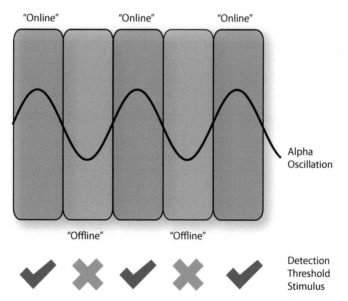

FIGURE 19.4   The phase of the alpha oscillation switches the brain between an "online" and an "offline" state. During the online state, the sensory detection threshold is low and threshold stimuli are successfully detected. During the offline state, the threshold is high and the stimuli are missed.

was more anterior than in the tasks that typically show modulation of sensory thresholds by alpha oscillation in posterior areas. These findings suggest that alpha oscillations represent a general gate to sensory processing not limited to low-level sensory perception.

We conclude our discussion of alpha oscillation as a sensory gating mechanism by emphasizing that the findings discussed earlier are all based on correlations. Correlations do not demonstrate the presence of a causal relationship. Several strategies can be used to modulate alpha oscillations in humans to more directly investigate the causal relationship between features of alpha oscillations and perceptual threshold. We will revisit these strategies when we discuss the manipulation of alpha oscillations by noninvasive brain stimulation later in this chapter.

Working Memory, Attention, and Creativity

Next, we will review what we know about the role of alpha oscillations in more sophisticated cognitive processes beyond the gating of sensory stimuli. For example, alpha oscillations are modulated during working memory tasks. Working memory requires (1) processing of external sensory stimuli, (2) subsequent internal maintenance and processing of these inputs during the retention interval, and (3) successful retrieval during the presentation of the test stimulus. For instance, during the Sternberg working memory task, a sequence of letters is presented and after a retention interval the participants need to indicate whether a test letter was previously presented in that trial (more details on this task in chapter: Theta Oscillations). The power of alpha oscillations during this task increases with the difficulty of the task, such that in trials where more letters have to be memorized, a more pronounced enhancement of alpha oscillations occurs [7].

Attention refers to the ability to allocate sensory processing areas to a certain subspace of sensory input. Examples are selective attention directed to a specific spatial location or a specific sensory modality. Aside from working memory, alpha oscillations may also serve as a mechanism for attentional selection by suppression of neural representation of the stimulus that needs to be ignored [8]. *Covert attention* denotes allocation of attention without overt behavioral output such as head or eye movement. Directing covert attention to a specific location increases sensory processing abilities for this area (such as detection of weak stimuli) at the expense of transiently impaired sensory processing of other locations. Typical attention paradigms use a cue that indicates what aspect of the sensory input needs to be focused on. If a cue points to the right hemifield in an attention task, alpha oscillation power increases in the mismatched hemisphere (ipsilateral to cued hemifield). This makes sense when we remember that the right visual field is processed by the left hemisphere, and vice versa. Such mismatch in alpha power between the two hemispheres is an effective predictor of task performance in spatial attention paradigms, and therefore is a possible candidate mechanism for spatial attention [9]. A similar modulation of alpha oscillations also occurs for other, nonspatial forms of attention such as intermodal (requiring paying attention to one of multiple sensory modalities) and feature-based visual attention. Object-based attention also selectively regulates alpha oscillations. For example, the balance between alpha in the right and left hemisphere is differentially regulated if attention is directed to words or faces (two processes known to be lateralized to the left and the right hemispheres, respectively [10]).

It is quite likely that the role of alpha oscillations in attention is more complex than the previously discussed findings would imply. First, alpha oscillations may serve different functional roles in different brain areas. For example, alpha oscillations in the inferotemporal cortex exhibited a positive correlation with behavioral performance contrasted with lower visual areas V2 and V4, which showed a negative correlation [11]. Second, faster oscillations (see chapter: Gamma Oscillations) are also associated with attentional processing. In the simplest model, alpha oscillations are reduced by the onset of visual stimuli, at which point gamma oscillations are increased. However, there could also be a more functional dissociation that cannot be isolated in the classical design, where stimulus onset co-occurs with the presence of the information being paid attention to. Indeed, in a task where only after seconds of visual stimulation the relevant information appears, alpha oscillations are maintained until onset of the relevant information [12]. Interestingly, the modulation of alpha oscillations correlates with how predictable the information is. In contrast, gamma oscillations may be recruited in the case of more surprising information. Together, alpha oscillation may therefore be a mechanism for providing the evidence based on expectation ("top-down"), where gamma oscillation is differentially recruited for unexpected stimuli that are likely of higher importance and therefore require more substantial visual processing.

Alpha oscillations have also been associated with creativity [13], which is defined as the ability to create novel and useful work. Specifically, creativity is linked to increases in alpha oscillations in both frontal and parietal locations. Tasks that require higher levels of creative thinking recruit more alpha oscillations than control tasks. Also individuals with higher creativity levels exhibit stronger alpha oscillations. Differential recruitment of alpha oscillations during creative ideation fits with the previously discussed role of alpha oscillations in attention. Creativity can be interpreted as "internally oriented attention" [13], where sensory input is suppressed to protect the internal thought process. This model is consistent with the notion of alpha oscillations carrying top-down information and suppressing bottom-up information flow.

Emotions

Alpha oscillations have also been implicated as a marker for emotional processing. One prevalent theory of emotion regulation posits that left frontal areas are the substrate of outward-directed, goal-directed behaviors, while the right frontal areas are the substrate of inward-directed, inhibited behaviors. In this theory, overall emotional state is a function of the relative neuronal activity levels in the frontal areas of the two hemispheres. This model can be tested with EEG, since alpha oscillations are typically assumed to correspond to a *hypoactivation* in terms of neuronal firing. Indeed, frontal alpha EEG power, often calculated as the logarithmic ratio of left and right frontal alpha power measured at EEG locations F3 (left) and F4 (right), supports this model [14]. Dominant alpha oscillations in the left hemisphere correspond to a reduced activation of the brain areas responsible for outward-directed, goal-directed behaviors. Indeed, left hemispheric alpha dominance has been associated with heightened stress response, heightened negative affective responses to negative stimuli, increased risk of mood disorders (such as major depressive disorder and premenstrual dysphoric disorder), overall lack of physical well-being, and even decreased immune function. In other words, lack of neuronal activity (ie, increased alpha oscillations) in the left hemisphere correlates with negative emotional disturbances. As it has turned out, such alpha asymmetry reflects not only a stable trait (ie, a biological fingerprint) that may have early development origin but also correlates with behavioral state that can rapidly change, even during the course of a typical study session, which includes questionnaires and electrophysiological measurements. It is intriguing to note (and further discussed in the chapter: Major Depressive Disorder) that repetitive transcranial magnetic stimulation (rTMS) for depression uses a 10-Hz stimulation paradigm [15] and that there are other emerging noninvasive treatment modalities that also target alpha oscillations [16].

PROBING ALPHA OSCILLATIONS WITH NONINVASIVE BRAIN STIMULATION

Alpha oscillations have become one of the prime targets of noninvasive brain stimulation (chapter: Noninvasive Brain Stimulation). Although originally designed to create outlasting effects of stimulation without consideration of cortical oscillations, rTMS at 10 Hz has been demonstrated to entrain alpha oscillations [17], thereby providing a means to noninvasively enhance alpha oscillations to directly probe their functional role. More recently, transcranial alternating current stimulation (tACS) at 10 Hz (10 Hz-tACS) or matched to the individual alpha peak frequency (IAF-tACS) was shown to enhance alpha oscillations during and immediately after stimulation [18]. Therefore such stimulation can be used to probe the functional role of alpha oscillations. The basic question in all of these studies is whether 10 Hz-tACS improves behavioral performance in a task that recruits alpha oscillations. In such studies, stimulation at a different frequency outside the alpha frequency band needs to be performed as a control condition. This enables us to determine if the observed effects of stimulation are frequency specific and not simply a general "electric stimulation" effect. The detailed mechanism of enhancement alpha oscillations by tACS remains debated. On the one hand, neuronal activity could entrain to the stimulation. Entrainment is defined as the temporal alignment of the endogenous oscillation(s) to the external periodic stimulus provided by tACS. According to this model, the phases of many endogenous alpha oscillations become aligned and thereby increase the amplitude of the overall oscillation measured by EEG. On the other hand, stimulation could induce synaptic plasticity that favors the occurrence of alpha oscillations. In the case of entrainment, we expect that neuronal activity is entrained at the stimulation frequency and that the alpha phase is stable across trials. In case of plastic changes of synaptic connectivity, changes in functional connectivity strength can be expected as a result of synaptic plasticity. Evidence for both types of mechanisms has been found [19,20], so the mechanism(s) by which periodic brain stimulation modulates alpha oscillations remains to be fully elucidated. Also the effect of tACS appears to be state dependent such that only stimulation in the "eyes open" (but not in the "eye closed") condition enhances alpha oscillations [21].

Independent of the mechanism of action, modulation of alpha oscillations with either tACS or TMS alters sensory detection performance in a phase-dependent way [19,22]. In extension of the model of alpha oscillations as a sensory gate, they also provide a temporal reference framework that determines the occurrence of sensory processing illusions such as the *sound-induced double-flash illusion*. In this illusion, the presentation of two beeps in rapid succession (typically less than 100 ms apart) leads to the perception of a simultaneously presented, single light flash as two flashes. The integration time window for this illusion correlates with the individual alpha period (inverse of IAF). Modulating the alpha oscillation by applying tACS with stimulation frequencies lower and higher than the individual frequency accordingly shrank or enlarged the relevant time interval in which the illusion occured [23].

Brain stimulation that targets alpha oscillations is not limited to mechanisms of sensory processing. In agreement with findings from EEG studies, 10 Hz-tACS of both frontal hemispheres significantly enhanced creative ideation compared to sham stimulation in a blinded, crossover study in healthy participants. Importantly, no such effect was found for 40 Hz-tACS, suggesting a frequency-specific engagement of alpha oscillations (instead of a nonspecific, general "electric effect") as the mechanism of creativity enhancement [24].

SUMMARY AND OUTLOOK

In this chapter, we discussed alpha oscillations, which have been undergoing a quite remarkable renaissance over the last few years. Originally dismissed as an uninteresting signature of the cortex idling, alpha oscillations have been reconceptualized as a mechanism that regulates functional interactions between brain areas that need to cooperate for the successful performance of behavioral tasks. Alpha oscillations regulate basic sensory detection performance all the way to higher-order cognitive capabilities such as creativity, presumably by an activity deactivating process that would diminish task performance. Alpha oscillations may particularly provide a top-down signal that informs sensory processing areas about how novel or attention-worthy specific input is as a function of how unexpected it is. Thereby alpha oscillations would ensure optimal neuronal resource allocation for the processing of behaviorally relevant sensory input. Alpha oscillations have emerged as a hotbed for the development of noninvasive brain stimulation modalities for both the demonstration of a causal role of network oscillations in cognition and behavior and for the treatment of psychiatric illnesses.

References

[1] Berger H. Uber das Elektrenkephalogramm des Menschen. Arch für Psychiatr Nervenkrankh 1929;87(1):527–70.
[2] da Silva FH, et al. Organization of thalamic and cortical alpha rhythms: spectra and coherences. Electroencephalogr Clin Neurophysiol 1973;35(6):627–39.
[3] Hughes SW, Crunelli V. Thalamic mechanisms of EEG alpha rhythms and their pathological implications. Neuroscientist 2005;11(4):357–72.
[4] Hughes SW, et al. Synchronized oscillations at alpha and theta frequencies in the lateral geniculate nucleus. Neuron 2004;42(2):253–68.
[5] Busch NA, Dubois J, VanRullen R. The phase of ongoing EEG oscillations predicts visual perception. J Neurosci 2009;29(24):7869–76.
[6] Strauss A, et al. Alpha phase determines successful lexical decision in noise. J Neurosci 2015;35(7):3256–62.
[7] Jensen O, et al. Oscillations in the alpha band (9-12 Hz) increase with memory load during retention in a short-term memory task. Cereb Cortex 2002;12(8):877–82.
[8] Foxe JJ, Snyder AC. The role of alpha-band brain oscillations as a sensory suppression mechanism during selective attention. Front Psychol 2011;2:154.
[9] Worden MS, et al. Anticipatory biasing of visuospatial attention indexed by retinotopically specific alpha-band electroencephalography increases over occipital cortex. J Neurosci 2000;20(6):RC63.

[10] Knakker B, Weiss B, Vidnyanszky Z. Object-based attentional selection modulates anticipatory alpha oscillations. Front Hum Neurosci 2014;8:1048.

[11] Bollimunta A, et al. Neuronal mechanisms of cortical alpha oscillations in awake-behaving macaques. J Neurosci 2008;28(40):9976–88.

[12] Bauer M, et al. Attentional modulation of alpha/beta and gamma oscillations reflect functionally distinct processes. J Neurosci 2014;34(48):16117–25.

[13] Fink A, Benedek M. EEG alpha power and creative ideation. Neurosci Biobehav Rev 2014;44:111–23.

[14] Davidson RJ. Anterior cerebral asymmetry and the nature of emotion. Brain Cogn 1992;20(1):125–51.

[15] George MS, et al. Daily repetitive transcranial magnetic stimulation (rTMS) improves mood in depression. Neuroreport 1995;6(14):1853–6.

[16] Leuchter AF, et al. Efficacy and safety of low-field synchronized transcranial magnetic stimulation (sTMS) for treatment of major depression. Brain Stimul 2015;8(4):787–94.

[17] Thut G, et al. Rhythmic TMS causes local entrainment of natural oscillatory signatures. Curr Biol 2011;21(14):1176–85.

[18] Zaehle T, Rach S, Herrmann CS. Transcranial alternating current stimulation enhances individual alpha activity in human EEG. PLoS One 2010;5(11):e13766.

[19] Helfrich RF, et al. Entrainment of brain oscillations by transcranial alternating current stimulation. Curr Biol 2014;24(3):333–9.

[20] Vossen A, Gross J, Thut G. Alpha power increase after transcranial alternating current stimulation at alpha frequency (α-tACS) reflects plastic changes rather than entrainment. Brain Stimul 2014;8(3):499–508.

[21] Neuling T, Rach S, Herrmann CS. Orchestrating neuronal networks: sustained after-effects of transcranial alternating current stimulation depend upon brain states. Front Hum Neurosci 2013;7:161.

[22] Romei V, et al. Causal implication by rhythmic transcranial magnetic stimulation of alpha frequency in feature-based local vs. global attention. Eur J Neurosci 2012;35(6):968–74.

[23] Cecere R, Rees G, Romei V. Individual differences in alpha frequency drive crossmodal illusory perception. Curr Biol 2015;25(2):231–5.

[24] Lustenberger C, et al. Functional role of frontal alpha oscillations in creativity. Cortex 2015;67:74–82.

Beta Oscillations

Segregation of the frequency spectrum into "classical frequency" bands is as old as the discovery of the electroencephalogram (EEG) by Hans Berger. The resulting nomenclature of Greek letters for the different frequency bands has proven useful, but it also has the potential to confuse because of a lack of commonly agreed upon definitions. This is particularly true for the beta frequency band discussed in this chapter. Tucked in between the alpha oscillations (see chapter: Alpha Oscillations), which can be easily detected in recordings from participants with closed eyes, and the gamma oscillations, which are fundamental for cognition (see chapter: Gamma Oscillations), beta oscillations have remained a poorly delineated, vaguely defined frequency band. As a result, we lack a clear definition of which frequencies belong in the beta band (typically 12–30 Hz). Most of what we know about the beta oscillation stems from research of the motor system. Beta oscillations are a crucial puzzle piece for understanding motor disorders. The chapter "Parkinson's Disease" will elaborate on beta oscillations in this disease state.

TERMINOLOGY

Beta oscillations have been recorded across the entire sensorimotor system: motor cortex, premotor cortex, spinal cord, muscles, dorsal root ganglion, deep cerebellar nuclei, striatum, posterior parietal cortex, and somatosensory cortex. The most prominent feature of beta oscillations is their transient suppression during movement; they are therefore referred to as an *antikinetic rhythm*. As we will see, beta oscillations are typically elevated in the absence of movement. This is referred to as beta *synchronization*. This term refers to the fact that macroscopic oscillations arise from the synchronization of neuronal populations, which leads to detectable EEG signals by superposition of the endogenous electric fields generated by individual neurons (see chapter: LFP and EEG). Accordingly, movement-induced suppression of beta oscillations is often referred to as *desynchronization* (Fig. 20.1).

Early work found that beta oscillations localized to precentral areas, and that the central fissure (separation between frontal and parietal cortices) delineates the brain areas associated with alpha and beta oscillations, respectively, although we now know that this separation is not so simple. Classically, the precentral beta oscillation was interpreted as "idling activity"

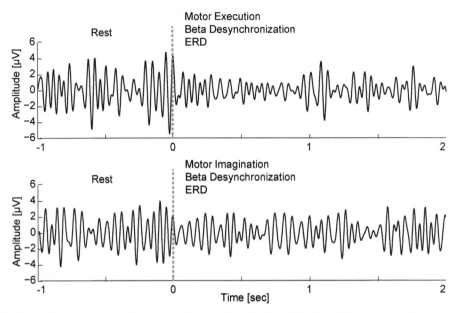

FIGURE 20.1   Both motor execution (*top*) and motor imagination (*bottom*) lead to a decrease in beta oscillation power. This phenomenon is referred to as beta desynchronization or event-related desynchronization (ERD).

of the motor cortex corresponding to resting muscle groups. As it has turned out, however, beta oscillations undergo more sophisticated modulation beyond movement-induced desynchronization, and may play more refined functional roles.

MECHANISMS OF BETA OSCILLATIONS

In terms of the mechanisms that generate beta oscillations, one of the strongest hints comes from the observation that administration of benzodiazepines (see chapter: Synaptic Transmission) increases the power and decreases the frequency of beta oscillations in the sensorimotor cortex [1,2]. Computational modeling has confirmed that increasing the $GABA_A$ conductance in a cortical network model can replicate the changes observed in resting-state beta oscillations in response to benzodiazepine administration. Increasing the conductance of inhibitory synapses on interneurons replicated the main changes in the beta peak in the spectrum, whereas changing the conductances of the inhibitory synapses on the excitatory pyramidal cells had little effect on the beta peak frequency. Therefore the inhibitory coupling of interneurons may be crucial for the genesis of resting-state beta oscillations, such that the inhibitory subnetwork defines the time periods during which excitatory cells can fire.

Beta oscillations share a reliance on the inhibitory subnetwork for their genesis with gamma oscillations. However, the two frequency bands are not only opposed in terms of their occurrence before, during, and after movements, but also play two distinct roles in providing a carrier frequency for functionally connecting (cortical) areas during behavior. Transitions from gamma to beta frequency activity (at least in vitro) are marked by a continued background gamma oscillation generated by the interneuron subnetwork, during which pyramidal cells fail to activate on every cycle. As a result, beta oscillations at half the frequency of the gamma oscillation appear as "beat skipping," which arises from increased slow potassium conductances that slow down pyramidal cell firing and increased excitatory coupling of pyramidal cells that supports the synchronous activity required for the emergence of a network oscillation. Thus beta oscillations can emerge as a *subharmonic* (frequency at $1/n$ of the fundamental frequency, here $n = 2$) of the gamma oscillation. A key functional advantage of the slower beta frequency is its ability to synchronize in the presence of longer conduction delays that are ubiquitous in connections between different cortical areas. In computer simulations, beta oscillations can support synchronization with up to 20 ms propagation delays, whereas gamma oscillations require propagation delays less than 10 ms [4].

Both alpha and beta band oscillations are often most prominent in deep (ie, infragranular) layers. Correspondingly, in vitro studies demonstrated that beta oscillations can emerge from pyramidal cells in layer V of the neocortex. Activation of acute cortical slices with kainate caused the emergence of gamma frequency band oscillations in superficial layers, beta frequency oscillations in deep layers, and a combination of the two frequency bands in layer IV. Physical separation of the deep layers from the superficial layers did not abolish the beta oscillations. Therefore neither network interactions with superficial layers nor apical dendrites of infragranular cells are likely to contribute to the rhythmogenesis in the beta band. Blocking AMPA or NMDA receptors equally failed to reduce the beta oscillations (but had a deleterious effect on the superficial gamma oscillation). Intracellular recordings showed that intrinsically bursting cells in layer V exhibited activity in the beta frequency band [5], mostly *spikelets* suggestive of a gap junction-mediated effect (Fig. 20.2). Spikelets are low-amplitude deflections in membrane voltage whose waveforms resemble action potentials. Indeed, application of gap junction blockers octanol and carbenoxolone mostly suppressed the beta activity in the slice. Together, these data suggest an infragranular locus of beta oscillation genesis in the neocortex, with a pronounced role of network interactions that do not rely on chemical synapses.

Mu rhythms in the beta band may arise through the dynamic interaction of two 10-Hz inputs to pyramidal cells [6]. In this model, the first input is a 10-Hz afferent (feedforward) thalamocortical signal. The second input is a 10-Hz signal that corresponds to intracortical signals that impinge on distal dendrites of layer II/III pyramidal cells. The dynamic interaction of these two inputs causes the emergence of mu oscillations in the alpha and beta frequency bands as a function of the relative strength and timing of the two inputs. The feedforward input causes intracellular current flow up the dendrites; the feedback input causes current flow down the dendrite toward the soma. The stronger the feedback input and the more precise its timing, the more pronounced are rhythms in

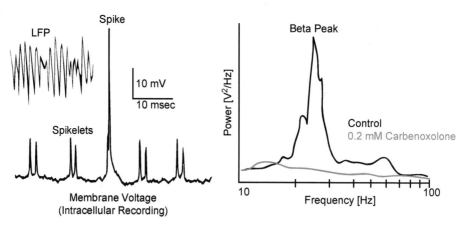

FIGURE 20.2   Beta oscillations in vitro. *Left*: Local field potential (LFP, *inset*) and membrane voltage time series with spikelets and spikes. *Right*: Spectrum shows that application of carbenoxolone suppresses the spectral peak in the beta frequency band. *Adapted from Roopun AK, et al. A beta2-frequency (20–30 Hz) oscillation in nonsynaptic networks of somatosensory cortex. Proc Natl Acad Sci USA 2006;103(42):15646–50.*

the beta band. Conversely, if the afferent input dominates, alpha oscillations are more prominent.

FUNCTIONAL ROLES OF BETA OSCILLATIONS

Dynamic Modulation of Beta Oscillations

A straightforward approach to understanding the functional role of beta oscillations is to delineate the circumstances (ie, experimental conditions) that modulate the power of oscillations in the beta band. Beta oscillations in the contralateral hemisphere of activated muscle groups are suppressed during movement execution (the previously discussed *event-related desynchronization*) and recover (and overshoot) after the movement is completed (*beta rebound*). More generally, no active movement is required for a decrease in beta oscillations—motor imagery (eg, participant is instructed to imagine manipulation of a displayed object), movement observation, passive movement (someone else moves a body part), and even kinesthetic illusions also reduce beta power. These dynamics make the case for an antagonism between movement and beta oscillations in the motor cortex. In contrast, beta oscillations are increased during postural maintenance such as stable object hold or lever press. Beta rebound does not require the previous occurrence of a movement. If a planned movement is not performed, for example, in response to a no-go cue (where a signal indicates to the participant *not* to execute a movement), beta oscillations also transiently increase. The beta rebound (*event-related synchronization*) remains poorly understood in terms of both its underlying mechanism and its functional role. Beta rebounds not only occur briefly after movement, but also after motor imagery and in response to somatosensory stimulation. The beta rebound is spatially localized to the motor cortex territory that corresponds to the previously activated muscle groups (eg, "mouth," "left toe," "left finger").

Motor System

The functional role of beta oscillations, that is, *why* the cortex generates beta oscillations, remains unclear. Beta oscillations could serve the simple purpose of maintaining stable posture by inhibiting other, new movements. In agreement with this model, beta oscillations are increased during instructed delay periods prior to initiation of movements. Alternatively, increased beta oscillations before movement initiation, typically during the period when the participant waits for a cue to start the movement, can be interpreted as an attentional mechanism. Experimental evidence for such a role of the beta oscillation in attention has been found in other cortical areas, particularly in the case of visual attention (see later for beta oscillations in nonsensorimotor areas). Similarly to how the alpha oscillation has been reinterpreted as an active mechanism for information routing (instead of a mere idling of visual thalamocortical circuits), the beta oscillation may also represent more than just an idling in the absence of voluntary movement, in contrast to the classical view outlined earlier. Beta oscillations may play a role in stabilizing ongoing motor activity, and may act as a protectant against recruitment of new, subsequent, or alternative motor (and perhaps) cognitive programs [7]. This theory is in agreement with the finding that initiation of new motor commands is less successful during periods of spontaneously increased beta oscillations. In contrast, executing correction of ongoing motor activity is facilitated by the presence of beta oscillations, again in agreement with the proposed role of beta oscillations to stabilize and facilitate maintenance of the current state.

To distinguish between a simple "hold" function and a more specific attentional mechanism, a task was devised where the participant needed to wait before performing a directed movement until several sequential cues were presented, where some of the cues were instructional (and therefore should be attended to) and others were noninstructional and had to be ignored (Fig. 20.3, [8]). The designation "instructional" refers to providing input that the participant needed to process to perform a given trial of the task correctly. In such a paradigm, beta oscillations were selectively enhanced for instructional cues, supporting the

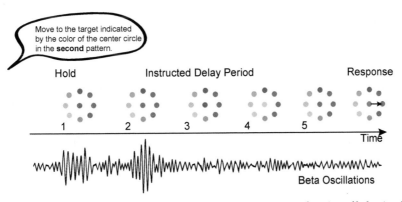

FIGURE 20.3 Modulation of beta oscillations in the human motor cortex as a function of behavioral relevance (ie, instructional value) of a visual cue. *Adapted from Saleh M, et al. Fast and slow oscillations in human primary motor cortex predict oncoming behaviorally relevant cues. Neuron 2010;65(4):461–71.*

more sophisticated role of beta oscillations to enable preferred processing of relevant stimuli. Noninstructional cues did not elicit beta oscillations, in conflict with the model of beta oscillations preventing movements.

Closing the Sensorimotor Loop

Beta band activity in the EEG and in the electromyograph (EMG, measurement of electric activity of the muscle) of the target muscles is temporally bound together—typically, there is significant coherence (see chapter: Network Interactions) between beta band EEG activity in the motor cortex and the EMG of the contralateral muscle [9]. The cortical network exhibiting/generating the beta oscillations includes the pyramidal tract neurons, the descending output neurons that send their axons to the brain stem (*corticobulbar tract*) and to the spinal cord (*corticospinal tract*). Not only are beta oscillations increased above baseline during periods of sustained muscle contractions, but also the beta coherence between muscles and the sensorimotor cortex is modulated. For example, when participants had to apply constant force to levers, corticomuscular beta coherence was predominantly related to the force/displacement ratio (so-called *compliance* of the levers). In all conditions, participants had to apply the same force, but the more compliant the condition was, the more the levers moved in response to the force (in essence like a spring), whereas for the *isotonic* (no displacement) condition, the levers were locked in place. The higher the compliance was, the stronger the beta coherence. Thus beta oscillations may be an essential component of sensorimotor feedback loops and therefore represent much more than an idling of the motor cortex. Indeed, recordings across precentral (motor) and postcentral cortical (somatosensory) territories in nonhuman primates revealed functional connectivity in the beta frequency band [10]. Specifically, Granger causality (see chapter: Network Interactions) detected effective connectivity from primary somatosensory areas to both the motor cortex and the posterior parietal cortex. This finding supports a functional role of beta oscillations in enabling somatosensory feedback during sustained hand-lever press. This is in agreement with the observation that in the absence of somatosensory input, maintenance of sustained motor output is impaired. However, notably, the fact that motor imagery modulates beta oscillation power in a qualitatively similar way as actual movement has been used as an argument against a role of beta oscillations in sensory processing.

Beyond the Sensorimotor Cortex

Beta oscillations are not limited to the sensorimotor system, but are also recruited in other cortical areas during cognitive tasks. Fluctuations in the beta band are prominent in tasks that require top-down attention, such as search tasks [11]. For example, in a task that required monkeys to perform a saccade (targeted eye movement) to a previously presented target stimulus, beta coherence between the prefrontal cortex and the parietal cortex was selectively enhanced if the target stimulus closely resembled other distractor stimuli. In contrast, trials where the target stimulus was clearly different from the distractor stimuli, gamma coherence was selectively enhanced. These findings point to a top-down control function of beta coherence in cognitive tasks. A delayed match-to-sample task that probes working memory also demonstrated pronounced coherence in the beta band between parietal and prefrontal areas

[12]. Beta coherence decreased during presentation of the sample stimulus and gradually increased in the delay period (no stimulus shown) such that beta coherence peaked at the onset of the match stimulus. Granger causality in the frequency domain favored posterior parietal cortex as the "sender" and prefrontal cortex as the receiver of signals in the beta frequency band, demonstrating that oscillatory activity in the beta band can organize widespread cortical networks and is not limited to a top-down attentional signal. At a more local scale, successful performance of a delayed match-to-sample task correlated with beta phase synchronization in the temporal lobe. Thus beta oscillations may also orchestrate neuronal activity at a more local scale during cognitive tasks.

PROBING BETA OSCILLATIONS WITH NONINVASIVE BRAIN STIMULATION

Noninvasive brain stimulation offers the opportunity to study the nervous system by application of well-defined perturbations to network-level targets (see chapter: Noninvasive Brain Stimulation). As we have already seen in the previous chapters, this approach has been very fruitful for modulating specific cortical activity patterns and studying the induced changes in network dynamics and behavioral performance. Thereby, the causal role of the modulated activity patterns can be deduced. In its simplest form, transcranial magnetic stimulation (TMS) can be used to stimulate a cortical target with a very brief pulse for the study of what engineering refers to as the impulse response—essentially the fundamental response of the system to an infinitely short perturbation. Motor-evoked potentials (MEPs) can be recorded by EMG as an objective measure of motor cortex output in the periphery. MEPs elicited by TMS depend on ongoing beta oscillations, similar to how alpha oscillation power and phase gate TMS-induced phosphenes in the visual cortex. Both the amplitude and the phase of the beta oscillation modulate MEP amplitude. Moreover, TMS during beta rebound after voluntary movement caused reduced MEPs, and therefore delineates beta rebound as a period of reduced excitability. Combining EEG with TMS provides insights into cortical response to the stimulation. Single-pulse TMS over different cortical sites revealed that central areas exhibit endogenous resonant frequencies in the beta band [13], contrasted to more frontal areas (gamma frequency band) and more posterior areas (alpha frequency band). It is important to recognize that strong focal activation of the motor cortex with TMS does not necessarily reflect the sophisticated motor signal endogenously created for the activation of the same muscle group. As a result, changes in "excitability" come with the important caveat that they are determined with a distinctly artificial perturbation.

Going beyond probing the system with single pulses, both magnetic and electric stimulation can be used to modulate beta oscillations by periodically patterning the stimulation waveform. In the case of TMS, a regular series of pulses (called a *pulse train*) can be applied (repetitive TMS). In the case of electric stimulation, a sine-wave electric current is applied to the scalp (transcranial alternating current stimulation, tACS). The chapter "Noninvasive Brain Stimulation" discusses the mechanism of action of these approaches and the evidence for these paradigms to modulate cortical oscillations in a frequency-dependent way. In agreement with the basic model of beta oscillations as an antikinetic rhythm, tACS at 20 Hz (beta frequency band) but not at 5 Hz reduced movement velocity in a motor task. Specifically, in this task a joystick had to be moved to track the motion of a cursor on the screen after an initial hold period. Importantly, 20 Hz-tACS selectively increased corticomuscular coherence in the beta frequency band in the epoch before onset of movement [14].

MU RHYTHM

Adding to the complexity of dissecting the functional role(s) of beta oscillations is the presence of a second cortical rhythm of similar frequency, one that is also dynamically regulated in relation to movement. The *sensory-motor rhythm* (also referred to as *mu rhythm*) denotes oscillatory activity in the upper alpha and lower beta bands. There is no agreement on the exact frequency range for the mu band. The mu rhythm overlaps with the alpha frequency band, therefore it is tempting to consider the mu rhythm to be a more anterior manifestation of alpha oscillations. The distinction arises from the fact that there are fundamentally different ways to define cortical oscillations. The first approach is a simple labeling of different sections of the spectrum as a function of frequency, and argues that the mu rhythm is the same as the alpha oscillation. The second, more-refined approach is based on classifying different oscillations based on a set of properties, including sources of origin, functional role, and network and cellular mechanisms. Following this second approach enables mechanistic studies of cortical oscillations. However, it comes at the price that a large number of different activity patterns remain poorly characterized and understood. Nevertheless, in case of the alpha versus mu dilemma, there are clear functional differences. As we have seen in the chapter "Alpha Oscillations," the main hallmark of alpha oscillations is their decrease upon opening of the eyes (*alpha blocking*). This is not the case with the mu rhythm, which decreases during voluntary (and imagined) movement and somatosensory stimulation (*mu blocking*), but not during "eyes open." Also source localization by high-density MEG points to distinct dipole sources that generate the two types of rhythms. Interestingly, the mu rhythm is present during rapid eye movement sleep, during which the cortex is activated but motor function is inhibited. Also it has been proposed that the mu rhythm is linked to the so-called mirror neuron system (located in the premotor area), which is activated when observing actions performed by others. Specifically, mu oscillations are suppressed in conditions where actions of others are observed. In agreement with the hypothesis that the mirror neuron system is impaired in people with autism, some studies showed that mu suppression when watching others is less prominent in this population in comparison to healthy controls [15].

BETA OSCILLATIONS IN BRAIN–COMPUTER INTERFACES

Brain–computer interfaces (BCIs) aim to provide assistance to patients who have limited neuromuscular function, such as people with spinal cord injuries or amyotrophic lateral sclerosis. The main idea is to record and decipher neuronal activity patterns and to use the extracted signals to communicate with the surrounding environment, such as controlling a computer cursor. BCI research has made great progress, and systems that provide brain-based control of sophisticated neural prosthetics are under development. Different neural signals have been used as input signals in BCI systems, ranging from scalp EEG signals all the way to neuronal spiking activity recorded with implanted electrode arrays in the human motor cortex [16,17]. As it turns out, employing beta oscillations recorded with EEG over the sensorimotor cortex is a successful strategy, probably because beta oscillations are easily modulated by motor imagery [18]. Importantly (and often overlooked), successful BCI operation requires the patient to learn how to generate specific neurophysiological activity patterns that are then decoded into actions by the BCI. Several systems are based on modulation of beta (and mu) rhythms over the sensorimotor cortex. In early training sessions, active effort by the patient is required, who typically learns how to control a cursor by motor imagery. After completion of several training sessions, the same actions are then performed much more effortlessly, similar to natural motor commands in healthy controls.

SUMMARY AND OUTLOOK

In this chapter, we discussed beta oscillations. This cortical rhythm is most closely related to the absence of movement and could therefore be conceptualized as the "idling rhythm" of the motor system. However, recent studies have found that beta oscillations may play an additional, more refined role in cognitive tasks. Overall, both the mechanisms and the functions of beta oscillations have remained very poorly understood. But clinically, beta oscillations have become of great importance, particularly with regard to disorders of the motor system. Pathological beta oscillations in the motor cortex may be a promising target for brain stimulation for the treatment of Parkinson's disease (see chapter: Parkinson's Disease). Also the robust relationship between movements and modulation of beta oscillations make this rhythm a valuable readout for BCIs for restoring motor function with brain-controlled prosthetics.

References

[1] Baker MR, Baker SN. The effect of diazepam on motor cortical oscillations and corticomuscular coherence studied in man. J Physiol 2003;546(Pt 3):931—42.

[2] Jensen O, et al. On the human sensorimotor-cortex beta rhythm: sources and modeling. NeuroImage 2005;26(2):347—55.

[3] Rangaswamy M, et al. Beta power in the EEG of alcoholics. Biol Psychiatry 2002;52(8):831—42.

[4] Kopell N, et al. Gamma rhythms and beta rhythms have different synchronization properties. Proc Natl Acad Sci USA 2000;97(4):1867—72.

[5] Roopun AK, et al. A beta2-frequency (20—30 Hz) oscillation in nonsynaptic networks of somatosensory cortex. Proc Natl Acad Sci USA 2006;103(42):15646—50.

[6] Jones SR, et al. Quantitative analysis and biophysically realistic neural modeling of the MEG mu rhythm: rhythmogenesis and modulation of sensory-evoked responses. J Neurophysiol 2009;102(6):3554—72.

[7] Engel AK, Fries P. Beta-band oscillations—signalling the status quo? Curr Opin Neurobiol 2010;20(2):156—65.

[8] Saleh M, et al. Fast and slow oscillations in human primary motor cortex predict oncoming behaviorally relevant cues. Neuron 2010;65(4):461—71.

[9] Aumann TD, Prut Y. Do sensorimotor β-oscillations maintain muscle synergy representations in primary motor cortex? Trends Neurosci 2015;38(2):77—85.

[10] Brovelli A, et al. Beta oscillations in a large-scale sensorimotor cortical network: directional influences revealed by Granger causality. Proc Natl Acad Sci USA 2004;101(26):9849—54.

[11] Buschman TJ, Miller EK. Top-down versus bottom-up control of attention in the prefrontal and posterior parietal cortices. Science 2007;315(5820):1860—2.

[12] Salazar RF, et al. Content-specific fronto-parietal synchronization during visual working memory. Science 2012;338(6110):1097—100.

[13] Rosanova M, et al. Natural frequencies of human corticothalamic circuits. J Neurosci 2009;29(24):7679—85.

[14] Pogosyan A, et al. Boosting cortical activity at beta-band frequencies slows movement in humans. Curr Biol 2009;19(19):1637—41.

[15] Pineda JA. The functional significance of mu rhythms: translating "seeing" and "hearing" into "doing". Brain Res Brain Res Rev 2005;50(1):57—68.

[16] Wolpaw JR, et al. Brain-computer interfaces for communication and control. Clin Neurophysiol 2002;113(6): 767—91.

[17] Hochberg LR, et al. Neuronal ensemble control of prosthetic devices by a human with tetraplegia. Nature 2006;442(7099):164—71.

[18] Wolpaw JR, McFarland DJ. Control of a two-dimensional movement signal by a noninvasive brain-computer interface in humans. Proc Natl Acad Sci USA 2004;101(51):17849—54.

21

Gamma Oscillations

Gamma oscillations refer to fast rhythmic network activity at frequencies typically higher than 30 Hz. Compared to the other oscillations discussed in this unit, gamma oscillations are relatively well understood in terms of the cellular and network mechanisms that give rise to them. Seminal findings about gamma oscillations in the visual system of cats have triggered widespread interest because of the proposed role of gamma oscillations in integration of visual information into coherent percepts. These studies proposed that gamma oscillations are a putative mechanism for the "hard problem"—how firing of neurons together can give rise to perception. The original enthusiasm was tempered by the fact that gamma oscillations are a ubiquitous phenomenon found to varying degrees in most experiments that study sensory processing. Nevertheless, findings of selectively impaired gamma oscillations in diseases, such as schizophrenia and autism have confirmed the importance of understanding this network oscillation. We will first disentangle the different definitions of gamma oscillation, dissect the mechanistic underpinnings of gamma oscillations, and then review the key findings about the functional role of gamma oscillations in both animal models and human recordings. We will also discuss how gamma oscillations relate to lower and higher frequency oscillations ("high gamma"). We conclude the chapter with a note on the role of the gamma oscillation during the development of the nervous system. Pathological changes in the gamma frequency band are covered in the chapters "Autism Spectrum Disorders" and "Schizophrenia."

TERMINOLOGY

As we have already seen with the other frequency bands, care needs to be taken in terms of how a given frequency band is defined. This caveat also applies to gamma oscillation. Often, experiments show a diffuse increase in higher frequency activity in the gamma band (>30 Hz). This modulation of spectral power typically fails to exhibit pronounced frequency peaks. This contrasts with, for example, the modulation of alpha oscillations as a function of whether the eyes of the participant are open or closed. Furthermore, no agreement exists with regard to the upper frequency limit of the gamma band. Often, the distinction between *low gamma* and *high gamma* is made ("low" and "high" referring to the frequency) with no agreement about the cut-off frequency that separates them [1]. Clearly, not all oscillations above 30 Hz represent a monolithic unit caused by a single mechanism. Oscillations with frequencies higher than ~100 Hz exhibit distinct characteristics and are discussed in the subsequent chapter "High-Frequency Oscillations." When discussing gamma oscillations recorded with scalp electroencephalogram (EEG) electrodes, it is important to remember that such recordings are challenging because of muscle artifacts that exhibit a similar frequency spectrum. In particular, microsaccades can cause stimulus-evoked artifacts that are easily mistaken for gamma oscillations [2].

Gamma oscillations are often used as a surrogate marker for action potential firing, for example, in electrocorticography recordings where action potentials cannot be detected with standard clinical electrodes. In addition, gamma oscillations correlate with the hemodynamic signal measured by magnetic resonance imaging [3].

CELLULAR AND NETWORK MECHANISMS OF GAMMA OSCILLATIONS

The time scale of gamma oscillations (a period of ~25 ms) is of a similar order of magnitude as the time constant of $GABA_A$ synaptic inhibition (see chapter: Synaptic Transmission). Indeed, substantial evidence points to a prominent role of synaptic inhibition in generating gamma oscillations. As a result, two major models of how gamma oscillations are generated have emerged that distinguish themselves by the relative importance of excitatory (pyramidal) cells in the network (Fig. 21.1, [4,5]).

In the *ING* (or *I−I*) *model*, gamma oscillations arise from the inhibitory interaction of interneurons and no excitatory cells are required to sustain the gamma oscillation. In the presence of tonic (constant, noise-free) depolarization, inhibitory interneurons fire, inhibit each other, and fire again when released from the synaptic inhibition. As a result, such a network of inhibitory interneurons rapidly synchronizes at a frequency determined by the decay time constant of the inhibitory postsynaptic potential. In this model, the inhibitory interneurons fire like a clock at the network frequency. In the presence of a noisy input, the averaged spiking activity of all inhibitory interneurons still exhibits a frequency peak in the gamma band, but individual neurons exhibit a more random firing pattern. In the *PING* (or *E−I*) *model*, excitatory cells drive inhibitory interneurons that fire with a delay. Once the excitatory cells recover from the synaptic inhibition they triggered, they fire again, and the next cycle of the oscillation commences. The key strategies to establish the relative importance of phasic (ie, temporally patterned) excitation as required by the PING model are (1) in vitro (slice) experiments and (2) computational models. Most in vitro studies of gamma oscillation used the hippocampal slice preparation. Without additional experimental manipulations, this preparation typically does not exhibit structured network activity such as gamma oscillations. Several different manipulations have been used for network activation, and perhaps not so surprisingly the proposed underlying mechanisms of gamma oscillations determined from these different approaches also diverge. Common strategies for network activation for the study of the gamma oscillation include electric stimulation, bath application of kainite receptor agonists, application of the cholinergic agonist carbachol, and increases in extracellular potassium concentration [6]. Dominance of the ING or PING mechanism is assessed by comparing how detrimental blocking AMPA versus GABA receptors is for the occurrence of (transient) gamma oscillations in the slice. For example, in hippocampal slice preparation,

ING: Mutually Connected Inhibitory Interneurons

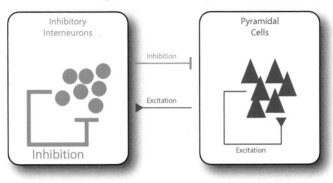

PING: Reciprocally Connected Pyramidal Cells and
Inhibitory Interneurons

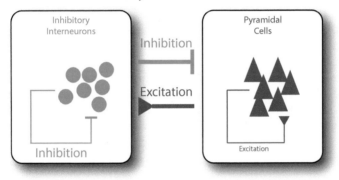

FIGURE 21.1   Models of how gamma oscillations are generated in local cortical circuits. *Top*: ING (mutually connected inhibitory interneurons) model. Mutual inhibition of interneurons generates oscillations and no synaptic excitation is required. Inhibitory projections recruit pyramidal cells. *Bottom*: PING (reciprocally connected pyramidal cells and inhibitory interneurons). Excitatory input to interneurons from pyramidal cells drive interneurons, which in turn inhibit pyramidal cells. Oscillation depends on synaptic excitation that drives the reciprocal interaction between pyramidal cells and inhibitory interneurons.

brief transients of gamma oscillations can be evoked with tetanic (high-frequency) stimulation in the presence of antagonists of fast glutamate receptors [7]. Application of metabotropic glutamate receptor antagonists abolished the oscillation; these receptors do not mediate phasic depolarization and therefore solely provide the required depolarization or increased excitability for network activation. In contrast, GABAergic antagonists block occurrence of the oscillation. Together, these experiments favor the ING model, in which no temporal patterning of excitatory input is required for oscillation genesis. Other experiments, especially the ones using carbachol, came to somewhat different conclusions and found that blocking AMPA receptors also impaired the gamma oscillation activity [8]. Computer simulations provide additional insights, as they allow the systematic variation of model parameters and, by the choice of model complexity, the identification of the minimally required elements for the emergence of the gamma oscillation.

So evidence exists for both the ING and the PING models, and it is likely that cooperation between the two proposed mechanisms may be what drives gamma oscillations in the intact brain. Independent of which model better reflects the circuit mechanisms that enable the occurrence of gamma oscillations, it is clear that fast-spiking [parvalbumin (PV)-positive] cells play a pivotal role. Their electrophysiological properties make them ideal candidates for sustaining gamma oscillations. In addition, pulsed optogenetic stimulation at a broad range of frequencies from below 10 Hz all the way to 200 Hz revealed that the strongest enhancement of rhythmic activity measured from the local field potential occurred around 40 Hz, when channelrhodopsin was expressed in PV-positive cells [9]. In contrast, stimulation of pyramidal cells that expressed channelrhodopsin showed the most pronounced enhancement of low frequencies and no gamma peak. These findings further confirm the role of soma-targeting, PV-positive cells that typically fire an action potential per gamma cycle that is phase locked to the gamma oscillation. Given the importance of inhibitory coupling between fast-spiking (FS) cells to the genesis of the gamma oscillation, in particular in the ING model, it is worth asking about the detailed properties of these inhibitory synapses. In terms of anatomy, basket cells (the most typical FS cell type) have a high connection probability among each other, an organizational principle that applies to many interneuron classes. In terms of physiology, the inhibitory synapses on these cells contrast with the inhibitory synapses of FS cells onto excitatory (pyramidal) cells. These inhibitory synapses that connect the FS cells are fast and strong [6]. Decay times of inhibitory postsynaptic currents are on the order of a few milliseconds, and synaptic currents are maintained to a large degree even under high-frequency activation. In addition, gap junctions are likely to contribute to the genesis of the gamma oscillation as suggested by impairment of gamma oscillations in the connexin-36 knockout mouse [10].

FUNCTIONAL ROLES OF GAMMA OSCILLATIONS

Organizing Network Activity into Cell Assemblies

Gamma oscillations define a time scale at which neurons interact. Activity on a given cycle may represent an "information package" or a *cell assembly* of cells that interact to perform a specific function or computation [11]. In the simplest model, such coincident activation of presynaptic cells (within one cycle of the gamma oscillation) will favor summation of synaptic inputs on shared postsynaptic cells. Although the lower-frequency oscillations discussed in the previous chapters enable similar grouping of active cells for each oscillation cycle, the gamma oscillation stands out since it enables formation of dynamic cell assemblies that are shorter in duration because of the higher frequency of the gamma oscillation. The sequence of action potential firing within a single cycle of the gamma oscillation may also be of functional importance (Fig. 21.2).

Conceptually similar to *theta precession* (see chapter: Theta Oscillations), pyramidal cells that receive strong excitatory input fire at an earlier phase of the gamma oscillation, since they can overcome a larger amount of synaptic inhibition provided by the network oscillation at that earlier phase. Pyramidal cells that receive less excitation can only fire later in the cycle,

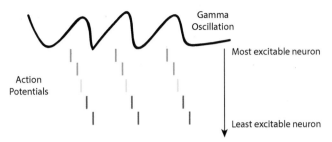

FIGURE 21.2 Each cycle of the gamma oscillation groups together action potentials to form assemblies of neurons. The more excitable a neuron (or the more excitatory input it receives), the earlier a neuron fires on a given cycle of the gamma oscillation. Note the similarities to theta precession in the place cells in the rodent hippocampus.

once more of the inhibition has died out. Together, gamma oscillations enable the transformation of relative input strength among pyramidal cells into a temporal code [12]. Neurons that receive strong input will fire before neurons that receive weaker input. This mechanism of establishing a temporal code also provides noise suppression since the pyramidal cells with weakest input will not get the chance to fire at all. Overall, gamma oscillations are therefore an ideal candidate mechanism for neuronal networks that encode information by spike timing. Despite the ongoing debate between how much the brain relies on average firing rates (*rate code*) and how much the brain relies on the precise timing of the activity of individual neurons (*spike-timing code*), understanding the potential underlying mechanisms is of fundamental importance. In particular, synchronization of neuronal activity enhances the effect on activity in downstream structures because all of the synaptic input arrives within a sufficiently short time window to interact.

Sensory Perception

Gamma oscillations were first studied in the visual cortex. Both the awake and the anesthetized animal respond to the presentation of visual stimuli with rhythmic organization of neuronal activity in the gamma frequency band. Gamma oscillations have been found in a wide variety of species, including invertebrates, and in a range of different brain areas under different task conditions. Here, we focus on the role of gamma oscillations in visual processing and perception. The original excitement about gamma oscillations in the visual system stems from the discovery that gamma oscillations may enable perception by binding together activity of cells that represent different features of what together form a unified percept (the *binding by synchronization hypothesis*, also referred to as the *temporal correlation hypothesis* [13,14]). Detection and recognition of a visual scene requires the segregation of the scene into a large number of features followed by integration into a unified percept. Individual neurons early in the visual stream respond to low-level visual features, such as oriented lines at specific locations in space. The gamma oscillation has emerged as a candidate neurophysiological mechanism that integrates and combines these individual activity patterns such that a unified percept can arise from the dynamic interaction of a large number of cells that represent individual features. In other words, the timing of activity of these individual cells relative to each other—the synchronization—represents the binding mechanism.

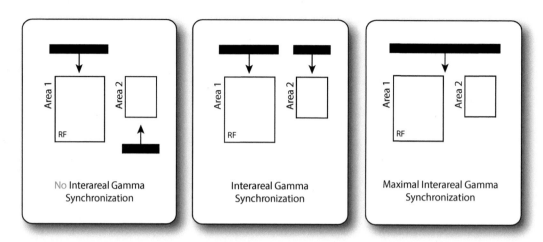

FIGURE 21.3   Interareal gamma synchronization for three different visual stimulation paradigms. *Left*: When the two receptive fields were stimulated with bars that moved in the opposite direction, no gamma synchronization was found. *Middle*: If the bars moved in the same direction, interareal gamma synchronization was found. *Right*: Interareal gamma synchronization was maximal for one joint bar. These results suggest that gamma band synchronization is selectively evoked when the individual stimuli can be combined to a unified percept, here a moving bar. Details in [15]. *RF*, Receptive field.

Specifically, comparison of three different visual stimulation conditions has revealed a potential role of synchronization in the gamma frequency band as a mechanism for temporal binding (Fig. 21.3).

In this type of experiment, neuronal spiking activity from two neurons in two different areas of the visual system is simultaneously recorded [15]. All three visual stimulation conditions use visual stimuli that optimally drive the responses of the two neurons. Specifically, an oriented line that matches the shared orientation tuning and the receptive field of the two cells is used. Note that neurons in the visual cortex maximally respond to oriented lines (hence orientation tuning) in a specific location in space (receptive field). In the first condition, the two bars (one per receptive field) are moved in opposite directions. In this case, no gamma-band synchronization of spiking activity is observed. In the second case, the two bars are moved in the same direction. This causes a moderate synchronization of the neuronal responses. The third condition used one continuous instead of two disjointed bars; in this condition gamma synchronization was most prominent. In other words, if the visual input to the two cells encoded aspects of one unified visual object, gamma synchronization was maximal. Such temporal organization for these three types of stimuli was found for both pairs of neurons in the primary visual cortex and pairs of neurons from two different visual areas. Therefore gamma oscillations may play a key role in imposing shared structure on neuronal activity as a function of global visual properties.

Attention and Working Memory

Gamma oscillations are selectively modulated by higher order cognitive functions, such as attention and memory [16]. Attention is the process by which neuronal resources are selectively allocated to stimuli of behavioral relevance. At the neuronal level, neurons that represent the attended sensory input dominate in terms of their neural activity over neurons that represent unattended visual input. Gamma oscillations appear to enable such dynamic selection of neurons. Specifically, neurons that represent the attended stimulus synchronize their firing in the gamma-frequency range. As a result, these neurons have a stronger impact on downstream neurons, since their postsynaptic input will arrive within the integration time window defined by the time constant of the postsynaptic neuron. Furthermore, attended stimuli trigger larger gamma oscillations than unattended stimuli. Gamma oscillations are also associated with the retention of sensory information in short-term memory and working memory. The role of the gamma oscillation in memory appears to be very similar to the role of theta oscillations in memory encoding. Indeed, we will see next that theta and gamma oscillations are often coupled together.

CROSS-FREQUENCY COUPLING AND GROUPING OF GAMMA OSCILLATIONS

The occurrence of transient gamma oscillations is, among others, gated by lower-frequency oscillations. In particular, theta oscillations appear to impose a temporal structure on the occurrence of gamma oscillations. This cross-frequency coupling between theta and gamma oscillations manifests itself as a modulation of the gamma power as a function of the theta phase. Theta oscillations in the hippocampus of rodents are highest during active exploration and rapid eye movement sleep (see chapter: Theta Oscillations). Both single-unit firing and gamma activity in the neocortex are phase locked to hippocampal theta oscillations [17]. Determining such interaction is an experimentally challenging task because it requires demonstration that the gamma recorded in the neocortex indeed is locally generated and does not reflect hippocampal gamma activity that can be detected because of volume conduction (see chapter: LFP and EEG). Such findings have suggested that theta oscillations enable the sequential structuring of information encoded by cell assemblies that form at every cycle of the gamma oscillation [17a].

Functionally, theta–gamma coupling has been proposed to serve as a mechanism to encode messages that consist of multiple items in a specific order. A theta cycle defines the message. Within that cycle, there are several gamma cycles, each of which corresponds to the selective activation of a specific dynamic cell assembly (group of cells) that encodes one of the items of the overall message. Typically, four to eight gamma cycles fit into a single theta cycle that allows for the encoding of messages with the corresponding number of items.

This theta–gamma hypothesis was first advanced by studies of navigation in the rodent hippocampus. As we have seen in the chapter "Theta Oscillations," the theta phase of the firing of place cells encoded the position of the animal within the place field (theta precession). In addition, place cells that respond to very close spatial locations fire on the same gamma cycle. Similar theta–gamma coupling was also found during other cognitive tasks. Both in rats and humans, theta–gamma coupling is associated with successful long-term memory encoding. Theta–gamma coupling has also been associated with working memory. Attempts have been made to link the theta–gamma neuronal code to behavioral performance in working memory tasks. Typically, the capacity of working memory (also called *span*) is seven items with a standard deviation of two items. One study showed that the span of an individual correlates with the number of gamma cycles within a theta cycle. However, methodological concerns about the initial study require further study of this phenomenon. Finally, demonstration of the functional relevance of theta—gamma coupling will require manipulation of this coupling combined with an assessment of behavioral consequences.

FAST GAMMA OSCILLATIONS ("HIGH GAMMA")

Although it remains unclear if higher frequency gamma oscillations represent a separate network activity pattern, it is worth considering their unique aspects [1]. In the hippocampus, low and high gamma oscillations occupy distinct frequency ranges (low gamma: 25—50 Hz; high gamma: 65—140 Hz). Interestingly, these two activity modes also differ in terms of functional connectivity within the hippocampal system. During low-frequency gamma oscillations, coherence between CA1 and CA3 is pronounced. During high-frequency gamma oscillations, however, CA1 is driven by the entorhinal cortex [18]. Similarly, low and high gamma oscillations are differentially modulated in the neocortex during cognitive tasks, further supporting the distinct role of the two types of gamma oscillation. Electrocorticography (ECoG) recordings routinely show a modulation of high gamma activity during tasks; often high gamma is used as a surrogate of network activation, that is, neuronal spiking. This approach helps to overcome the limitation of the commonly used macroelectrodes in ECoG, which cannot detect extracellular action potentials.

Multiple mechanisms for the genesis of high gamma oscillations have been proposed. In vitro, low, and high gamma oscillations appear to be distinct rhythms [19]. Low gamma oscillations were associated with a current sink—source pair (see chapter: LFP and EEG) in layer V. In contrast, high gamma oscillations were accompanied by a sink—source pair in layer III. Pharmacological manipulations showed that both low and high gamma oscillations depended on GABAergic inhibition, AMPAergic excitation, and gap junctions. Conceptually, several other mechanisms can lead to high gamma oscillations. First, if a population of neurons is divided into subpopulations that each oscillate at a low gamma frequency but with different (stable) phase offsets between them, the overall network activity pattern can exhibit peak frequencies at integer multiples of the oscillation frequency of any given subpopulation. Second, an individual cell (population) that rhythmically fires in the high gamma frequency could drive (ie, entrain) an entire downstream neuronal population. Third, if two lower-frequency signals are combined in a nonlinear way, that is, they interact in a way different from summation (eg, multiplication), higher frequency signals can arise (called *nonlinear mixing*).

DEVELOPMENT

Gamma oscillations also occur in the immature nervous system, both spontaneously and evoked by sensory activity [20]. But the underlying mechanism and possibly their functional role are quite distinct. In neonatal rats (up to postnatal day 7), whisker deflections trigger a brief epoch of gamma oscillations originating in the granular layer of the somatosensory cortex (layer IV, cortical input layer, see chapter: Microcircuits of the Neocortex). The localization, determined by current source density, differs from the more superficial localization of cortical gamma oscillations in the adult animal. In further contrast, these early gamma oscillations are not abolished by GABA antagonists applied to the cortex, since they are driven by excitatory postsynaptic potentials from thalamic afferents. In addition, gamma oscillations during early development are time locked to the whisker deflection, whereas gamma oscillations in the adult brain in response to sensory input are induced, that is, not time locked to stimulus onset. Therefore network activity in the same frequency band as adult gamma oscillations may play a fundamentally different role in the immature brain, namely, the alignment of thalamocortical connections instead of horizontal communication across cortical columns. Of note, such gamma bursts during development are preferentially triggered by single-whisker stimulation and coincide in terms of their occurrence during development with the critical period for the formation of the thalamocortical map of whisker presentation (cortical fields responding to individual whiskers are referred to as barrels based on their visual appearance). As a note of caution, it is not yet clear if human infants exhibit such bursts of gamma oscillation in sensory systems during development. Given the localization to layer IV, detecting such activity from scalp-level electrodes in humans remains a technical challenge because the dipoles may not be strong enough to be detectable at the surface.

SUMMARY AND OUTLOOK

The discovery of the gamma oscillation in the visual cortex has triggered the development of an entire field within neuroscience that focuses on rhythmic activity patterns in the brain. In comparison to the other cortical rhythms, the gamma oscillation is relatively well understood in terms of the underlying cellular and synaptic mechanisms, although the differential role of synaptic excitation and inhibition appears to differ in different experimental preparations. Despite (or because of) their omnipresence in the cortex, gamma oscillations have become a point of some contention in terms of their functional relevance. Research continues and the gamma oscillation has remained a central puzzle piece in understanding both physiological and pathological brain function.

References

[1] Uhlhaas PJ, et al. A new look at gamma? High- (>60 Hz) gamma-band activity in cortical networks: function, mechanisms and impairment. Prog Biophys Mol Biol 2011;105(1−2):14−28.
[2] Yuval-Greenberg S, et al. Transient induced gamma-band response in EEG as a manifestation of miniature saccades. Neuron 2008;58(3):429−41.
[3] Niessing J, et al. Hemodynamic signals correlate tightly with synchronized gamma oscillations. Science 2005;309(5736):948−51.
[4] Tiesinga P, Sejnowski TJ. Cortical enlightenment: are attentional gamma oscillations driven by ING or PING? Neuron 2009;63(6):727−32.
[5] Buzsaki G, Wang XJ. Mechanisms of gamma oscillations. Annu Rev Neurosci 2012;35:203−25.
[6] Bartos M, Vida I, Jonas P. Synaptic mechanisms of synchronized gamma oscillations in inhibitory interneuron networks. Nat Rev Neurosci 2007;8(1):45−56.

[7] Whittington MA, Traub RD, Jefferys JG. Synchronized oscillations in interneuron networks driven by metabotropic glutamate receptor activation. Nature 1995;373(6515):612—5.

[8] Fisahn A, et al. Cholinergic induction of network oscillations at 40 Hz in the hippocampus in vitro. Nature 1998;394(6689):186—9.

[9] Cardin JA, et al. Driving fast-spiking cells induces gamma rhythm and controls sensory responses. Nature 2009;459(7247):663—7.

[10] Buhl DL, et al. Selective impairment of hippocampal gamma oscillations in connexin-36 knock-out mouse in vivo. J Neurosci 2003;23(3):1013—8.

[11] Harris KD, et al. Organization of cell assemblies in the hippocampus. Nature 2003;424(6948):552—6.

[12] Fries P, Nikolic D, Singer W. The gamma cycle. Trends Neurosci 2007;30(7):309—16.

[13] Engel AK, et al. Temporal binding, binocular rivalry, and consciousness. Conscious Cogn 1999;8(2):128—51.

[14] Singer W, Gray CM. Visual feature integration and the temporal correlation hypothesis. Annu Rev Neurosci 1995;18:555—86.

[15] Engel AK, et al. Synchronization of oscillatory neuronal responses between striate and extrastriate visual cortical areas of the cat. Proc Natl Acad Sci USA 1991;88(14):6048—52.

[16] Jensen O, Kaiser J, Lachaux JP. Human gamma-frequency oscillations associated with attention and memory. Trends Neurosci 2007;30(7):317—24.

[17] Sirota A, et al. Entrainment of neocortical neurons and gamma oscillations by the hippocampal theta rhythm. Neuron 2008;60(4):683—97.

   [a] John E, Lisman, Ole Jensen. The theta-gamma neural code. Neuron March 20, 2013;77(6):1002—16.

[18] Colgin LL, et al. Frequency of gamma oscillations routes flow of information in the hippocampus. Nature 2009;462(7271):353—7.

[19] Oke OO, et al. High-frequency gamma oscillations coexist with low-frequency gamma oscillations in the rat visual cortex in vitro. Eur J Neurosci 2010;31(8):1435—45.

[20] Minlebaev M, et al. Early gamma oscillations synchronize developing thalamus and cortex. Science 2011;334(6053):226—9.

High-Frequency Oscillations

In this last chapter of this unit on cortical oscillations, we focus on oscillations with frequencies higher than the ones subsumed in the gamma frequency band. Unfortunately, there is no commonly agreed-upon frequency range that would precisely define *high-frequency oscillations* (HFOs). As a rule of thumb, oscillations with frequencies greater than 100 Hz are termed HFOs. Overall, very little is known about HFOs and it is likely that there are several very distinct types of HFO. In this chapter, we will discuss three types of HFO. First, hippocampal circuitry generates *ripples*, short bouts of HFO, during non-rapid eye movement (NREM) sleep and quiet waking. Ripples play a central role in "replaying" previously generated activity patterns and are therefore an attractive candidate for the network mechanism of memory consolidation. Second, the somatosensory cortex generates brief periods of HFOs with very high frequencies. These fast HFOs have been found in both humans and animal models. Their functional role remains mostly unclear. Third, *fast ripples* are associated with epilepsy. We will revisit pathological high-frequency activity signatures in the chapter "Epilepsy" in the unit Network Disorders. Despite the differences in their functional roles, all of these activity patterns probably require similar cellular and synaptic mechanisms for such millisecond synchronization.

TERMINOLOGY

The detection and study of HFOs is challenging. It is difficult to separate extracellular action potentials and local field potential (LFP) signatures of HFO spikes, because they occupy the same frequency band. Standard low-pass filtering (below 300 Hz) for isolating the LFP and high-pass filtering for isolating multiunit spiking activity (above 300 Hz) does not work, because some of these fast, transient oscillations exhibit peak frequencies right at the artificial but commonly used separation point of spiking activity and LFP.

HIGH-FREQUENCY OSCILLATIONS IN THE HIPPOCAMPUS

Mechanisms

The best-understood HFOs are the ripples in the rat hippocampus (see chapter: Microcircuits of the Hippocampus) during sleep and quiet waking [1]. Ripples are short epochs of fast oscillatory activity at 100–300 Hz. These physiological signals occur predominantly during quiet waking (as opposed to exploratory behaviors) and slow-wave sleep. Hippocampal ripples occur together with *sharp waves*, which are pronounced deflections of the LFP signal recorded in the striatum radiatum of the hippocampus (up to 100-ms long). The combined event, the *sharp-wave/ripple complex* (SWR), is the result of highly synchronized neuronal firing. Nevertheless, not all action potentials within an SWR occur exactly at the same time. Instead, the relative timing of the individual action potentials reflects the temporal sequence of firing of the same neurons during waking. Similarities of sharp waves with responses to electric stimulation of the Schaffer collaterals (axons from CA3 pyramidal cells that project to CA1) suggest that sharp waves are the result of synchronized bursting in CA3. Indeed, slices of hippocampal CA3 can generate events that resemble sharp waves. The strong recurrent excitation in the CA3 network represents the ideal network substrate for the genesis of such synchronized events. This synchronized bursting is considered to be the endogenous ("default") activity pattern of CA3. Cholinergic input, for example, suppresses recurrent excitation and thereby reduces the occurrence of sharp waves. Synaptic inhibition may play a counterintuitive role in the genesis of sharp waves, because a release from synchronized inhibition facilitates the emergence of a sharp wave. Ripples are local events in CA1 triggered by sharp waves. Notably, other depolarizing input can also trigger ripples in CA1. For example, direct activation of a few pyramidal cells with optogenetic stimulation can trigger ripples [1a]. The ripple LFP waveform is generated both by superimposed extracellular action potentials and synaptic potentials. Multiple mechanisms support the genesis of ripples. Basically, ripples are the result of local synaptic interaction between CA1 pyramidal cell and inhibitory interneurons that target the somata of pyramidal cells. For example, parvalbumin-positive basket cells are synchronized to ripples in their firing and thus provide rhythmic inhibition to the pyramidal cells. Gap junctions (see chapter: Synaptic Transmission) theoretically provide a further mechanism for synchronization of fast network activity. However, the evidence for the presence of such a mechanism is mixed. In some experiments, pharmacological blockade of gap junctions reduced ripples. But these compounds are notorious for their effects on other systems, such as GABAergic synaptic inhibition. Ripples are also unchanged in knockout mice that lack the genes for the connexin proteins that form gap junctions. Alternatively, ectopic spikes—spikes that are spontaneously triggered in axons—could travel to the soma and thereby generate an antidromic spike. Through hypothesized axo-axonal gap junctions, this mechanism would generate spikelets that have indeed been observed in recordings of the neuronal membrane voltage. The origin and nature of spikelets is under debate, however. Given the lack of demonstration of a structural basis for axo-axonal gap junction coupling, its existence and potential role in generating ripples remains under investigation. Finally, coupling of neurons via the endogenous electric field is also likely to contribute to the synchronization of SWR (see chapter: Neuronal Communication Beyond Synapses).

Functions

SWR events may enable the transfer of information between the hippocampus and the neocortex during sleep. In fact, there are complex, bidirectional interactions between the two main thalamocortical sleep rhythms, low-frequency oscillations and sleep spindles, and SWRs. For example, the transition from a Down to an Up state can trigger an SWR. However, the occurrence of SWRs can also trigger delta waves and spindles in the neocortex. The entorhinal cortex, as the gateway to the hippocampus, plays a central role in this interaction.

SWR events are involved in memory consolidation. Memories are divided into *declarative* and *procedural* (ie, nondeclarative) memories. Declarative memories encompass both episodic memories (unique personal experiences) and semantic memories (facts), and depend on the hippocampal system. Sleep and memory have long been associated; there is convincing evidence that sleep promotes declarative memory. Overall, NREM sleep (see chapter: Low-Frequency Oscillations) appears to be particularly important. But which exact activity pattern is responsible for consolidation of declarative memories remains an open question, since low-frequency oscillations, sleep spindles, and SWRs are all linked together. The role of SWRs in consolidation of declarative memories is multifold. First, parameters of SWRs (such as their rate of occurrence) are correlated with declarative memory consolidation. Second, targeted interruption of SWRs by brain stimulation impairs declarative memory consolidation. Third, sequences of activation of neurons during exploration and learning are "replayed" during SWRs (Fig. 22.1, [2]). For example, place fields are activated in a specific order when a rat runs on a track. During phases of immobility, the same spike sequences are replayed during SWRs. During replay, the sequences are compressed in time. Intriguingly, replay can be both forward (same order as during exploration) or reverse (backward).

EVOKED HIGH-FREQUENCY OSCILLATIONS

Human Neocortex

HFOs with very high frequencies are also observed in the neocortex in the absence of any pathological condition. The human somatosensory cortex responds to electric stimulation of the median nerve with a burst of HFOs [3]. The median nerve provides both motor and sensory function to parts of the arm and hand. Electric stimulation causes an evoked response, which is typically determined by averaging the responses to multiple stimulation pulses. This so-called *somatosensory-evoked potential* (SEP) exhibits a stereotypical waveform with several troughs and peaks denoted with the letters N (for "negative") or P (for "positive"), accordingly, followed by a number that indicates the typical delay in milliseconds after onset of the stimulation pulse. The initial cortical response to median nerve stimulation manifests itself either as a P20 (frontal) or N20 (parietal) response. A very fast oscillation (about 600 Hz) rides on top of this initial response. These signals are detected both in EEG and MEG studies and are also referred to as *sigma bursts* (σ-bursts). Sigma bursts can be recorded in both the awake and the sleeping human participant. However, the presence of this HFO is state dependent, since its amplitude is reduced as a function of the sleep stage (see chapter: Low-Frequency Oscillations). The deeper the NREM sleep, the less HFOs are detected. In contrast, during REM sleep, sigma bursts are again (partially) present. The suppression of these fast evoked oscillations during NREM sleep is remarkable, as ripples are more present during sleep than awake.

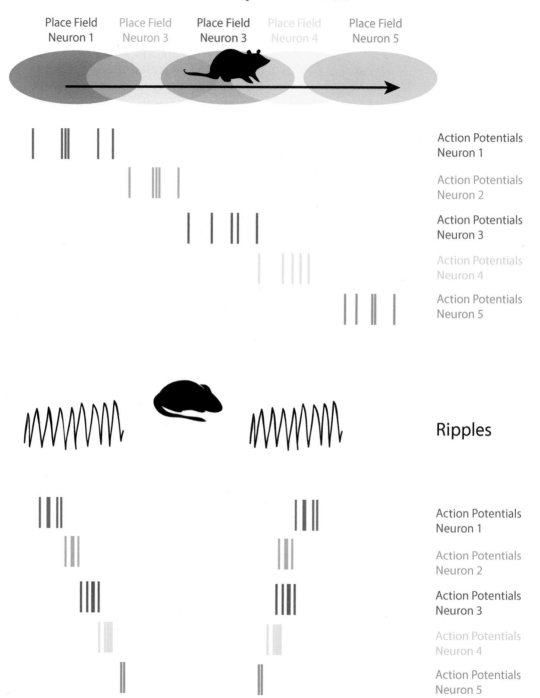

FIGURE 22.1 *Top*: Sequential activation of place cells in the rodent hippocampus during active exploration. *Bottom*: When the animal rests or sleeps, the same sequences of neurons are reactivated (both forward and backward) during ripples.

The sigma bursts are a unique signal since they appear to reflect highly synchronized action potential firing. The superposition of time locked action potentials gives rise to an extracellular signal called the *population spike*. The sigma burst is therefore a series of population spikes. This contrasts with the fact that the main generator of EEG and MEG signals is synaptic activity and not action potentials. Source localization enables the determination of the source (location) of signals recorded from the scalp (chapter: LFP and EEG). In the case of SEPs, the N20 is the primary cortical response mediated by excitatory postsynaptic potentials in the apical dendrite of pyramidal cells in the postcentral gyrus (primary somatosensory cortex; Brodmann area 3b). Localization of sigma-burst activity turned out to be more difficult. Electric recordings (EEG, depth electrodes) point to a subcortical localization, in particular the thalamocortical tract. In contrast, magnetic recordings (MEG) indicated a cortical origin, based on the fact that the localizations for the cortical N20 and the sigma burst were closely correlated. The resolution of these apparently conflicting findings is that the sigma burst measured by EEG is longer than the one picked up by MEG. Thus the HFO picked up by electric recordings is likely a superimposed signal caused by multiple dipole sources. In contrast, the MEG only detects one of the contributing signal sources, the one in the cortex, because of the preference of MEG for tangential sources.

Animal Neocortex

Responses very similar to the sigma burst in humans can also be recorded in animal models. For example, the transient deflection of a whisker in the rat evokes a biphasic sensory potential that includes HFOs [4]. Intracellular recordings showed that the membrane potential of neurons in the somatosensory cortex follows the fast frequency of the HFOs, and that the occurrence of action potentials is gated by that oscillation. Regular-spiking cells fire action potentials on some cycles of the HFOs, but they only exhibit a small (subthreshold) membrane deflection on the other cycles. Fast-spiking (presumably interneurons) cells fire a burst of action potentials in response to the whisker stimulation. The individual spikes in this burst are time locked to the cycles of the HFOs. Mechanistically, the sigma burst and related high-frequency-evoked activity requires a substrate that can generate synchronized action potential firing with millisecond precision. Any more jitter would lead to cancellation of the signals and hinder the occurrence of a synchronized, macroscopic signal. Interestingly, a multitude of mechanistic elements could contribute to such signaling. The burst firing of thalamocortical relay cells at the level of the thalamus is one possible contributor. These burst discharges are not limited to sleep, but can also occur in the awake state. At the level of the neocortex, several cell types—including excitatory cells that can be classified as "bursting" or "chattering" based on their response to current injection and fast-spiking inhibitory interneurons—can fire high-frequency bursts that may contribute to the sigma burst. Animal model studies enable the combination of noninvasive recordings with local pharmacological manipulations by targeted drug injection into the brain. Such a combination of MEG and local injection of a synaptic blocker revealed that the early part of the HFO elicited by electric nerve stimulation was resistant to the blocking synaptic transmission, whereas the later component was abolished [5]. This experimental strategy thereby confirmed the presumed differentiation into different mechanisms of evoked HFOs as a function of time after stimulation onset.

PATHOLOGICAL HIGH-FREQUENCY OSCILLATIONS: FAST RIPPLES

High-Frequency Oscillations in Epilepsy

HFOs are also implicated in epilepsy. In addition to "physiological ripples," patients with epilepsy also exhibited higher frequency HFOs detected by depth recordings in or near the epileptic zone (Fig. 22.2). Since there are no invasive recordings of brain activity from healthy participants, an animal model of epilepsy that exhibits similar pathological high-frequency discharge patterns is instrumental for delineating physiological and pathological HFOs through comparison with healthy control animals [6]. Indeed, rats injected with kainic acid (KA) in the brain develop chronic, recurrent seizures several months after the injection. Recordings from the hippocampus and the cortical entorhinal cortex in those animals display physiological ripples indistinguishable from control animals. However, faster oscillations ("fast ripples") were also observed but only in proximity to the KA-induced lesion and in the dental

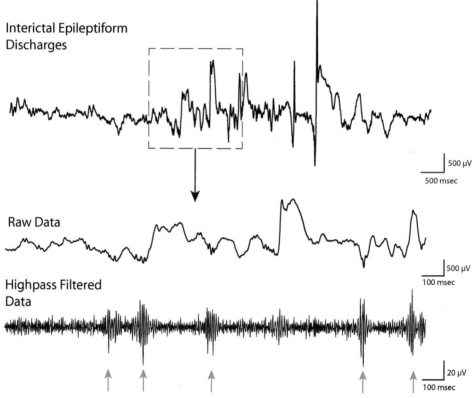

FIGURE 22.2 Bursts of high-frequency oscillations (HFOs) during epileptiform discharges in a patient with epilepsy. *Top*: Raw trace from depth electrode. *Middle*: Enlarged window from top trace. *Bottom*: High-pass filtered data to detect HFOs (*red arrows*).

gyrus and entorhinal cortex in the same hemisphere as the lesion. This coexistence of ripples (physiological) and fast ripples (pathological) is also found in depth recordings from patients with medication-refractory epilepsy. Therefore pathological HFOs are of great interest as a possible localizing signal to be used to delineate surgical resection sites for the treatment of medication-refractory epilepsy. Power in this higher frequency, pathological HFO band may also serve as a signal for seizure prediction, because it appears to increase before onset of ictal (seizure) activity.

Mechanisms

Fast, pathological ripples are similar to physiological ripples, which are slower. Here, we will look at the mechanism that increases the ripple frequency. Fast ripples originate in hippocampus CA3; their frequency is double the frequency of the physiological ripples. Therefore fast ripples are an oscillation at the first harmonic frequency of the physiological ripples. One approach to studying the mechanisms of fast ripples is to use an animal model that exhibits hippocampal seizures and examine the cellular and network properties in acute hippocampal slices of these animals. This approach has revealed that the fast ripple frequency is the result of less precise spike timing in CA3 [7]. Two main factors were identified that caused the reduced temporal precision of action potential firing. First, the extracellular space is enlarged because of cell loss. This increased extracellular volume fraction reduces the resistivity of the extracellular space. As a result, the endogenous electric fields are diminished. Second, there are larger fluctuations in synaptic input compared to control that also reduce spike timing reliability. Intriguingly, increasing synaptic inhibition by application of benzodiazepines switched fast ripples to "regular" ripples; this finding is in agreement with the prominent role of synaptic inhibition in mediating spike timing in the hippocampal circuit (see chapter: Microcircuits of the Hippocampus).

SUMMARY AND OUTLOOK

In this chapter, we have discussed fast cortical oscillations. We noted that there are three different scenarios under which oscillations with frequencies higher than ~100 Hz can occur. First, sharp waves followed by ripples appear to enable the central role of the hippocampus in memory consolidation. Second, afferent electric stimulation can trigger a transient HFO in both humans and animal models. Third, ripples can play a role in pathological network dynamics of epilepsy. No coherent model that brings together these different forms of HFO exists yet.

References

[1] Buzsaki G. Hippocampal sharp wave-ripple: a cognitive biomarker for episodic memory and planning. Hippocampus 2015;25(10):1073—188.
[1a] Stark E, Roux L, Eichler R, Senzai Y, Royer S, Buzsáki G. Pyramidal cell-interneuron interactions underlie hippocampal ripple oscillations. Neuron July 16, 2014;83(2):467—80.
[2] Diba K, Buzsaki G. Forward and reverse hippocampal place-cell sequences during ripples. Nat Neurosci 2007;10(10):1241—2.

[3] Curio G. Linking 600-Hz "spikelike" EEG/MEG wavelets ("sigma-bursts") to cellular substrates: concepts and caveats. J Clin Neurophysiol 2000;17(4):377–96.

[4] Jones MS, et al. Intracellular correlates of fast (>200 Hz) electrical oscillations in rat somatosensory cortex. J Neurophysiol 2000;84(3):1505–18.

[5] Ikeda H, et al. Synchronized spikes of thalamocortical axonal terminals and cortical neurons are detectable outside the pig brain with MEG. J Neurophysiol 2002;87(1):626–30.

[6] Bragin A, et al. Hippocampal and entorhinal cortex high-frequency oscillations (100–500 Hz) in human epileptic brain and in kainic acid–treated rats with chronic seizures. Epilepsia 1999;40(2):127–37.

[7] Foffani G, et al. Reduced spike-timing reliability correlates with the emergence of fast ripples in the rat epileptic hippocampus. Neuron 2007;55(6):930–41.

NETWORK DISORDERS

Disorders of the central nervous system (CNS) are complex, poorly understood, and difficult to treat. The hope and premise of this book is that network neuroscience can contribute to bringing relief to affected patients and their families by enabling discoveries that will lead to new treatment options. In the previous units of this book, we have learned about different strategies to study the structure, dynamics, and functions of brain networks. We then focused our attention on cortical oscillations of different frequencies that have been found to be likely mediators of a broad range of brain functions. In this unit, we learn where we stand today in the application of these methods and concepts to complex CNS disorders. We limit ourselves to a few neurological and psychiatric illnesses for which we have reached at least some level of understanding of the network-level pathologies. Many other illnesses are not covered in this unit because our insights into their network dynamics and (dys-)function are still rudimentary.

The first two chapters are dedicated to disorders classically defined as neurological disorders. First, we discuss "Parkinson's Disease" (PD). PD is of unique interest since we have a good understanding of the associated network-level pathology, namely, a dysregulation of activity in loops that connect the neocortex, basal ganglia, and thalamus. We have previously encountered PD in the chapter "Deep Brain Stimulation" (DBS). PD (together with essential tremor) is the only diagnosis for which DBS is routinely used. Then, we will learn about epilepsy. This disorder is very heterogeneous and our mechanistic understanding remains quite limited. Historically, epilepsy is an illness that is clinically defined by pathological electric activity patterns in the brain. Therefore the study of epilepsy calls for different research strategies than those applied in disorders where the associated pathological network activity patterns are unknown.

The next three chapters are dedicated to psychiatric illnesses. The origin of research on the biological mechanisms of these disorders is quite recent. Nevertheless, tremendous progress has been made, although the field is riddled with seemingly contradictory results. This demonstrates the urgent need for more research with the type of interdisciplinary strategies presented in this book. Schizophrenia is perhaps the psychiatric disorder for which we have the most comprehensive (and yet still very limited) understanding of the pathological changes in network structure and dynamics. Autism may exhibit similarities with schizophrenia. Nevertheless, very little is known about the network neuroscience of autism. Lastly, depression is a common psychiatric disorder for which there are many different models and theories. Recent attempts to conceptualize depression as a network disorder that can be treated with brain stimulation has led to both encouraging successes and humbling setbacks.

Together, these chapters aim to provide the starting point for translational research on CNS disorders that leverages the methods and concepts covered in this book. This unit on network disorders concludes the units with regular chapters. The unit with toolboxes follows; they provide background as needed to work through the main units of the book.

Parkinson's Disease

Parkinson's disease (PD) is a progressive, neurodegenerative disorder of aging that affects both motor and cognitive function. The etiology of PD is mostly unknown, but it likely involves both genetic and environmental factors. In contrast, the network-level pathology of PD is reasonably well understood. There are quite effective medications to treat the motor symptoms of PD. In addition, medication-resistant PD is routinely treated with deep brain stimulation (DBS). Despite the clinical success of DBS in PD, the mechanisms of action of DBS are poorly understood. In this chapter, we will first review the basic clinical aspects of PD, and then we will consider the network-level pathology. Finally, we will discuss the effect of DBS on the diseased networks. More details on the fundamentals on DBS are provided in the chapter "Deep Brain Stimulation."

SYMPTOMS

PD is a progressive disorder with a mean onset age of 55 years. The primary pathology of PD is the loss of dopaminergic neurons in the basal ganglia, specifically in the substantia nigra (SN) (see later). The main motor symptoms in PD are tremor, rigidity, *bradykinesia* (slowing of movements), *hypokinesia* (reduced movement range), and *akinesia* (lack of conscious movements). Tremor is most pronounced during rest. Rigidity is found for passive movement of the patients' limbs. Clinical diagnosis is established based on these motor symptoms. Gold-standard diagnosis, by establishing the presence of the associated neurodegeneration, is only possible postmortem. In addition to the motor symptoms, patients exhibit cognitive deficits, particularly in executive functions, such as attention and working memory. People with PD are also at elevated risk for depression and dementia. It is increasingly recognized that the cognitive symptoms of PD may emerge before the motor symptoms, and that they are an important contributor to the disability caused by PD. The early-disease course of PD can be structured into the following phases: the *prephysiological* (no signs or symptoms, but possibly at genetic risk), the *preclinical* (still no signs or symptoms but presence of biomarkers), the *premotor* (only nonmotor symptoms), and the *motor* phase (classically defined PD).

PATHOLOGY AND MEDICATION TREATMENT

The etiology of the disease remains unknown in the vast majority of PD cases. Epidemiological studies point toward a possible role of environment factors, such as exposure to herbicides and pesticides that have similar effects to a toxin that was found to induce a PD-like state in primates (see later). There is also an infrequent familial form of PD, the genetic basis of which has been discovered. The main pathological findings in PD are the loss of dopaminergic neurons in the basal ganglia and the presence of Lewy bodies (abnormal protein aggregates) in the cytosol of neurons. Although the degeneration of the dopaminergic system is pronounced and can explain the main symptoms of PD, it is interesting to note that the neurodegeneration is not limited to dopaminergic neurons but also affects other neuromodulatory systems and brain networks. First-line treatment is the medication L-dopa, a precursor of dopamine. Over time, the L-dopa dose needs to be increased, and pronounced motor system symptoms eventually resurface. DBS for the basal ganglia has become a standard treatment for this later, medication-resistant phase. All currently available treatments, both medication and brain stimulation, are symptomatic and do not alter the time course of the degeneration of dopaminergic neurons.

ANIMAL MODELS

Several types of animal models are used for the study of PD [1]. In neurotoxin models, toxins (6-hydroxydopamine, abbreviated as 6-OHDA; 1-methyl-4-phenyl-1,2,3,6-tetrahydropyridine, abbreviated as MPTP) are administered that cause loss of dopaminergic cells. 6-OHDA is injected directly in the basal ganglia and produces loss of dopamine neurons. The discovery of MPTP as a PD model goes back to the accidental intravenous injection of MPTP by a drug user, who in response rapidly developed PD-like symptoms. MPTP kills dopamine neurons and causes irreversible effects that closely reflect symptoms in human PD patients. The similarity of this model to human PD is more substantial in the case of primates than in the case of mice. Both of these neurotoxin models demonstrate that environmental toxins (at sufficient exposure levels) can cause PD in laboratory animals. However, the PD-like state is acutely induced and these models lack the typical progressive disease time course of PD in humans. In addition, most models lack the presence of Lewy bodies. In addition to neurotoxins, inflammatory processes can also lead to PD-like neurodegeneration by loss of dopaminergic neurons in experimental animals. Similarly, no Lewy bodies are present in these models. Genetic models based on inducing the rare mutations associated with familial PD have mostly failed to exhibit PD-like symptoms in animals.

NETWORK PATHOLOGIES

The study of PD is both humbling and exciting since it requires an integrated understanding of multiple cortical and subcortical brain areas and their interconnections: the neocortex, thalamus, and basal ganglia. We will start with a rather static view of the circuit that considers which brain area is connected to which other brain area and whether individual connections between brain areas are excitatory (glutamatergic) or inhibitory (GABAergic). First, we will go over the individual components of the circuit (Fig. 23.1). The neocortex has been previously introduced (see chapter: Microcircuits of the Neocortex). Here, we focus on the *basal ganglia*, a set of distinct subcortical nuclei comprising the *striatum*, the *pallidum*, the *substantia nigra*, and the *subthalamic nucleus*. The striatum is divided by the *internal capsule*, a large white matter tract, into the *caudate* and the *putamen*. Most neurons in the striatum are

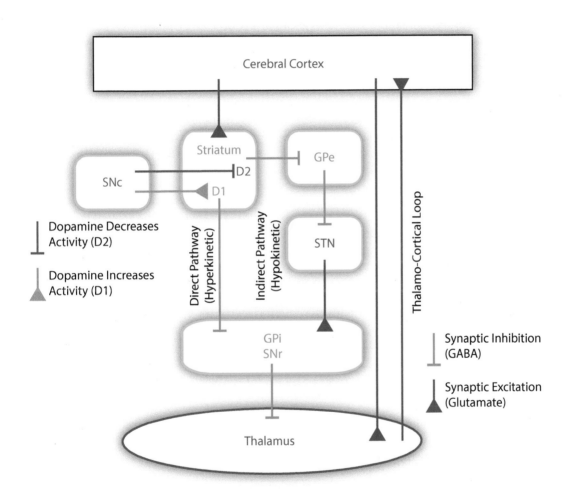

FIGURE 23.1   The basal ganglia (*yellow*) connect the cortex (*black*) with the thalamus (*green*), in addition to the direct thalamocortical connections. The direct pathway is hyperkinetic and the indirect pathway is hypokinetic. Dopamine released by the SNc differentially modulates the striatal projections to the GPi/SNr and to the GPe. By increasing the activity of MSNs of the direct pathway, dopamine has a prokinetic action since it suppresses activity in the GPi/SNr. Although dopamine inhibits the MSNs of the indirect pathway, its action is also prokinetic since it reduces activity in this hypokinetic pathway. *SNc*, Substantia nigra compacta; *GPi*, globus pallidus internal segment; *SNr*, substantia nigra reticulate; *GPe*, globus pallidus external segment; *MSN*, medium spiny neuron.

medium spiny neurons that are GABAergic and therefore inhibit their targets. Medium spiny neurons receive excitatory input from the cortex (and the thalamus, not shown). The pallidum is composed of the *globus pallidus* (itself further divided into the *external* and *internal segments*, GPe and GPi, respectively) and the *ventral pallidum* (limbic area, not further considered here). The pallidum also mostly contains GABAergic neurons. The GPe receives input from the striatum and projects to the subthalamic nucleus. The GPi, in contrast, receives input from both the so-called *direct* and *indirect* pathways. The SN is divided into the *SN reticulate* (SNr) and the *SN compacta* (SNc). Functionally, the GPi and the SNr are grouped together. The SNc produces the dopamine that is central to understanding PD. The subthalamic nucleus (STN) contains glutamatergic neurons and is therefore the only excitatory nucleus of the basal ganglia. The STN receives input from the GPe and projects to the GPi and the SNr.

The *direct pathway* is a *hyperkinetic* (ie, promoting movement) pathway since its activation increases activity in the motor cortex. In essence, the cortex excites the striatum, which inhibits the SNr and the GPi, which impose reduced inhibition on their target, the thalamus. Therefore thalamic activity increases and the excitatory thalamocortical projections excite the neocortex. The *indirect pathway* is *hypokinetic* (ie, hindering movement), and has the opposite effect on activity in the motor cortex. In this pathway, neurons in the striatum project to and inhibit GPe, which in turn reduces its inhibition of STN. As a result, the increased activity in STN excites the SNr and the GPi, which both inhibit the thalamus such that thalamic activity and therefore excitatory input to the cortex is diminished.

This entire circuit that connects the cortex via the basal ganglia to the thalamus, which in turn closes the loop back to the cortex, is under the control of dopamine. There are two distinct types of *medium spiny neurons* (MSNs) in the striatum. Those that project to the SNr and GPi express substance P and dynorphin as well as D1 dopamine receptors (see chapter: Neuromodulators). Those that project to GPe express enkephalin and D2 dopamine receptors. So, the two types of neurons have opposite responses to dopamine. In response to dopamine, the MSNs of the direct pathway will increase their activity and thereby suppress activity in the SNr and the GPi. As a result, the inhibitory input provided by SNr and GPi to the thalamus will be reduced, and activation of the thalamocortical circuit will be increased. The MSN of the indirect pathway will decrease their activity in response to dopamine, and accordingly their inhibitory input to the GPe is reduced. The resulting increased level of activity in the GPe then inhibits the STN. The reduced activity in the STN then reduces the activity in its targets, the SNr and the GPi. This reduced activity results in a decreased inhibition of the thalamus and therefore also increases activity in the thalamus and the neocortex. Thus an increase in dopaminergic tone in the striatum acts synergistically on both the direct and the indirect pathways such that overall there is a *prokinetic* action of the basal ganglia on the thalamocortical system.

A pathological reduction in dopaminergic input from the SNc to the striatum, as occurs in PD, reduces the strength of this prokinetic action. In this circuit model, one would expect STN to exhibit pathologically increased activity levels. Indeed, STN inactivation or lesion improves the symptoms associated with PD (more details later, where we discuss the STN as the primary target for DBS in PD). Notably, this entire model of the direct and indirect pathways and their synergistic pathological upregulation of synaptic inhibition onto the thalamocortical system via increased activity in the basal ganglia output SNr and GPi rests on a static model of firing rates. In this *rate-based model*, excitatory and inhibitory synaptic input increases and decreases neuronal activity, respectively. This model inspired surgical lesion approaches in which GPi and the upstream STN were lesioned as a treatment to reduce pathologically increased inhibitory output of the basal ganglia to the thalamus. Clinical success with these lesions then inspired DBS of the STN, which is also highly effective for the treatment of motor symptoms. The outcomes of these procedures are in strong agreement with the rate-based model. However, several neurophysiological findings are inconsistent with this rate-based model of PD. For example, the elegant separation of D1 and D2 receptor action on the direct and indirect pathways did not hold up in studies where D1 and D2 antagonists were used.

NETWORK OSCILLATIONS

In addition to firing rates, the temporal fine structure of neuronal activity, such as oscillations is of importance (see unit: Cortical Oscillations). Indeed, loss of dopaminergic neurons also alters firing patterns of neurons in the basal ganglia; typically, the firing patterns become more prone to bursting [2].

In anesthetized animals, neurons in the STN follow the slow rhythmic firing of the cortex. Following a 6-OHDA lesion, this coupling is increased and neurons in the GPe also exhibit similar slow rhythmic activity [3]. Therefore it is likely that low-frequency oscillations in the cortex drive the low-frequency oscillations in the basal ganglia in models of PD. MSNs typically do not fire many action potentials and therefore gate (block) the propagation of slow rhythmic cortical activity into the basal ganglia. In dopamine-depleted animals, the MSNs in the striatum are more depolarized, and fire action potentials following the slow rhythm of the cortex [4]. Note that these findings are based on recordings in anesthetized animals (in which it is easier to achieve intracellular recordings of the membrane voltage). Nevertheless, such enhanced coupling between rhythms in the striatum and the cortex has also been found in awake PD patients [5]. Both animal models and human PD patients exhibit pathological bursting and synchronization [6]. In summary, the findings on enhanced rhythmicity in the basal ganglia loop of PD animal models and people with PD propose an additional, refined model in which loss of dopaminergic tone decreases the ability of the striatal network to filter out rhythmic activity from the cortex. According to this model, movement is impeded by pathologically synchronized activity in the basal ganglia.

DEEP BRAIN STIMULATION

High-frequency stimulation of the GPi or the STN is highly effective for the treatment of the motor symptoms in PD. Overall, stimulation of STN outperforms stimulation of GPi, which appears less consistent in terms of effectiveness and also does not allow for a reduction in L-dopa dose. Given the limitations of pharmacological treatment with L-dopa, especially the need to increase the dose over time to maintain effectiveness, DBS is a highly valuable treatment strategy. DBS targets evolved from the beneficial effects of surgical lesions of these sites and from primate studies that demonstrated a positive effect on motor symptoms.

It should be noted that DBS in PD can also have undesired side effects, such as cognitive impairment and psychiatric symptoms (eg, hallucinations and mood symptoms). Overall, the actions of DBS on the circuit are poorly understood. This gap in knowledge hinders the rational choice of stimulation parameters (instead of today's trial-and-error procedure used by the treating physician) and thus the development of even better stimulation paradigms that are more effective and have fewer negative side effects.

As an interesting side product, DBS procedures enable the recording of neurophysiological signals in the target areas (to verify correct electrode positioning). Such recordings have demonstrated elevated synchronization between the basal ganglia and the cortex in the beta frequency band in people with PD (see chapter: Beta Oscillations). Both L-dopa and DBS reduce this synchronization. These results support the hypothesis that pathological synchrony could be the network-level mechanism of motor impairment in PD.

SUMMARY AND OUTLOOK

In this chapter, we discussed PD, a neurodegenerative disorder that has ushered in invasive brain stimulation as a standard treatment for a CNS disorder. Strikingly, the main motor deficits of PD can be explained by tracing a loop of multiple brain areas that involves the motor cortex, the basal ganglia, and the thalamus. Some of the main symptoms can be explained by straightforward consideration of synaptic excitation and inhibition of the different areas involved in that loop. However, despite the impressive clinical benefits of DBS for PD, this simple rate-coding model of the neurobiological substrate of PD cannot explain the mechanism of action of DBS. Instead, it appears that the more complex temporal structure of neuronal activity, such as oscillations and their synchronization will need to be included in any model that explains how DBS treats PD symptoms. With this challenge and opportunity, PD remains at the forefront of network neuroscience research.

References

[1] Dauer W, Przedborski S. Parkinson's disease: mechanisms and models. Neuron 2003;39(6):889—909.
[2] Bevan MD, et al. Move to the rhythm: oscillations in the subthalamic nucleus-external globus pallidus network. Trends Neurosci 2002;25(10):525—31.
[3] Magill PJ, Bolam JP, Bevan MD. Dopamine regulates the impact of the cerebral cortex on the subthalamic nucleus-globus pallidus network. Neuroscience 2001;106(2):313—30.
[4] Tseng KY, et al. Cortical slow oscillatory activity is reflected in the membrane potential and spike trains of striatal neurons in rats with chronic nigrostriatal lesions. J Neurosci 2001;21(16):6430—9.
[5] Shimamoto S, et al. Subthalamic nucleus neurons are synchronized to primary motor cortex local field potentials in Parkinson's disease. J Neurosci 2013;33(17):7220—33.
[6] Hammond C, Bergman H, Brown P. Pathological synchronization in Parkinson's disease: networks, models and treatments. Trends Neurosci 2007;30(7):357—64.

Epilepsy

Epilepsy is a serious neurological illness defined by the occurrence of abnormal electrical activity in the brain. In contrast to other disorders of the CNS, the network-level pathology of epilepsy is well known and in fact serves as the key diagnostic marker. Clinicians who treat people with epilepsy use both noninvasive and invasive electrophysiology as their most important diagnostic tool. The historically strong link between epilepsy and neurophysiology research has helped to advance both fields [1]. Nevertheless, key questions about the mechanisms of epileptic seizures have remained unanswered or have become a matter of debate. One example is the commonly held notion that an epileptic seizure represents a bout of *hypersynchronous* (overly synchronized) activity in the brain. Experimental recordings during seizures have raised questions about this model. In this chapter, we will first introduce the basic clinical aspects of *epileptology* (the branch of neurology concerned with epilepsy), followed by a discussion of medication, surgical, and brain stimulation treatment for epilepsy. Finally, we will review the key advances in terms of understanding the dynamics and their underlying mechanisms in the epileptic brain.

EPILEPTOLOGY

There is often confusion about the basic terms of epilepsy and seizures. A *seizure* is defined as a transient period of abnormal excessive or synchronous neuronal activity in the brain (Fig. 24.1).

A large number of conditions can lead to seizures; epilepsy is one of them. *Epilepsy* is defined by the number and frequency of unprovoked seizures. There is a history of revisions of definitions and a continued effort to improve classification schemes of epileptic seizures. Originally, classification of epilepsy was performed by distinguishing between *primary* (cause unknown) and *secondary* (cause known) seizures, and by distinguishing between *partial* (clear onset zone in the brain) and *generalized* (no clear onset zone). In addition, the distinction between *simple* (no loss of consciousness) and *complex* (loss of consciousness) was made. Initial revisions then replaced the terms "primary" and "secondary" with *idiopathic* (no known cause except for possible genetic risk factors) and *symptomatic* (epilepsy being the consequence of another disorder of the brain, such as neurodevelopmental disorders). Today, classification of different types of epilepsy is based on the notion of certain electrographic (results

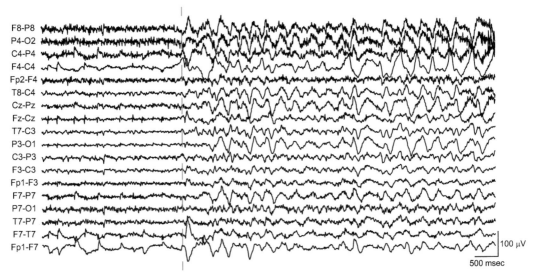

FIGURE 24.1   Electroenchephalogram (EEG) of an epileptic seizure. *Red dashed line* indicates seizure onset.

from electrophysiological studies) and clinical features occurring together and enabling the definition of *epilepsy syndromes*. Key diagnostic labels (features) of epileptic seizures are summarized in Table 24.1.

In addition to the seizures (called *ictal activity*), there are also pathological discharges in the time periods between seizures. These typically brief discharges are called *interictal spikes* because of their waveform (Fig. 24.2). We discuss interictal spikes and their underlying mechanisms in more detail later in this chapter.

TABLE 24.1   Terminology of Epileptic Seizures (International League Against Epilepsy)

Motor	Any involvement of musculature in seizure (increase or decrease in motor activity)
Tonic	Sustained increase in motor activity (milliseconds to minutes)
Atonic	Sudden loss of motor activity
Myoclonus/ myoclonic	Sudden, brief involuntary muscle contraction
Clonic	Repetitive, brief involuntary muscle contractions (typically 2–3 Hz)
Automatism	Coordinated, repetitive motor activity during state of cognitive impairment (eg, chewing or licking)
Dyscognitive	Disturbance of cognitive function (perception, attention, emotion, memory, executive function)

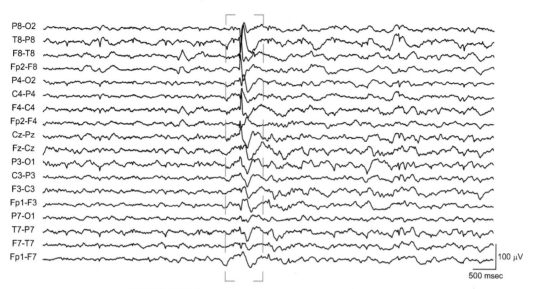

FIGURE 24.2 Interictal spike (*red-dashed box*).

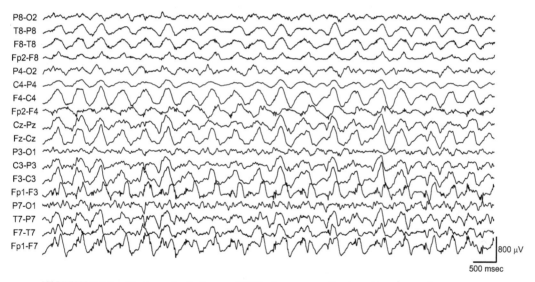

FIGURE 24.3 Electroenchephalogram (EEG) of a status epilepticus (time excerpt).

Not all seizures terminate on their own. Status epilepticus is defined as continued seizure activity for more than 5 min (Fig. 24.3). Status epilepticus is a medical emergency and requires acute medical intervention.

The occurrence of a single, unprovoked seizure does not warrant the diagnosis of epilepsy. The risk of a second seizure (leading to the diagnosis of epilepsy) is about 40%. After the occurrence of a second seizure, the likelihood of further unprovoked seizures is close to

100%. Of note, seizures with an identified cause (such as high fever) are not included in the seizure count for diagnosis, since they do not count as unprovoked. At the point of two unprovoked seizures, the diagnosis of epilepsy is established and *antiepileptic drugs* (AEDs) are prescribed. About 50% of patients become seizure free with the first AED prescribed. If seizure control is not achieved with the first AED, the likelihood of complete seizure control is dramatically reduced and is only about 5% for the third or later AED trial. *Drug-resistant epilepsy* is defined as the failure of two AEDs to achieve seizure control. Overall, about 30% of epilepsy patients exhibit drug-resistant epilepsy, and about 10% are considered truly intractable. This subset of patients represents candidates for a therapeutic approach different from AEDs, such as surgery or invasive brain stimulation.

We will now review some of the main features and clinical manifestations of epileptic syndromes in preparation for the later discussion of cellular and network-level mechanisms that drive epileptic seizures. The beginning of an epileptic seizure may include an initial phase of maintained consciousness and abnormal sensory perception called *aura*. If the aura only affects one sensory modality (eg, vision), that specificity can help to localize the seizure onset zone. Auras are not limited to sensory modalities but can also manifest as emotional auras (often manifested as fear) or psychic auras (wrong interpretation of current experience). Psychic auras include illusion of a familiar memory (*déjà vu*) or misinterpretation of a familiar visual scene as unfamiliar (*jamais vu*). An aura is different from the *prodromal symptoms* that can occur hours to days before seizure onset. Prodromal symptoms are less perceptual experiences than general symptoms of nervousness, anxiety, dizziness, and headache.

Generalized motor seizures (GMS) are one of the most frequent and most debilitating types of epileptic seizures (Fig. 24.4). GMS manifest themselves as pathological activity that

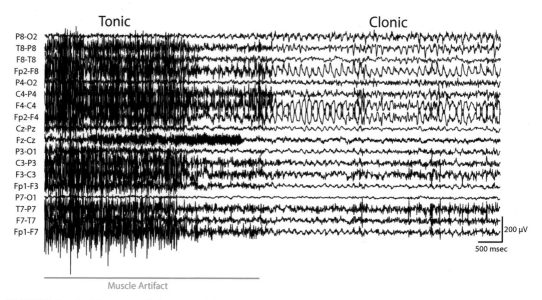

FIGURE 24.4   Electroencephalogram (EEG) of generalized motor seizure with tonic and clonic phase. EEG is contaminated with muscle artifacts during the tonic phase (*red*).

includes both hemispheres at all levels of the brain from the neocortex to the brainstem. GMS are *tonic–clonic seizures* (*grand mal*). The *tonic phase* is associated with fast rhythmic spikes in the EEG, and is marked by pathological muscle contraction that can also lead to impaired tissue oxygenation caused by respiratory suppression. The *clonic phase* is associated with slower but higher-amplitude EEG spikes, and is marked by rhythmic jerking movements. The spikes are the result of synchronous action potentials from a large group of neurons and should not be confused with the term spike used to describe individual action potentials.

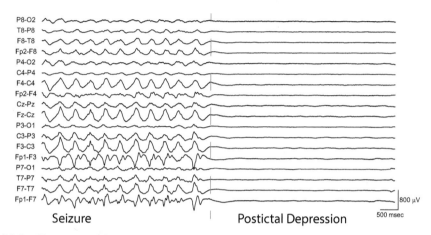

FIGURE 24.5   Electroencephalogram (EEG) of seizure ending followed by postictal depression (flat EEG).

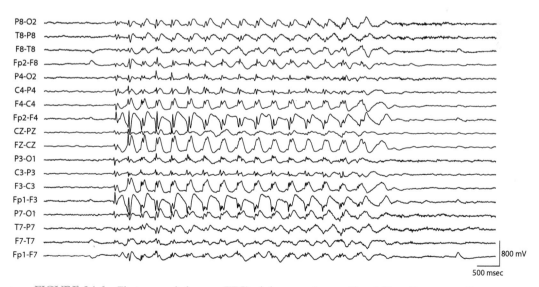

FIGURE 24.6   Electroencephalogram (EEG) of absence seizure with ∼3-Hz spike-wave pattern.

Tonic—clonic seizures are followed by an unresponsive state (*postictal depression*) character-ized by a suppression of background activity in the EEG (Fig. 24.5).

Absence seizures are very different from GMS in their clinical manifestation. They are defined by an abrupt cessation of ongoing behavioral activity, accompanied by a blank stare and an equally sudden return of the behavioral activity from before the seizure onset. The EEG of absence seizures is dominated by spike-and-wave discharges at approximately 3 Hz (see later for discussion of underlying mechanisms) (Fig. 24.6). Atypical absence seizures are often associated with other disruptions of the EEG between seizures (*interictal periods*). Without medication treatment, absence seizures can be frequent (several times per hour) and short (around 15 s). Absence seizures are a feature of many epileptic syndromes, including childhood absence epilepsy. We will revisit absence seizures in the section on mechanisms since they are associated with a pathological oscillation in the thalamocortical system that has been extensively studied in animal models.

ANTIEPILEPTIC DRUGS

AEDs aim to reduce brain excitability and have four main types of targets: (1) intrinsic voltage-gated ion (sodium, potassium, and calcium) channels, (2) GABAergic inhibition, (3) glutamatergic excitation, and (4) neurotransmitter release. Sodium channels enable action potentials (see chapter: Dynamics of the Action Potential) and are targeted by sodium channel blockers, such as lamotrigine. These drugs preferentially suppress high firing rates (in a use-dependent way) and thereby interfere more with pathological neuronal activity and less with physiological activity. Low-threshold calcium channels (T-type) in neurons of the thalamus and the cortex are critically important for thalamic bursting associated with states of reduced vigilance. Similar thalamic bursting occurs during absence seizures, which are quite successfully treated with ethosuximide, which blocks T-type calcium channels. Potassium channels reduce neuronal excitability and come in many different flavors. In the context of epilepsy, the family of Kv7 potassium channels that mediate the so-called M-current—a potassium current activated by membrane voltages close to the threshold of ac-tion potential generation—is of particular interest. Because of their activation profile, Kv7 channels are particularly important for spike-frequency adaption and thus negative feedback in the presence of elevated activity. The development of AEDs that target Kv7 is not straight-forward since one member, KV7.1, is predominantly expressed in the heart. The AED ezogabine [2] is a positive modulator of all Kv channel types expressed in the brain but not in the heart (Kv7.2—Kv7.5). Enhancing synaptic inhibition is theoretically a straightfor-ward way to reduce neuronal excitability. Indeed, positive allosteric modulators of $GABA_A$ receptors, such as barbiturates and benzodiazepines exhibit broad therapeutic effects. Two other successful targets of the GABAergic system are the GAT-1 GABA transporter tiagabine, which increases extracellular GABA, and the GABA transaminase vigabatrin, which blocks metabolic breakdown of GABA. Two targets associated with synaptic release mechanisms are the membrane protein SV2A (targeted by levetiracetam) and the alpha2delta-1 calcium channel accessory subunit (targeted by gabapentin and pregabalin). Given the importance of α-amino-3-hydroxy-5-methyl-4-isoxazolepropionic acid (AMPA)ergic transmission for network function, it is surprising that a selective AMPA receptor antagonist works as a clinically approved AED (perampanel). Notably, there are several highly effective AEDs with unknown or complex mechanisms of actions that defy the aforementioned classification of mechanisms, such as valproate.

SURGICAL TREATMENT OF EPILEPSY

Drug-resistant epilepsy can be treated by surgical intervention to either resect (remove) or isolate the networks where seizures originate. Clinically, such procedures have a favorable risk–benefit profile when performed in carefully selected patients. From a network neuroscience perspective, epilepsy surgery poses a set of interesting questions and scientific opportunities. First, epilepsy surgery planning typically includes multimodal structural and functional assessments [EEG, magnetic resonance imaging (MRI), functional MRI (fMRI), single-photon emission computed tomography (SPECT), magnetoencephalography (MEG), etc.]. Integration of the results from these different modalities and the subsequent choice of surgical strategy represent a formidable challenge of analyzing, understanding, and predicting the structure–function relationship of neuronal networks. Second, implantation of electrodes for invasive electrophysiology is typically performed during the hospital stay leading up to the surgery. These patients provide a unique window into human brain function that has given rise to an entire research community that performs research studies (unrelated to epilepsy). Importantly, electrode positioning in these patients is strictly defined by clinical needs, with no considerations made regarding potential research studies. Invasive electrophysiology in human epilepsy patients is also discussed in the chapter "LFP and EEG."

Localizing Seizures by Electrophysiology

MRI (see chapter: Imaging Structural Networks With MRI) has revolutionized the planning of epilepsy surgery since it enables the direct visualization of epileptogenic lesions. That localization then allows the assessment of the position of *eloquent cortex* (brain tissue that would lead to clear neurological impairment after removal) relative to the lesion targeted by surgery. However, not every patient with drug-resistant epilepsy exhibits a lesion in a routine structural MR scan. These patients are prime candidates for noninvasive and invasive monitoring of brain activity. Typical noninvasive monitoring includes a combination of regular scalp EEG with videography. Under close medical supervision, different manipulations are used to increase the number of observed seizures [eg, sleep deprivation, stimulation with strobe light (ie, photic stimulation), medication withdrawal]. Videography and EEG are used in conjunction to identify and localize seizure origins. MEG is another noninvasive technique aside from EEG that measures neuronal activity with high temporal resolution (see chapter: LFP and EEG). Magnetic fields pass through all tissues unhindered, including the skull. Thus MEG outperforms EEG in terms of signal localization. However, for an MEG scan, the patient needs to lie still (motion correction is available only for small motions), and therefore only interictal recordings are possible. Interictal periods often exhibit pathological activity signatures, such as interictal spikes that can be used as a proxy signal for epileptic seizures for localization purposes. Interestingly, source localization based on EEG has not yet been broadly adopted in the clinic because of the high number of electrodes needed and the large number of assumptions required about electric properties of different tissues. In contrast, source localization based on MEG is routine. The best clinical outcomes have been found when EEG and the MEG data colocalized the seizure onset zone. Source localization with MEG is particular helpful in epilepsy when there are structural defects in the skull (eg, head injury) that distort source localization with EEG. Overall, MEG is quite valuable for localization and differentiation of the seizure onset zone and the eloquent cortex.

Localizing Seizures by Imaging

Diffusion tensor imaging (DTI, see chapter: Imaging Structural Networks With MRI) measures the directionality of water diffusion and enables tractography, the identification and tracking of major myelinated fiber bundles in the brain. Although not yet routinely used clinically, DTI may become an important tool for surgery planning. Both *positron emission tomography* (PET) and SPECT can be used to localize seizures. Both techniques are based on administration of radioactively tagged probes, such as glucose. PET enables assessment of glucose consumption, cerebral blood flow, neurotransmitter synthesis, and ligand binding. SPECT is primarily used to measure perfusion (delivery of blood) to the brain. The longer half-life of the SPECT ligands allows injection during a seizure and subsequent imaging. Both techniques typically display reduced perfusion (*hypoperfusion*) in the epileptic focus during the interictal state. Given these changes in perfusion, fMRI (see chapter: Imaging Structural Networks With MRI) potentially also provides important information on seizure localization. Given the challenges of acquiring (f)MRI during seizures that occur at hard-to-predict moments in time and may have a significant motor component, fMRI is particularly appropriate for the study of interictal activity. Advances in artifact suppression strategies allow for simultaneous EEG and fMRI acquisition such that the BOLD signal for specific pathological EEG features can be investigated. The clinical benefit from such an approach has not yet been demonstrated.

Localizing Seizures by Invasive Electrophysiology

The main indication for invasive electrophysiology recordings before epilepsy surgery is the lack of clear targeting information about where to perform the surgery. This applies to the following scenarios: absence of structural lesions on the MR scan, unclear findings from multimodal imaging, and nearby or overlapping eloquent cortex. Two main methods are employed for direct cortical recordings: subdural grid recordings and so-called depth electrodes (also referred to as stereo EEG). Subdural grid electrodes are commonly used in the United States. Several grid shapes with different numbers of contact sites are available to match the specific needs of a given case. The grids are typically inserted through either open craniotomies or burr holes. The main advantage of grids is the resulting number of locations both for recording and stimulation. However, areas in sulci and in deep structures are not adequately sampled by these grids. In contrast, depth electrodes (stereo EEG) allow for accurate sampling of deep brain structures. The main advance (and challenge) of this method is the requirement for a detailed determination of the hypothesized seizure network for planning of the electrode targeting strategy. Electrodes are inserted through small drill holes and removed at the end of the hospital stay. Depth electrodes are particularly indicated if subdural grid monitoring has failed (eg, for deep structures, such as the mesial temporal lobe). The main disadvantage of depth recordings (in addition to the complexity of the planning) is the limited spatial sampling, which does not allow as clean a functional mapping of eloquent cortex as grid electrodes do. Depth monitoring is more frequently performed in Europe. In the United States, grid electrodes are often combined with depth electrodes. Independent of the type of electrode used, invasive electrophysiology is primarily used to identify onset zones of seizure and to localize eloquent cortex. The latter is achieved by electric high-frequency stimulation, which creates a transient "virtual lesion." If such stimulation impairs speech or cognition, the corresponding electrode site is designated as eloquent and is excluded from surgical resection.

BRAIN STIMULATION TREATMENT

Brain stimulation has emerged as a new treatment modality for epilepsy. Contrasted with surgery, pathological brain areas are not removed, but are controlled by electric stimulation delivered through chronically implanted electrodes and stimulation devices. Technically, this approach closely follows the successful use of deep brain stimulation in movement disorders. Basically, electric stimulation is used to target specific activity patterns associated with development of epileptic seizures. Earlier brain stimulation approaches had looked promising until they were suggested not to be efficacious based on carefully controlled (albeit very small) clinical trials (chronic cerebellar stimulation and stimulation of the centromedian nucleus of the thalamus). Two more recent advances of brain stimulation approaches stand out: stimulation of the anterior nucleus of the thalamus (pursued by Medtronic) and responsive neurostimulation (pursued by Neuropace). In the SANTE study (funded by Medtronic), it was found that stimulation of the anterior nucleus of the thalamus was more effective than no stimulation (device turned off) for drug-resistant partial and secondarily generalized seizures [3]. Neuropace pursued a different approach based on feedback stimulation of neocortical targets in patients with drug-resistant focal epilepsy. Compared to the sham group, the stimulation group experienced significantly reduced seizure frequency [4]. The Neuropace technology combines recording and stimulation of brain activity. Optimization of stimulation parameters is based on processing the data collected from the device and is interpreted in the framework of all data available from all patients implanted with the device. Importantly, such a feedback stimulation approach relies on the early and reliable detection of activity leading to seizures, that is, *seizure prediction*.

MECHANISMS OF EPILEPTIC SEIZURES

Our understanding of the mechanisms that underlie epileptiform activity remains incomplete [5,6]. Most research has focused on the neocortex, thalamus, and hippocampus. These brain structures appear to be particularly likely to exhibit pathological electric activity patterns. This vulnerability may be caused by the strong excitatory feedback that can amplify signals, and plasticity mechanisms that can remodel and strengthen neuronal connections.

In terms of cellular electrophysiology, the basic assumption is that seizures are periods of hyperactivity caused by a disruption of the balance between excitatory and inhibitory synaptic transmission and an increase in intrinsic excitability that favors burst firing of action potentials. Indeed, blocking synaptic inhibition both in vitro and in vivo reliably induces electric activity patterns that resemble epileptic seizures. It was therefore quite a surprise when it became clear from electrophysiological recordings that inhibitory interneurons continue to fire action potentials during (experimental) epileptic seizures. These findings required a conceptual revision of the synaptic mechanisms of seizures.

At the level of network dynamics, an often-repeated assumption is that epileptiform activity is characterized by pathologically increased levels of synchrony. In this model, "hypersynchronization" is considered to be the driver of "runaway" dynamics that would eventually lead to sufficient depolarization that terminates the seizure by inducing depolarization block through inactivation of sodium channels. In reality, however, many neurophysiological recordings of epileptiform activity fail to confirm this hypothesis [7]. Rather, it appears that synchronization may in fact be the mechanism that leads to seizure termination. Computational models have demonstrated the plausibility of such seizure termination by a shift toward excitation, which caused the emergence of synchronized bursting, which in turn fostered termination of the pathological activity [8].

In addition to this shift in thinking about the role of neuronal synchronization in epileptic seizures, two further aspects have been consistently overlooked because of technical limitations and conceptual barriers. First, both ictal and interictal electric activity include high-frequency oscillations (see chapter: High-Frequency Oscillations) that were not discovered until recently because of the traditional convention of considering only EEG signals below 100 Hz in epileptology. Second, the traditional use of comparably large invasive electrodes in humans (macroelectrodes) delayed the detection of small clusters of neurons that exhibit pathologically synchronized activity, called *microseizures*, which are only detected by microelectrodes. The clinical relevance of these recent discoveries remains to be established.

The mechanisms of epileptiform activity have been and are studied with a wide variety of methods and model systems. *Epileptogenesis* is defined as the process that leads to development of epilepsy after an initial insult. In animal models, there are multiple routes to trigger epileptogenesis. For example, repeated (daily) electric stimulation of certain brain areas permanently lowers the threshold for seizures, that is, it makes the occurrence of seizures more likely. This process is called *kindling*. Other insults that increase the risk of epileptic seizures and are used in animal models include local administration of toxins, such as tetanus toxin and trauma. Of note, AEDs have very limited impact on epileptogenesis, and given the window of opportunity between insult and overt epileptic seizures, a better understanding of this latent period is instrumental for the development of new treatments for epilepsy following traumatic brain injury. *Posttraumatic epilepsy* is common after head injury (eg, combat injuries in veterans, car accidents) and the process of pathological reorganization of neuronal circuits is poorly understood. Aberrant synaptic plasticity has been proposed as the underlying mechanism.

Epileptic Spikes

Acute manipulations that induce seizure-like activity are used to study the dynamics of pathological electric activity patterns in model systems. Pharmacological reduction of GABAergic synaptic inhibition and of potassium currents are some of the most frequently used experimental strategies to induce such activity. The interictal spike is the network-level event associated with epilepsy that is easiest to study. It is a brief, sharp deflection in the EEG signal that occurs in between epileptic seizures. Not everyone with interictal spikes in the EEG has epilepsy, but these transient events have proven to be a fruitful model system to study intrinsic and synaptic mechanisms of such network activity signals. In intracellular recordings, these spikes are associated with a massive depolarization that causes action potential firing, and in some cases depolarization block. This phenomenon is called *paroxysmal depolarizing shift* (PDS), where "paroxysmal" refers to the sudden recurrence of these transient events. Certain pyramidal cells in both the hippocampus and the neocortex (in particular in layer V) can fire rapid bursts of action potentials caused by complex interaction of intrinsic ion channels and cell morphology. The bursting cells are coupled by excitatory synapses that rapidly activate large networks of excitatory cells to fire in near-synchrony. Two nonsynaptic mechanisms may provide additional positive feedback that further boosts neuronal activity during PDSs. First, excessive neuronal activity causes an increase in extracellular potassium concentration that weakens the strength of potassium currents and thus prevents repolarization of neurons. In addition, synchronized neuronal firing causes an increase in the endogenous electric field, which in turn further boosts synchronization (see chapter: Neuronal Communication Beyond Synapses). These events are limited in time because of mechanisms that counteract the positive feedback process that encourages the occurrence of PDS. First, activity-dependent currents that hyperpolarize neurons, such as the BK channels that mediate a potassium current activated by intracellular calcium concentration, probably play an important role in terminating epileptiform bursts. Recruitment of synaptic inhibition further helps to terminate such pathological bursts.

Spike Waves

During seizures, the most prominent EEG activity is the rhythmic occurrence of the "spike-and-wave" pattern at a frequency of few cycles per second. This activity signature is named for its appearance. Spike waves are associated with absence epilepsy. Intracellular recordings of the membrane voltage have shown that the "spike" phase corresponds to the firing of action potentials and that the "wave" phase corresponds to a hyperpolarization. Similar mechanisms for the interictal spike also likely contribute to the genesis of the spike wave. Of note, spike waves are thought to involve the thalamocortical system. In fact, spike waves may arise from pathological hijacking of the mechanisms that generate sleep spindles (see chapters: Low-Frequency Oscillations and Alpha Oscillations). For the thalamocortical system to generate spindles, thalamocortical cells need to be relatively hyperpolarized. In this case, inhibition provided by cells in the reticular nucleus sufficiently hyperpolarize thalamocortical neurons to generate a rebound burst of action potentials mediated by the low-threshold calcium current I_T. A strengthening of the activation of reticular neurons (eg, by loss of inhibition between reticular cells or by an increase in excitatory corticothalamic afferents) causes a reduction of the frequency of the spindle oscillation and leads to a ~3-Hz spike-wave activity pattern. Similar to what we have discussed for physiological rhythms of the thalamocortical system, spike waves can therefore arise through complex interaction of intrinsic and synaptic properties in this system. It is particularly intriguing that modulators that enhance GABAergic transmission, such as barbiturates, in fact *promote* seizures in the case of absence epilepsy.

SUMMARY AND OUTLOOK

In this chapter, we discussed epilepsy, a neurological disorder with obvious pathological activity patterns in brain networks. Historically, epilepsy research has been tightly integrated with basic neurophysiology work, since seizure-like events can be easily elicited in both in vivo and in vitro preparations. We first reviewed the main clinical features and treatment strategies of epilepsy. The subset of patients who require surgical intervention not only present a unique challenge in terms of linking electrophysiology and imaging results to localize the seizure locus, but they also provide a unique opportunity to collect local electrophysiological signals from the human brain with invasive recording methods. This new convergence of basic science with epileptology may become a model for how basic and clinical research can work more hand in hand. Recordings of human brain activity are poised to become much more common given the first clinical success of chronic recording and stimulation paradigms for the treatment of epilepsy. Despite this clinical progress, it is humbling that most of the underlying neuronal and synaptic mechanisms of epileptic seizures remain to be discovered.

References

[1] Niedermeyer E, Lopes da Silva FH. Electroencephalography: basic principles, clinical applications, and related fields. 5th ed. Philadelphia: Lippincott Williams & Wilkins; 2005. xiii. p. 1309.
[2] Brodie MJ, et al. Efficacy and safety of adjunctive ezogabine (retigabine) in refractory partial epilepsy. Neurology 2010;75(20):1817–24.
[3] Fisher R, et al. Electrical stimulation of the anterior nucleus of thalamus for treatment of refractory epilepsy. Epilepsia 2010;51(5):899–908.
[4] Morrell MJ, RNS System in Epilepsy Study Group. Responsive cortical stimulation for the treatment of medically intractable partial epilepsy. Neurology 2011;77(13):1295–304.
[5] McCormick DA, Contreras D. On the cellular and network bases of epileptic seizures. Annu Rev Physiol 2001;63:815–46.
[6] Jefferys JG. Advances in understanding basic mechanisms of epilepsy and seizures. Seizure 2010;19(10):638–46.
[7] Jiruska P, de Curtis M, Jefferys JG, Schevon CA, Schiff SJ, Schindler K. Synchronization and desynchronization in epilepsy: controversies and hypotheses. J Physiol February 15, 2013;591(4):787–97. http://dx.doi.org/10.1113/jphysiol.2012.239590. Epub 2012 November 26.
[8] Fröhlich F, Bazhenov M, Timofeev I, Sejnowski TJ. Maintenance and termination of neocortical oscillations by dynamic modulation of intrinsic and synaptic excitability. Thalamus Relat Syst June 1, 2005;3(2):147–56.

Schizophrenia

Schizophrenia is a severe mental illness that affects about 1% of the population. Symptoms typically emerge in late adolescence and persist for a lifetime. Pharmacological treatment helps improve some of the symptoms, but it fails to fully mitigate them in most cases. No single cause of schizophrenia has been identified. There is clearly a genetic component, and early subclinical differences in brain structure and function can be detected in children at high familial risk. As a result, it has been proposed that schizophrenia is a neurodevelopmental disorder. Brain imaging approaches, in particular (functional) magnetic resonance imaging [(f)MRI] and noninvasive electrophysiology [electroencephalography (EEG) and magnetoencephalography (MEG)], have given rise to an intriguing view that conceptualizes schizophrenia as a *network disorder*. In this chapter, we will first review the main clinical features of schizophrenia. Then, we will consider the network impairments associated with schizophrenia, spanning from basic synaptic transmission deficits all the way to impaired organization of large-scale brain activity patterns. The toolbox "Psychiatry" provides helpful background about psychiatry in general.

SYMPTOMS OF SCHIZOPHRENIA

Schizophrenia symptoms can be broadly clustered into *positive*, *negative*, and *cognitive symptoms*. These symptoms can also co-occur with major depressive or manic episodes, a condition referred to as *schizoaffective disorder*.

Positive and Negative Symptoms

The diagnosis of schizophrenia is established with a structured clinical interview that excludes other disorders that present similarly but are of known other cause (such as hallucinations caused by drug abuse). Symptom severity is often assessed with the *Positive and Negative Syndrome Scale*, which probes three main dimensions: positive symptoms, negative symptoms, and general psychopathology. Positive symptoms are symptoms "added" to the experience of a person and include delusions, conceptual disorganization, hallucinatory behavior, excitement, grandiosity, suspiciousness, and hostility. Negative symptoms are features of the experience of a person that are "taken away," and include blunted affect, emotional withdrawal, poor rapport, social withdrawal, difficulty in abstract thinking, lack of spontaneity and flow

Network Neuroscience
http://dx.doi.org/10.1016/B978-0-12-801560-5.00025-2

of conversation, and stereotyped thinking. General psychopathology refers to other, more difficult-to-classify symptoms, such as disorientation, poor attention, and lack of judgment.

The difficulty with these behavioral measures is that they can suffer from limited reliability, caused by several factors. First, these measures rely on patients being able and willing to cooperate, which is not always the case. Second, different experts assessing the same person may come to different scores on these scales since there is no truly objective way to rate answers (an issue of interrater reliability). In addition, quantification of symptoms does not provide any mechanistic insight into the disease process. This limits the ability to develop treatments that target the underlying pathologies of neurobiological processes. In addition, from a social, economic, and political perspective, demonstration of objective biological impairment of the central nervous system in patients with mental illnesses, such as schizophrenia plays an important role in reducing stigma and improving the care and quality of life of affected patients. Many of these limitations of assessing symptoms can be overcome by performing measurements of brain network activity.

Auditory Hallucinations

Auditory hallucinations are a common and highly distressing positive symptom in patients with schizophrenia. Often, auditory hallucinations present as voices that comment on or criticize the behavior of the patient. Such *verbal auditory hallucinations* are defined as vocal perceptions in the absence of the corresponding, appropriate external stimulus. Their underlying neuropathology is only partially understood. Originally, imaging studies focused on localized changes in gray matter volume in auditory and speech-related areas, such as the left superior and middle temporal gyrus and the Heschl gyrus. Indeed, symptom severity of verbal auditory hallucinations correlates with volume loss of gray matter in some of those areas [1]. Correspondingly, verbal hallucinations activate auditory and language-related cortices and also more widespread frontotemporal locations. These findings implicate auditory areas in verbal auditory hallucinations, but fail to explain conceptually and mechanistically how they arise. In that regard, one leading theory proposes impaired top-down control of auditory areas by frontal areas (pathological disconnectivity) as the core of the pathology at the network level. Such defective communication could lead to thoughts being perceived as voices (*inner speech model*). Several lines of evidence support this model. EEG coherence in the theta band between left lateral frontal and posterior temporal sites is significantly higher in healthy controls during speaking rather than listening to simple sentences that are common during auditory hallucinations [2]. This functional connectivity may provide auditory areas with the required signal to indicate that the auditory input is generated by one's own speech. This difference in functional connectivity between listening and speaking is not present in patients with auditory hallucinations (and only to some extent in patients with schizophrenia who do not experience auditory hallucinations). In addition, functional connectivity between the temporoparietal junction and the right inferior frontal language production area (Broca's area) is impaired in patients with schizophrenia who experience verbal auditory hallucinations. This finding is in further agreement with a reduced top-down instructional signal, also referred to as *corollary signal*. Indeed, inhibitory stimulation of the temporoparietal junction with 1 Hz transcranial magnetic stimulation(TMS) improved symptoms. Reduced functional connectivity between language production (Broca's) and language perception (Wernicke's) areas predicted the response magnitude to brain stimulation. Together, these results suggest impaired top-down instruction from frontal areas to signal auditory areas that the spoken words are self-generated. However, diffusion tensor imaging (DTI) studies have yet to consistently demonstrate impaired structural connectivity between frontal and temporoparietal areas. Another proposed explanation of how auditory hallucinations arise is the *central auditory processing deficit* model, often associated with pathological functional connectivity between left and right auditory cortices [3].

Cognitive Function in Schizophrenia

Cognitive impairment is ubiquitous in patients with schizophrenia. Cognitive symptoms do not respond to established treatments, and they are largely independent of positive symptoms. Cognitive impairment predicts reduced functioning of patients. Impaired aspects of cognition include attention, verbal memory, visual memory, problem solving, speed of processing, working memory, and social cognition. Strikingly, cognitive impairment is also measurable in children and adolescents who go on to develop schizophrenia later in life.

Working memory is a key cognitive function impaired in schizophrenia and for which there is no established treatment. Working memory is defined as the ability to transiently store and manipulate information for subsequent goal-directed action. Working memory relies on the dorsolateral prefrontal cortex. *Memory cells* in that area exhibit sustained increases in neuronal firing during the retention period, presumably encoding the information to be stored during working memory. At the level of BOLD signal measured by fMRI in humans, patients with schizophrenia show a mixed pattern of both increased and decreased activation during working memory tasks in comparison to healthy controls (task-related hypo- and hyperfrontality, respectively). These discrepancies across studies are not yet understood. Pharmacological treatment has so far failed to deliver a clear and consistent benefit for treatment of working memory deficits. Of particular note, neither enhancement of N-methyl-D-aspartate (NMDA) signaling by glycine (an allosteric modulator of the NMDA receptor) or D-cycloserine (a partial NMDA agonist) nor enhancement of GABAergic signaling by MK-0777 (a selective GABA$_A$ receptor agonist) have provided improvements despite their well-motivated pharmacological rationale. *Cognitive remediation training* (such as specialized computer games to improve attention and working memory) provides moderate improvement of overall cognition, including working memory. Noninvasive brain stimulation has become a tool of interest because of its ability to modulate neuronal excitability and possibly induce long-lasting changes by recruiting neuroplasticity (see chapter: Noninvasive Brain Stimulation). More research is needed to determine its efficacy for the treatment of working memory and other cognitive functions.

BRAIN NETWORKS IN SCHIZOPHRENIA

Sensory Gating: The P50 Response

A key cognitive symptom in patients with schizophrenia is *impaired sensory gating*, defined as reduced ability to suppress processing of irrelevant and uninformative sensory input.

The so-called *auditory dual-click task* is commonly used to assess sensory gating. In this task, two consecutive, brief click sounds (spaced by 500 ms) are played, and the sensory response to the tones is measured by EEG. When the responses from a large number of trials of such click pairs are averaged, an evoked potential with a positive-going waveform at about 50 ms after the onset of each click is found (and thus denoted as P50). In this notation, "P" stands for the positive sign of the deflection in the evoked potential and "50" denotes the typical delay from stimulus onset in milliseconds (see also chapter: LFP and EEG for background). The amplitude of the response to the second click is typically lower (A2) than the amplitude of the response to the first click (A1). The ratio A2/A1 thus assumes a value smaller than 1. This ratio is termed the *P50 gating ratio*. Patients with schizophrenia exhibit larger ratios, since they have less reduction in response to the second click [4]. These larger A2/A1 values are indicative of the inability to suppress the response to the second click. Nevertheless, there are limitations to the P50 gating ratio. For example, since the determination requires the division of the two numbers A1 and A2, the measure is particularly sensitive to poor signal-to-noise ratio, since small differences in signal measurements of P50 responses can be amplified by the division. Still, the P50 gating ratio may be one of today's best *trait deficits* or *endophenotypes*, defined by (1) being associated with an illness of interest, in this case schizophrenia, (2) state independence by being present even if illness is not active, (3) reliability and stability, and (4) evidence of heritability.

Cognitive Processing: The P300 Response

A second EEG potential that has been associated with deficits in cortical function in patients with schizophrenia is the *P300 event-related potential* (ERP). Typically, P300 is determined from EEG recordings (at midline locations Pz or Cz) during an "oddball" paradigm where a frequent (auditory) stimulus (*standard*) is randomly interleaved with an infrequent stimulus (*target*) and the participant is asked to pay attention to the targets (eg, by counting the number of target occurrences). The P300 amplitude is determined as the peak amplitude measured as the difference between the prestimulus baseline and the largest positive-going deflection of the ERP waveform in a time window of 250–500 ms after the onset of the stimulus. P300 is considered a marker of cognitive processing, particularly in relation to working memory requirements, allocation of attentional resources, and updating of mental models of the environment [5]. Both the amplitude and the timing of the P300 signal have been associated with cognitive function. P300 amplitude is higher in the case of superior memory performance. P300 latency inversely scales with cognitive function, so it increases with cognitive decline.

The underlying neurobiological processes that give rise to the macroscopic P300 signal remain mostly unclear. Nevertheless, many factors are known to modulate the P300 response. For example, higher body temperature and higher heart rate are both associated with shorter P300 latencies. Alcohol consumption (similar to fatigue) reduces the amplitude and increases the latency. On the other hand, caffeine consumption decreases latency. As a result, P300 values not only depend on the specifics of the task used, but are also state dependent. In patients with schizophrenia, P300 amplitude is reduced and P300 latency is longer. Proposed explanations (with varying degrees of supporting evidence) are reduced effort or motivation (although P300 amplitude is still reduced even in tasks that require no effort), reduced attention, and more limited allocation of resources to the task. P300 is both a trait that marks disease (vulnerability) and a marker of clinical state. For example, P300 amplitude is negatively correlated with illness duration and so appears to track disease progression.

Auditory Steady-State Response

The third EEG response pattern uses auditory stimuli that exhibit a frequency structure matched to the frequency range of cortical oscillations. Either auditory click trains or amplitude-modulated tones at different frequencies, typically 20, 30, and 40 Hz, elicit rhythmic network activity that can be readily measured by EEG. The most straightforward analysis strategy reflects the previously used approach of trial-averaged sensory responses (*auditory steady-state response*, ASSR). Subsequent determination of the frequency spectrum of the responses at different frequencies shows the relative impact of stimulation at different frequencies. Alternatively, phase locking to the stimulus can also be computed. Intriguingly, healthy, control participants display more pronounced rhythmic responses to 40 Hz than to other click train frequencies. But patients with schizophrenia exhibit a specific impairment of such rhythmic responses to 40 Hz click trains [5a]. This result is particularly intriguing given the growing evidence of impaired gamma oscillations in patients with schizophrenia.

Cortical Oscillations

The deficit in gamma-frequency ASSR dovetails with a growing number of studies that found frequency-specific alterations of cortical network dynamics of patients with schizophrenia in the gamma frequency band. It has been proposed that schizophrenia is a condition of disconnectivity characterized by a failure to adequately integrate and process information in brain networks. Cortical oscillations, particularly in the gamma frequency band (see chapter: Gamma Oscillations), appear to enable such integration and are therefore a likely candidate for impaired network function in patients with schizophrenia. Such integrative or feature-binding qualities of gamma oscillations have first been proposed in the visual system, and indeed tasks that require spatial integration for perceptual decisions recruit gamma oscillations in the visual system. For example, in a task that requires the presence of an illusionary square (*Kanisza square*, Fig. 25.1) to be indicated by a button press, there is strong intertrial phase coherence in the gamma band (see chapter: Network Interactions) only in trials with an illusionary square present [6].

In people with schizophrenia, no such response was found for either stimulus. Furthermore, presentation of either stimulus transiently increased the number of functional connections between EEG electrode sites as measured by phase coherence in healthy participants, a feature that was mostly absent in patients with schizophrenia.

These deficits in recruitment of gamma oscillations and functional connectivity in the gamma band are not limited to perceptual tasks. Pathological changes in gamma oscillations have also been found in tasks that require higher-order cognitive processes, such as *cognitive control*, defined as the ability to flexibly adjust cognition and behavior to match intentions and goals. One challenge in studying recruitment of gamma oscillations in such tasks is that the occurrence is less time locked to the stimulus (referred to as evoked by the stimulus) but instead is induced (increased by the task but not time locked). Specifically, when participants

Illusionary Square   Control Stimulus

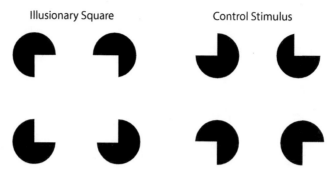

FIGURE 25.1   Visual stimuli for the study of gamma oscillation in perception. *Left*: Illusionary square that evoked gamma band organization of brain activity in healthy controls but not in people with schizophrenia. *Right*: Control stimulus. Details in [6].

conducted a task that on some trials required them to overcome the natural inclination to follow the direction of an arrow presented on the screen, gamma oscillations were differentially increased when the participants had to exert cognitive control to overcome their habitual response. Patients with schizophrenia, however, mostly failed to display differential recruitment of (induced) gamma oscillations for the two conditions [6a].

Baseline gamma oscillations can be determined from resting state recordings where participants are instructed to remain still in the absence of any task or, alternatively, from the baseline period before trials during a behavioral task. Overall, despite some heterogeneity in the literature, baseline gamma oscillations have been found to be *elevated* in patients with schizophrenia [6b]. These findings demonstrate that the pathology in the gamma oscillations is an inability to dynamically regulate gamma oscillation rather than an inability to generate gamma oscillations.

Changes in meso- and macroscopic network dynamics in patients with schizophrenia are not limited to the gamma frequency band. For example, in a task where degraded pictures of faces (so-called Mooney faces) and similar visual patterns are presented and the participant has to indicate if a face is present in the visual stimulus, long-range communication between different EEG sensor locations is impaired in patients with schizophrenia. Interestingly, these deficits were in the beta frequency band (12−30 Hz) in terms of functional connectivity determined by phase locking of EEG signals from different spatial locations. This can be interpreted in terms of the known deficits in Gestalt perception where visual elements need to be combined for successful perception in patients with schizophrenia [7].

Most of these studies are potentially confounded by reduced attention, motivation, and cooperation by participating patients. Therefore direct probing of network responses in the absence of behavioral tasks provides important additional information, because this approach is more likely to be free of such confounding, state-dependent effects. For example, the application of TMS in combination with EEG showed that patients with schizophrenia exhibited a reduced and delayed evoked gamma response to a TMS pulse [8]. Therefore, noninvasive brain stimulation has the potential to become not only a treatment but also a diagnostic tool in the future.

Structural Connectivity

Noninvasive imaging of network connectivity with DTI (see chapter: Imaging Structural Network With MRI for more details on this method) provides a unique tool to probe for evidence of altered structural connectivity in patients with schizophrenia and cohorts at risk of developing schizophrenia. DTI enables the visualization and quantification of the major white matter fiber tracts. Briefly, fractional anisotropy (FA) quantifies the degree to which diffusion is directionally restricted. The higher the FA number for white matter, the higher the presumed integrity of the corresponding tract. FA is computed for every voxel (the equivalent of a pixel, but in three dimensions) and can subsequently be used for tractography, where the major tracts are reconstructed. DTI studies have consistently found impaired white matter tract integrity in patients with schizophrenia [9,10]. The most prevalent finding is an impairment (typically measured as a reduced FA) of the frontal (left, deep white matter) and temporal (left, deep white matter) regions, the corpus callosum, which connects the two cerebral hemispheres, and thalamocortical connectivity [11]. However, the specific locations vary from study to study, a common but unresolved problem in the human imaging field in general. Moreover, the effects of disease course (high-risk but before disease onset, first episode, chronic schizophrenia), medication status, and other confounds remain unclear. Furthermore, network structure analysis of functional connectivity based on fMRI does not show uniform reduction of connectivity, but instead both increases and decreases, possibly reflecting a compensatory mechanism of neuronal activity in response to impaired structural connectivity.

MECHANISMS AND CAUSES OF SCHIZOPHRENIA

Cellular and Synaptic Impairment

Animal models (also referred to as *preclinical models*) are a powerful tool for the mechanistic study of pathological changes associated with schizophrenia. The obvious drawback, however, is that since the cause(s) of schizophrenia are unknown, and since we do not have the ability to ask animals questions about their internal mental state, any proposed animal model can be questioned about to what extent it indeed represents schizophrenia.

A main class of animal models centers on the hypothesis that impaired NMDA receptor functioning is a core mechanism of schizophrenia. Indeed, administration of NMDA receptor antagonists, such as ketamine, MK-801, and phencyclidine causes not only behavioral changes resembling schizophrenia but also increases in resting-state gamma oscillations in both animal models and humans. Mice that express fewer NMDA receptors (reduced NMDAR1 subunit expression) exhibit electrophysiological, behavioral, and molecular deficits that resemble schizophrenia in humans. But despite the psychosis-like symptoms caused by dopaminergic agents, such as amphetamines, no clear associations with gamma oscillations have been found. Given the key role of parvalbumin (PV)-positive GABAergic inhibitory interneurons, it naturally follows to hypothesize the involvement of synaptic inhibition in the pathophysiology of aberrant network dynamics in schizophrenia [12]. Indeed, NMDA hypofunction has been associated with downstream GABAergic hypofunction. Some of the more direct evidence for this relationship is derived from mouse models with selective knockdown of the NMDAR1 subunit on PV-positive cells or complete lack of NMDAR expression [12a,b]. These mice exhibit behavioral and molecular deficits that resemble those seen in human patients. In addition, the mice exhibit an enhanced baseline gamma oscillation, impaired rhythmic activity in response to optogenetic stimulation of PV interneurons, and selective cognitive impairment.

Neurodevelopmental Origin

Despite disease onset typically occurring in young adulthood, schizophrenia is conceptualized as a neurodevelopmental disorder. Genome-wide association studies that sequence the genome of large numbers of patients and healthy controls (up to many thousands of participants) have found a large number of alleles, likely in the thousands, each of which has only a very small effect in terms of clinical diagnosis. These studies have shattered the hope of pinpointing a small number of causative genetic features that could explain schizophrenia. Clearly, genetic makeup alone does not explain schizophrenia, since twins with identical genetic makeup only have a 50% concordance rate. The role of environmental factors is similarly complex and multifaceted. Epidemiological studies have found that many factors contribute to an elevated risk of developing schizophrenia, such as prenatal *maternal immune activation*, perinatal hypoxia, adolescent stress, and several others. But none of these environmental factors represent a single causative factor that can explain disease emergence. As a result, schizophrenia is likely to be caused by a complex interaction of genetic and environmental factors that together cause neurodevelopmental pathology of neurons and networks of neurons that lead to overt symptom onset in the young adult.

Given the impaired gamma oscillations and the disease-mimicking effects of the perturbing NMDA receptor-mediated drive of PV-positive interneurons, the development of the GABAergic inhibitory system is an intriguing candidate target mechanism (see chapter: Microcircuits of the Neocortex). Gene expression levels in the postmortem tissue of patients with schizophrenia point toward genes implicated in development, particularly development of the GABAergic neurons and synapses. For example, reduction in expression of glutamic acid decarboxylase 1 (GAD1), one of the GABA synthesizing enzymes, is a common finding. The transient suppression of GAD1 expression in mouse models has a range of different effects depending on the time point of perturbation; early experimental suppression leads to deficits in neuronal migration, later suppression causes PV-positive cell-specific damage. In terms of environmental perturbations, maternal immune activation (see later) perturbs the development of GABAergic circuits in rodent models, particularly in PV-positive cells. It is important to remember that changes in gene expression can be caused by genetic or environmental factors, or gene—environment interaction. Adding to this complexity is the fact that perturbed GABAergic circuits are not unique to schizophrenia but are common to a large number of psychiatric illnesses, including autism, anxiety, depression, and many more.

One of the environmental factors associated with schizophrenia is activation of the maternal immune system during pregnancy. The prenatal period is highly sensitive to damaging perturbations and can negatively alter the developmental trajectory of the offspring. Retrospective epidemiological studies that survey exposure and outcomes at the population level pointed toward a negative effect of maternal infection during pregnancy for the offspring. Many infectious diseases during pregnancy, such as influenza, rubella, measles, and herpes simplex, are associated with increased risk of schizophrenia in the offspring. However, epidemiological studies can never reach the level of causal evidence. Therefore animal experiments play a key role, and have provided comprehensive evidence for persistent structural, functional, and behavioral abnormalities in the offspring of mothers who were subject to immune system activation. Initial studies exposed animals to the influenza virus. Today, noninfectious activation of the immune system has become the approach of choice. One commonly used approach is the injection of polyriboinosinic—polyribocytidylic acid, abbreviated as poly(I:C), a synthetic analog of double-stranded RNA, into the pregnant mother [13]. Double-stranded RNA is generated during viral infection and is recognized by the mammalian host as foreign. Systemic administration of poly(I:C) mimics acute viral infection and causes a broad set of abnormalities in the brain of the offspring. Interestingly, and in striking parallel to schizophrenia in humans, behavioral deficits only fully emerge in the adult offspring born to an immune-challenged mother. The cognitive and behavioral impairments include deficits in sensorimotor gating, selective attention, social behavior, exploratory behavior, working memory, and cognitive flexibility. Another approach, injection of the bacterial endotoxin lipopolysaccharide, models the acute response of the maternal host to a bacterial infection. Despite differences in the specific immune system activation patterns and detailed outcomes in the offspring, similarly broad negative consequences ensue for the development of the brain of the offspring. In further agreement with the emerging framework of complex interactions of genetic and environmental factors, animal models with genetic modifications to mimic schizophrenia (such as by mutations in the disrupted-in-schizophrenia 1 gene) are particularly susceptible to environmental perturbations.

SUMMARY AND OUTLOOK

In this chapter, we have learned about symptom clusters that define schizophrenia. We discussed in more detail auditory hallucinations, a positive symptom of schizophrenia. Auditory hallucinations are of particular interest since there are preliminary findings on pathologies in the functional interaction between different (cortical) brain areas, and there have been early attempts to correct these network deficits with noninvasive brain stimulation. Moving forward, auditory hallucinations have the potential to become a successful network-level target for novel treatment approaches. We also considered cognitive symptoms, which are a debilitating facet of schizophrenia symptoms with very little effective treatment. The study of the network basis of these symptoms provides a starting point to build (reverse) translational bridges to systems and cognitive neuroscience. We then reviewed different network-level markers of brain dysfunction in schizophrenia. There are classical, well-studied responses, such as the P50 and P300 evoked potentials, which are quite reliable but provide little mechanistic insight. More recent work has focused on induced oscillations, particularly (but not limited to) gamma oscillations, which show quite complex patterns of impairment. Clearly, simplistic models, such as the brain of patients with schizophrenia exhibiting fewer gamma oscillations, are wrong. We then turned our attention to the causes and mechanisms of these network deficits and the associated symptoms. Synaptic inhibition by the GABAergic system has emerged as a likely target. We looked at several types of insults that could modulate the early life development of this system, in particular models of maternal infection during pregnancy. Overall, schizophrenia is a complex network disorder, with the origin of the disorder to be eventually traced back to a cascade of insults early in development.

References

[1] Gaser C, et al. Neuroanatomy of "hearing voices": a frontotemporal brain structural abnormality associated with auditory hallucinations in schizophrenia. Cereb Cortex 2004;14(1):91–6.

[2] Ford JM, et al. Reduced communication between frontal and temporal lobes during talking in schizophrenia. Biol Psychiatry 2002;51(6):485–92.

[3] Jardri R, et al. Cortical activations during auditory verbal hallucinations in schizophrenia: a coordinate-based meta-analysis. Am J Psychiatry 2011;168(1):73–81.

[4] Bramon E, et al. Meta-analysis of the P300 and P50 waveforms in schizophrenia. Schizophr Res 2004;70(2–3): 315–29.

[5] Polich J, Kok A. Cognitive and biological determinants of P300: an integrative review. Biol Psychol 1995;41(2):103–46.

[5a] Kwon JS, et al. Gamma frequency-range abnormalities to auditory stimulation in schizophrenia. Arch Gen Psychiatr 1999;56:1001–5.

[6] Spencer KM, et al. Abnormal neural synchrony in schizophrenia. J Neurosci 2003;23(19):7407–11.

[6a] Minzenberg MJ, Firl AJ, Yoon JH, Gomes GC, Reinking C, Carter CS. Gamma oscillatory Power is impaired during cognitive control independent of medication status in first-episode Schizophrenia. Neuropsychopharmacology December 2010;35(13):2590–9. http://dx.doi.org/10.1038/npp.2010.150.

[6b] Spencer KM. Baseline gamma power during auditory steady-state stimulation in schizophrenia. Front Hum Neurosci January 13, 2012;5:190. http://dx.doi.org/10.3389/fnhum.2011.00190.

[7] Uhlhaas PJ, et al. Dysfunctional long-range coordination of neural activity during Gestalt perception in schizophrenia. J Neurosci 2006;26(31):8168–75.

[8] Ferrarelli F, et al. Reduced evoked gamma oscillations in the frontal cortex in schizophrenia patients: a TMS/EEG study. Am J Psychiatry 2008;165(8):996–1005.

[9] Fornito A, et al. Schizophrenia, neuroimaging and connectomics. NeuroImage 2012;62(4):2296–314.

[10] Wheeler AL, Voineskos AN. A review of structural neuroimaging in schizophrenia: from connectivity to connectomics. Front Hum Neurosci 2014;8:653.

[11] Ellison-Wright I, Bullmore E. Meta-analysis of diffusion tensor imaging studies in schizophrenia. Schizophr Res 2009;108(1–3):3–10.

[12] Gonzalez-Burgos G, Lewis DA. NMDA receptor hypofunction, parvalbumin-positive neurons, and cortical gamma oscillations in schizophrenia. Schizophr Bull 2012;38(5):950–7.

[12a] Mohn AR, Gainetdinov RR, Caron MG, Koller BH. Mice with reduced NMDA receptor expression display behaviors related to Schizophrenia. Cell August 20, 1999;98(4):427–36.

[12b] Korotkova T, Fuchs EC, Ponomarenko A, von Engelhardt J, Monyer H. NMDA receptor ablation on parvalbumin-positive interneurons impairs hippocampal synchrony, spatial representations, and working memory. Neuron November November 4, 2010;68(3):557–69.

[13] Meyer U. Prenatal poly(i:C) exposure and other developmental immune activation models in rodent systems. Biol Psychiatry 2014;75(4):307–15.

Autism Spectrum Disorders

Autism was originally defined as "an innate disturbance of affective contact" by Leo Kanner in 1943 and was included in the chapter on childhood schizophrenia in early versions of the Diagnostic and Statistical Manual of Mental Disorders (DSM) (see toolbox: Psychiatry). Today, autism is part of a new *autism spectrum disorders* (ASD) diagnostic label in the DSM-5 [1] that subsumes autism according to Kanner's definition, Asperger's syndrome, and other early-onset interaction and communication impairments. One of the most puzzling and concerning aspects is the 10-fold increase in the rate of ASD diagnoses over a 20-year interval. According to the Centers for Disease Control and Prevention, the prevalence of ASD was 1 in 68 children in 2010 [2]. It remains to be discovered what causes ASD. But progress has been made in understanding the neurobiology of ASD. In this chapter, we will first introduce the symptoms of ASD, then discuss ASD from the perspective of brain networks, and finally review what we know about the mechanisms and etiology of ASD. The toolbox "Psychiatry" provides helpful background about psychiatry in general.

SYMPTOMS OF AUTISM SPECTRUM DISORDERS

Early diagnosis of ASD is crucial, because intense behavioral interventions early in life substantially improve overall outcome. ASD symptoms center on difficulties with social interactions and communications, and repetitive behaviors. Many people with autism also have other medical problems, such as epileptic seizures. There is a wide range of severity of manifestations of symptoms of ASD. Although the diagnosis is usually made in children who are a few years old, more research has revealed early (precursor) symptoms that manifest as early as a few months of life. For example, children with autism fail to respond to their name, exhibit reduced interest in interpersonal interaction, and display limited early vocalizations, such as babbling. These are early signs of a general difficulty connecting with other people and understanding social clues, such as smiles and subtleties, such as sarcasm. Overall, people with autism appear to fail to understand the world through someone else's eyes. Repetitive behaviors, such as head banging or stereotyped hand movements are common in people with ASD. Such stereotypy is not limited to motor actions—patients with ASD also tend to obsess over (often very technical) subject matters with unrelenting focus. It has been noted that some of the classical behavioral symptoms such as obsessing over technical details render people

319

with ASD much more adept at solving certain very difficult tasks than typically developing (TD) people.

DYNAMICS AND STRUCTURE IN BRAIN NETWORKS IN AUTISM SPECTRUM DISORDERS

Macroscopic Network Dynamics: Electroencephalogram

The most obvious and common finding of resting-state electroencephalogram (EEG) in people with autism is the occurrence of epileptiform discharges, even in the absence of clinical seizures. For resting-state EEG, heterogeneous findings of increased and decreased power levels and alterations of coherence (see chapter: Network Interactions) have been found [2a]. For adults with ASD, decreased alpha power and increased beta and gamma power have been reported quite consistently. Overall, evidence points toward changes in functional connectivity as a function of spatial scale [3]. Local connections appear to be enhanced, whereas more global connections (between more distant electrodes) are reduced [4]. Coherence in the alpha frequency band is reduced, suggesting an impaired functional integration of the frontal lobe with the remainder of the cortex (long-range hypoconnectivity). In contrast, local connectivity in the theta band is increased, suggesting local hyperconnectivity. Little is known about cortical oscillations in young children with ASD (or at risk). Likely, the deficits in network activity structure and functional connectivity differ from the ones found in adults [4a].

Histology

This shift from global (long range) to local connections is also supported by histological and structural imaging findings [5,6]. At the microscopic level, the number of *minicolumns* is increased in the brain of people with ASD compared to TD humans. Minicolumns are radially oriented clusters of neurons that may represent a unitary functional module. Neuronal size is smaller in the brains of patients with autism. Together with the larger number of minicolumns, this biases cortical connectivity toward short-range connections, since smaller neurons are less able to maintain the energy demands of long-range connections. Therefore these microscopic findings support reduced long-range connectivity and pathologically enhanced short-distance connections. Patients with ASD exhibit a pathological head enlargement that may be caused by brain overgrowth early in life, which has been proposed to be associated with pathological local hyperconnectivity.

White matter structure can be separated into two relatively distinct aspects, the *superficial* and the *deep* white matter. Superficial white matter is composed of fibers that run orthogonal to the cortical surface. Axons that form the deep white matter travel in parallel to the cortical surface. This distinction is relevant to understanding ASD, because the fibers that form the superficial white matter are local connections, whereas the fibers that constitute the deep white matter provide long-range connections. In ASD, superficial white matter is pathologically increased and deep white matter pathologically reduced [7].

(Functional) Magnetic Resonance Imaging

Functional imaging studies with (functional) magnetic resonance imaging (fMRI) have shown conflicting evidence of hyper- and hypoconnectivity in ASD. It is likely that the connectivity differences between TD children and children with ASD exhibit a developmental time course. Also detailed methodological concerns have arisen that need to be taken into account for correct data interpretation. Most prominently, head motions (more prevalent in children with ASD) can introduce artifactual biases in functional connectivity measures based on fMRI data (see chapter: Imaging Functional Networks With MRI). Even small amounts of movement can introduce spurious functional connectivity with spatial structure. In addition, the choice of analysis strategy for extracting functional connectivity from BOLD fMRI data can lead to findings of hyper- or hypoconnectivity for the same dataset. For example, functional connectivity based on coactivation, which reflects online engagement of sensory-motor and cognitive processes, appears to bias data toward findings of hyperconnectivity, possibly because of the task engagement of neuronal structures. In contrast, intrinsic functional connectivity analysis extracts spontaneous BOLD fluctuations by bandpass filtering the BOLD time series and removing task-evoked activity by regression. This method biases the results toward findings of hypoconnectivity. Despite these limitations of fMRI [8], impairment of medium- and long-distance functional connectivity is a relatively common finding in fMRI studies of people with ASD [9]. In agreement with such impairment of functional long-distance connections, the corpus callosum appears different in patients with ASD. The corpus callosum is one of the main gateways of long-range projections between cortical hemispheres. Pathologies of the corpus callosum have been found in people with ASD, including findings of lower volume, increased diffusion, and lower fractional anisotropy (poorer organization of fibers, see chapter: Imaging Structural Networks with MRI).

Behavioral Analogies of Altered Network Connectivity

It can be speculated that this pattern of hypo- and hyperconnectivity correlates with the behavioral profile. Patients with autism usually surpass TD children in tasks that require local processing (such as sensory discrimination tasks), but exhibit reduced performance in tasks that require information integration across multiple systems to "see the big picture." Along this line of speculation is the behavioral hyperspecificity and inferior generalization observed in people with ASD. For example, the classical Wechsler IQ test includes tasks that require the participant to reconstruct a global pattern by arranging individual blocks that have red, white, or diagonally separated red−white faces. The test requires breaking up each design into local units followed by a target manipulation of individual blocks to reconstruct the design. The task involves both global processing (the percept of the overall pattern to be reproduced) and local processing, analyzing the individual elements (blocks) that contribute to the pattern. In cases where perception of the overall pattern conflicts with local processing, patients with ASD outperform TD participants [10]. If segmentation is provided to aid local processing of the figure, this superior performance is abolished. Therefore TD are impeded in solving the task by global processing, whereas it is likely that people with ASD fail to engage global processing, and so excel by only processing the local information about individual elements in the task [11].

MECHANISMS AND ETIOLOGY OF AUTISM SPECTRUM DISORDERS

Genes, Environment, and Multisystem Pathologies

ASD are neurodevelopmental disorders with onset in childhood. ASD are complex disorders probably caused by many common genetic variants acting in combination with unknown environmental risk factors [12]. Fragile X syndrome (FXS) has many similarities with ASD at the neurobiological and behavioral level. But the disorders differ in their etiology. FXS is a *monogenetic* disorder caused by an expansion of the fragile X mental retardation protein 1 gene. Both disorders are more prevalent in males. FSX is an inherited syndrome with an X-linked dominant pattern. This means that the mutated gene is located on the X chromosome. It is called "dominant" since girls (who have two X chromosomes) can develop the disorder with just one copy of the X chromosome affected. However, disease manifestation is more prevalent in boys than girls (by about a factor of two) and is typically more severe in boys. ASD is almost five times more common in boys than in girls—the reason is unknown.

It is quite clear that environmental factors (probably through interaction with genetic vulnerabilities) contribute to autism etiology [13]. The early study of environmental causes was marred by academic fraud—a fraudulent publication claimed a link between childhood vaccines and autism. Those findings were concocted by a fraudster and have been proven to be wrong in a long series of careful scientific studies. However, exposure to certain chemicals may increase the risk for autism. For example, valproic acid, an antiepileptic (anticonvulsant) drug, which is also prescribed as a mood stabilizer, increases the prevalence of autism in offspring when taken during pregnancy. The highest vulnerability occurs a few weeks after conception. Rat models of valproic acid exposure exhibit structural and behavioral phenotypes similar to people with autism. At the structural level, changes in dendritic organization agree with the local hyperconnectivity and long-range hypoconnectivity discussed earlier. At the behavioral level, exposed rats exhibit a range of symptoms reminiscent of ASD, such as reduced social interactions, increased sensitivity to sensory stimuli, and exaggerated fear responses and memories. Another example of an environmental factor is organophosphate insecticides, such as chlorpyrifos, which have been shown to be detrimental to neurodevelopmental outcomes in rats. Similarly, epidemiological studies have shown association between maternal pesticide exposure and autism prevalence.

Understanding ASD and its cause(s) will require a broad approach that considers the interaction of multiple complex systems. For example, ASD has been hypothesized to relate to changes in the gut microbiota. This aligns with the poorly understood but very common finding of gastrointestinal symptoms in patients with ASD. Interestingly, decades of studies have been performed on the link between the gut-associated immune system, enteric nervous system, and gut-based endocrine system. However, these findings had not penetrated the fields of neurology and psychiatry until recently [14]. One of the earlier fundamental observations was that germ-free mice display an exaggerated response to stress that was reduced by colonization of the gut with bifidobacteria. Overall, recolonization of germ-free mice with normal microbiota or probiotic treatment reverses both biochemical (tryptophan metabolism) and behavioral (anxiety, sociability) deficits. Probiotic treatment has positive behavioral effects. Interestingly, these effects are strain specific, with *Bifidobacterium* and *Lactobacillus* appearing to be the most beneficial genera in terms of their effect on anxiety and depression-like behavior in mice. Despite the promise of these findings, it remains to be seen if changes in the microbiome cause CNS dysfunction. Despite the psychiatric and cognitive side effects of antibiotic use, the neuronal effects of changes in the microbiome induced by antibiotics remain mostly unstudied.

Developmental Origin of Connectivity Deficits

Differences in connectivity schemes may be related to differential brain growth trajectory in children with ASD and TD children. A large fraction of children with ASD exhibit pathologically accelerated brain growth that is limited to the first 2 years of their lives. As discussed earlier, larger brains may have relatively higher proportions of short-range to long-range connectivity. These altered growth trajectories follow an anterior—posterior organization pattern in which frontal areas are the most affected while occipital areas are the least affected. White matter differences appear to be most prominent for intrahemispheric fibers immediately underneath frontal areas, such as the dorsolateral prefrontal cortex. In terms of developmental trajectories, differences in head size disappear later during development, although people with ASD maintain elevated gray matter volumes and only limited increase in white matter. This contrasts with TD individuals, where gray matter volume decreases and white matter increases quite dramatically during development. At the structural level, one of the most consistent findings is a poor differentiation of the junction between the white and the gray matter. This can be caused by excess superficial white matter or pathologically increased number of *interstitial neurons*, defined as neurons in the white matter. Such an increase in interstitial neurons can be caused by pathologically arrested migration or by excess remnants of neurons in the fetal *subplate*, from which cortical neurons originate. The subplate is a transient neuronal structure that is fully established in the second trimester of human pregnancy and dissolves by postnatal month 6. The subplate is not only the source of cortical neurons, but it also represents a transient intermediate relay station of cortical afferents that form synapses onto neurons in the subplate before advancing their axonal terminals into the newly developed cortical layers. Notably, interstitial neurons remain in place in TD humans into adult life, particularly in frontal and prefrontal areas.

SUMMARY AND OUTLOOK

In this chapter, we learned that ASD affects a rapidly growing number of people, and that the underlying cause is likely to be a complex interaction between genetic and environmental factors. We reviewed evidence from different methods and levels of analysis that can be summarized in an overall model of increased short-range connectivity and decreased long-range connectivity. It is interesting to note that this overall pattern of long-range hypoconnectivity and short-range hyperconnectivity has also been reported in numerous other disease states.

In particular, "frontal disconnection" has been reported in many studies, including studies of patients with schizophrenia, attention deficit disorder, dyslexia, Down syndrome, depression, and HIV/AIDS.

References

[1] American Psychiatric Association, DSM-5 Task Force. Diagnostic and statistical manual of mental disorders: DSM-5. 5th ed. Washington, DC: American Psychiatric Association; 2013. xliv. 947 p.

[2] Centers for Disease Control and Prevention. Prevalence of autism spectrum disorder among children aged 8 years — autism and developmental disabilities monitoring network, 11 sites, United States, 2010. Morbidity and Mortality Weekly Report 2014.

[2a] Wang J, Barstein J, Ethridge LE, Mosconi MW, Takarae Y, Sweeney JA. Resting state EEG abnormalities in autism spectrum disorders. J Neurodev Disord 2013;5:24.

[3] Murias M, et al. Resting state cortical connectivity reflected in EEG coherence in individuals with autism. Biol Psychiatry 2007;62(3):270—3.

[4] Coben R, et al. EEG power and coherence in autistic spectrum disorder. Clin Neurophysiol 2008;119(5):1002—9.

[4a] Orekhova EV, Elsabbagh M, Jones EJ, Dawson G, Charman T, Johnson MH. The BASIS Team. EEG hyperconnectivity in high-risk infants is associated with later autism. J Neurodev Disord 2014;6:40.

[5] Casanova MF, et al. Minicolumnar abnormalities in autism. Acta Neuropathol 2006;112(3):287—303.

[6] Zikopoulos B, Barbas H. Altered neural connectivity in excitatory and inhibitory cortical circuits in autism. Front Hum Neurosci 2013;7.

[7] McFadden K, Minshew NJ. Evidence for dysregulation of axonal growth and guidance in the etiology of ASD. Front Hum Neurosci 2013;7.

[8] Nair A, et al. Impact of methodological variables on functional connectivity findings in autism spectrum disorders. Hum Brain Mapp 2014;35(8):4035—48.

[9] Wass S. Distortions and disconnections: disrupted brain connectivity in autism. Brain Cogn 2011;75(1):18—28.

[10] Shah A, Frith U. Why do autistic individuals show superior performance on the block design task? J Child Psychol Psychiatry 1993;34(8):1351—64.

[11] Mottron L, et al. Enhanced perceptual functioning in autism: an update, and eight principles of autistic perception. J Autism Dev Disord 2006;36(1):27—43.

[12] Parellada M, et al. The neurobiology of autism spectrum disorders. Eur Psychiatry 2014;29(1):11—9.

[13] Landrigan PJ. What causes autism? Exploring the environmental contribution. Curr Opin Pediatr 2010;22(2):219—25.

[14] Mayer EA, et al. Gut microbes and the brain: paradigm shift in neuroscience. J Neurosci 2014;34(46):15490—6.

Major Depressive Disorder

Major depressive disorder (MDD), often referred to as *depression*, is a serious mood disorder. Depression affects about 10% of the world population, with a higher prevalence in women than in men. Depression warrants particular interest by the network neuroscientist, because several Food and Drug Administration (FDA)-approved and experimental approaches are available to target and modulate the hypothesized network pathologies of depression by brain stimulation. The etiology of depression is poorly understood—depression appears to be caused by a complex interaction of genetic and environmental factors. In this chapter, we will learn about the symptoms of MDD and how they are assessed, we will focus on what is known about dysfunction in brain networks in depression, we will review what we know about the mechanisms and causes of depression, and we will conclude by providing an overview of the brain stimulation approaches to target the dynamics in depression. The toolbox "Psychiatry" provides helpful background about psychiatry in general.

SYMPTOMS OF DEPRESSION

The main symptoms of depression are depressed mood, loss of pleasure, feelings of guilt, suicidal ideation and intent, insomnia, and physical symptoms, such as loss of energy and fatigue. The most tragic consequence of depression is suicide; yearly in the United States twice as many people die by suicide than by homicide. First-line medication treatment is most commonly based on selective serotonin reuptake inhibitors. Not everyone responds to medication treatment, and a significant fraction of patients have *treatment-resistant depression* (TRD), defined as an incomplete response to an adequate antidepressant treatment.

Diagnosis of MDD is based on the Diagnostic and Statistical Manual of Mental Disorders, Fifth Edition (DSM-5) ([1], see toolbox: Psychiatry). One of the two main symptoms (depressed mood and loss of interest or pleasure) is required along with the presence of at least a subset of additional symptoms. Often, the *Mini International Neuropsychiatric Interview* (MINI) is used to assess and track psychiatric diagnoses. The MINI is a structured interview in which patients are asked to answer questions "Yes" or "No" (eg, "Were you ever depressed or down, or felt sad, empty or hopeless most of the day, nearly every day, for two weeks?"). The MINI is designed to map onto diagnoses defined by the DSM-5. In research studies, the most common assessment tools to track symptoms are the *Hamilton Depression Rating Scale* (HAMD) and the *Montgomery—Asberg Depression Rating Scale.*

These tools enable the quantification of disease severity. For example, the HAMD rates depressed mood (sadness, hopeless, helpless, worthless) on a scale from 0 (absent) to 4 ("patient reports virtually only these feeling states in his/her spontaneous verbal and non-verbal communication"). Overall, the two tools to assess depression are very similar and the scores accordingly correlate. The HAMD has been more frequently used to assess the impact of the treatment of acute symptoms. The Beck Depression Inventory (BDI) is a *self-assessment tool* that patients fill in. As a result, the BDI has limitations, since patients are known to modulate their answers as a function of the context of administration of the test. *TRD* is typically defined on a numeric scale from Stage 1 to Stage 5 [2]. To meet the definition of TRD Stage 1, failure of an adequate trial of one class of major antidepressant is required. To meet the criteria of Stage 5, failure of four different classes of antidepressants (including a tricyclic and a monoamine oxidase inhibitor) as well as electroconvulsive therapy is required (*Thase–Rush staging model*).

BRAIN NETWORKS IN DEPRESSION

Alpha Oscillations

Studying the resting state (participant instructed to "not think of anything particular") seems to be particularly relevant for depression since the passive resting state is associated with spontaneous processing of internal information. In terms of resting-state electroencephalogram (EEG), two frequency bands have been most often implicated in depression (and related mood disorders): alpha and theta rhythms ([3,4], see chapters: Alpha Oscillations and Theta Oscillations). Alpha oscillations (8-12 Hz) have been associated with decreased neuronal activity. In the simplest model of alpha oscillations in depression, the left and the right hemispheres represent opposite motivational drives. The left hemisphere is responsible for *approach motivation* and the right hemisphere is responsible for *withdrawal*. Therefore, alpha power asymmetry with larger alpha power in the left hemisphere (typically calculated as the logarithmic ratio of the power in the two hemispheres) would indicate reduced approach motivation, and has indeed been associated with depression [5]. Although alpha asymmetry is a quite robust finding associated with depression, not all studies have confirmed the role of alpha oscillations as a biomarker for depression. One unresolved question is whether alpha asymmetry more closely resembles a *state* or a *trait*. Alpha asymmetry is found in patients with depression, even during an epoch without symptoms, but also in healthy adults who, for example, score high on depression inventories, identify as shy, and have high cortisol levels. Therefore, alpha asymmetry may be less of a marker for depression, but more of a marker for vulnerability to negative affect (withdrawal).

Theta Oscillations

Theta oscillations (4-8 Hz) have also been implicated in depression. As a limitation, EEG measures from sensor signals exhibit poor spatial specificity and, therefore, they are difficult to integrate with imaging data that provides superior spatial resolution at the cost of temporal resolution. Methods of source localizations applied to EEG data help to bridge this gap (see chapter: LFP and EEG). One particularly interesting case is the role of activity in the rostral anterior cingulate (rACC, Brodmann Areas 24a and b). Measures of elevated activity in rACC by functional magnetic resonance imaging (fMRI) before treatment predict response to antidepressant treatments (both medication and brain stimulation). Through source localization, it was found that increased theta oscillations in the rACC reflect the increased activity levels determined with imaging. Specifically, when source localization with low-resolution electromagnetic tomographic analysis was used, elevated theta oscillations in rACC were robustly associated with treatment response [5a]. The rACC is a hub of the default mode network (DMN), which also includes the ventromedial and dorsal prefrontal cortex, posterior cingulate, retrosplenial cortex, lateral parietal cortex, lateral temporal cortex, and the hippocampal formations (see chapter: Imaging Functional Networks With MRI). The DMN is active in the absence of external tasks (therefore, it is also sometimes called the task-negative network) and has been associated with introspection and rumination, which can both be positive (mindfulness) and negative (self-analytic focus that increases depression).

Network Dynamics of Sleep

Sleep disturbances are a highly prevalent symptom in patients with depression [6]. At the level of network dynamics, the main aspects of altered sleep patterns in depression are reduced slow-wave sleep (SWS) with a shift of delta activity from the first to the second sleep cycle, and pathologically increased rapid eye movement (REM) sleep (eg, shortened latency for occurrence of the first REM episode and increased density of REM). Increased REM density, that is, frequency of REMs during REM sleep, is of particular interest, because it does not change with age, in contrast to most other aspects of sleep architecture. Therefore, studying the network dynamics of sleep can potentially lead to a definition of the endophenotype (robust and measurable characteristic) of depression.

Hyperactivity of the hypothalamic–pituitary–adrenocortical (HPA) system may be an important mechanistic contributor to sleep disturbances in depression, since corticotrophin-releasing hormone promotes wakefulness and REM sleep. Indeed, renormalization of HPA activity in patients with depression improves sleep. Sleep–wake transitions are conceptualized as the interaction of two processes: the circadian process, which is entrained by external stimuli (such as light), and the homeostatic process, which promotes sleep after wakefulness. Basically, several neuromodulatory systems promote wakefulness through the activation system (see chapter: Neuromodulators). REM sleep is a high cholinergic state that arises from a shift of the balance between cholinergic neurons that promote REM sleep and serotonergic neurons that prevent REM sleep. During waking, adenosine accumulates. Adenosine is a by-product of the degradation of adenosine triphosphate, the main cellular energy source used to maintain ion concentration gradients during neuronal activity. Therefore, the amount of adenosine produced is a function of neuronal activity. In this model, high levels of adenosine trigger sleep, during which adenosine levels decrease to baseline levels. As adenosine starts to accumulate, the activity of cholinergic neurons is inhibited, which leads to disinhibition of sleep-promoting neurons. Sleep deprivation has a rapid antidepressant effect since the resulting increase in adenosine may reduce the stress-induced arousal response and may restore the cholinergic and monoaminergic balance by increasing serotonergic neurotransmission. Clinically, response to both medication and psychological treatments for depression appears to be a function of the level of abnormalities in the sleep network dynamics. Both the REM and SWS disturbances appear to be trait markers of vulnerability since they are present before disease onset and persist even when the patient gets well.

MECHANISMS AND CAUSES OF DEPRESSION

Pathologies of the Stress Axis

Stressful life events are a major environmental risk factor for depression. Deregulation of the stress response in terms of activation of the HPA axis leads to elevated glucocorticoids, which in some patients with depression fail to activate a stress-limiting negative feedback loop. This view of depression focuses on stress-induced modulation of (subcortical) reward pathways. The model is motivated by the fact that many symptoms of depression can be explained by a dysregulated valuation process of stimuli leading to a failure to value positive stimuli (*anhedonia*) and increased negative valuation. Findings in animal experiments showed that chronic stress reduced excitatory synaptic input to medium spiny neurons in the nucleus accumbens (NAc). As a result, the afferent drive of the NAc to the ventral tegmental area (VTA) is reduced. The VTA plays a key role in the reward system since its neurons release dopamine (see chapter: Neuromodulators). However, the antidepressant effect of VTA activation remains controversial because optogenetic activation in rodents has led to differing findings across studies [6a]. Other subcortical structures included in this model are the lateral habenula and the amygdala. The lateral habenula directly excites the VTA and indirectly inhibits it via excitation of the inhibitory mesopontine rostromedial tegmental nucleus. The lateral habenula is activated by lack of an expected reward or punishment. Spontaneous action potential firing in the lateral habenula is increased in models of learned helplessness. The amygdala is critical for fear learning and is increased in volume and neuronal activity in patients with depression.

Animal Models and Tests of Mood Disorders

As with other psychiatric disorders, studying depression in animal models is challenging. Nevertheless, animal models that reflect certain aspects of depression have been successfully used to gain a mechanistic understanding of the neurobiological substrate of the disease [7]. For example, in the forced-swim test and the tail-suspension test, rodents are brought into situations from which they cannot escape. After a while, the animals stop trying to escape from the situation, a behavior that is interpreted as despair—a main symptom of depression in humans. Treatment of rodents with antidepressant medication prolongs the duration of these *escape-oriented behaviors*. Inactivation of the infralimbic cortex in rodents (roughly the equivalent of Brodmann area 25 targeted by one of the prevalent deep brain stimulation (DBS) approaches, see chapter: Deep Brain Stimulation) caused an antidepressant-like response. This process of testing findings from human studies in preclinical animal models is referred to as *reverse translation*.

Learned helplessness is modeled in rodents by exposing them to electric shocks without giving them the opportunity to escape. When the animals are subjected to subsequent electroshocks in an environment where they *can* escape, they make fewer attempts to escape in comparison to the control animals, which always had the opportunity to escape. Antidepressant treatment reverses such learned helplessness. Chronic, nonpredictable exposure to mild stress leads to a depressive phenotype in rats that is only reversed with chronic antidepressant treatment. Of note, these tests and models have high *predictive validity* since they respond to antidepressant treatment. But they are of low *face validity*, because the behavioral phenotype does not easily map onto the symptoms of MDD in humans.

BRAIN STIMULATION TREATMENTS

From a brain stimulation perspective, depression stands out as a psychiatric disorder for which there are several different brain stimulation modalities both clinically used and under investigation. First and foremost, electroconvulsive therapy (ECT) effectively treats depression. In ECT, patients are anesthetized and receive electric stimulation that causes massive electric discharges in the brain that resemble epileptic seizures. ECT is very successful for the treatment of even the most severe cases of depression, but its use is limited by the potential for cognitive side effects in some patients. The stimulation amplitudes used in ECT are a multitude of the electric current strengths used in investigational noninvasive brain stimulation approaches, such as transcranial direct current stimulation (tDCS, see chapter: Noninvasive Brain Stimulation). Despite its superior effectiveness for the treatment of depression, the mechanism of action of ECT remains ununderstood. Conceptually, ECT is often likened to a "reset" of pathological activity or functional connectivity in the brain of people with depression.

The clinical success of ECT has paved the way for investigation of other brain stimulation modalities for the treatment of MDD. Most prominently, repetitive transcranial magnetic stimulation (rTMS) is FDA approved for the treatment of depression. Typically, rTMS treatment consists of several weeks of daily stimulation sessions in which rTMS is used to target the left dorsolateral prefrontal cortex [8]. The rationale for this stimulation location reflects the model of impaired left frontal functioning in depression (approach motivation; see our discussion earlier of alpha oscillations [8a] as a network signal). The stimulation frequency of 10 Hz was chosen with the goal of enhancing activity and causing plastic reorganization that leads to sustained clinical benefits after completion of the stimulation treatment. More recent evidence of rTMS at 10 Hz entraining alpha oscillations has provided grounds for speculation that 10-Hz rTMS may (partially) work by targeting or resetting alpha oscillations. Experimental evidence in support of this theory is currently lacking. Despite its noninvasive nature, rTMS does not match the convenience of swallowing a pill. Instead, rTMS still requires a quite substantial commitment from the patient. The large number of required treatment sessions can conflict with the daily lives of the patients. In addition, the TMS technology is quite costly and is typically administered in specialized clinical centers. These limitations have inspired the study of tDCS for depression (and other psychiatric diseases [9]). The underlying rationale of tDCS for depression closely resembles the one for rTMS. Anodal tDCS is applied to the left frontal cortex in the hope of selectively increasing the pathologically decreased activity levels in the left dorsolateral prefrontal cortex. This investigational brain stimulation modality has promise because of its low cost and the portable nature of the stimulation technology. However, the clinical effects are less well characterized, but appear to be inferior to rTMS. Other investigational treatment approaches that have shown promise in initial studies include low-field magnetic stimulation and synchronized transcranial magnetic stimulation, which employs weak, periodic magnetic fields to target the individualized alpha oscillation frequency [10,11]. DBS for depression has also shown promise [12], although randomized, controlled studies so far have failed to show consistent treatment benefits ([13], see chapter: Deep Brain Stimulation). Current research endeavors are focused on identifying the network-level reasons why some patients show near-miraculous clinical benefits but others appear to fail to benefit [14]. Which of these investigational treatments will become a viable treatment approach approved by the FDA is unclear at this point. Nevertheless, the emergence of these different stimulation modalities further accelerates the recasting of depression into a network disorder.

SUMMARY AND OUTLOOK

In this chapter, we introduced the symptoms and the assays used to diagnose and measure depression severity. We then focused on the network-level pathologies in depression. In particular, we discussed changes in the symmetry of alpha oscillations as a potential marker for a pathological imbalance between approach motivation (left hemisphere) and withdrawal (right hemisphere). Theta oscillations detected along the midline also relate to depression. Finally, we discussed specific changes in sleep network dynamics, such as changes in REM sleep. All of these markers have so far failed to meet the standard of a robust endophenotype or biomarker to predict treatment response. Nevertheless, the findings about pathological changes to these network-level signals collectively represent what we understand about brain dynamics during depression, and therefore, they are of both translational and (in the future) clinical importance. Little is known about the etiology of depression, and animal models only capture certain aspects of the disease. Depression is unique in the sense that there is a wide array of brain stimulation treatments that span a range from daily and used around the world all the way to the highly investigational.

References

[1] American Psychiatric Association, DSM-5 task Force. Diagnostic and statistical manual of mental disorders: DSM-5. 5th ed., Washington, DC: American Psychiatric Association; 2013. xliv. 947 p.

[2] Nemeroff CB. Prevalence and management of treatment-resistant depression. J Clin Psychiatry 2007;68(Suppl. 8): 17−25.

[3] Baskaran A, Milev R, McIntyre RS. The neurobiology of the EEG biomarker as a predictor of treatment response in depression. Neuropharmacology 2012;63(4):507−13.

[4] Fingelkurts AA, Fingelkurts AA. Altered structure of dynamic electroencephalogram oscillatory pattern in major depression. Biol Psychiatry 2015;77(12):1050−60.

[5] Jesulola E, et al. Frontal alpha asymmetry as a pathway to behavioural withdrawal in depression: research findings and issues. Behav Brain Res 2015;292:56−67.

[5a] Pizzagalli D, Pascual-Marqui RD, Nitschke JB, Oakes TR, Larson CL, Abercrombie HC, Schaefer SM, Koger JV, Benca RM, Davidson RJ. Anterior cingulate activity as a predictor of degree of treatment response in major depression: evidence from brain electrical tomography analysis. Am J Psychiatry March 2001;158(3):405—15.

[6] Palagini L, et al. REM sleep dysregulation in depression: state of the art. Sleep Med Rev 2013;17(5):377—90.

[6a] Deisseroth K. Circuit dynamics of adaptive and maladaptive behaviour. Nature January 16, 2014;505(7483): 309—17.

[7] Slattery DA, Cryan JF. The ups and downs of modelling mood disorders in rodents. ILAR J 2014;55(2):297—309.

[8] Lefaucheur J-P, et al. Evidence-based guidelines on the therapeutic use of repetitive transcranial magnetic stimulation (rTMS). Clin Neurophysiol 2014;125(11):2150—206.

[8a] Thut G, Veniero D, Romei V, Miniussi C, Schyns P, Gross J. Rhythmic TMS Causes Local Entrainment of Natural Oscillatory Signatures. Curr Biol July 26, 2011;21(14):1176—85. http://dx.doi.org/10.1016/j.cub.2011.05.049.

[9] Kuo MF, Paulus W, Nitsche MA. Therapeutic effects of non-invasive brain stimulation with direct currents (tDCS) in neuropsychiatric diseases. Neuroimage 2014;85(Pt 3):948—60.

[10] Leuchter AF, et al. Rhythms and blues: modulation of oscillatory synchrony and the mechanism of action of antidepressant treatments. Ann NY Acad Sci 2015;1344:78—91.

[11] Leuchter AF, et al. Efficacy and safety of low-field synchronized transcranial magnetic stimulation (sTMS) for treatment of major depression. Brain Stimul 2015;8(4):787—94.

[12] Mayberg HS, et al. Deep brain stimulation for treatment-resistant depression. Neuron 2005;45(5):651—60.

[13] Dougherty DD, et al. A randomized sham-controlled trial of deep brain stimulation of the ventral capsule/ventral striatum for chronic treatment-resistant depression. Biol Psychiatry 2015;78(4):240—8.

[14] Smart OL, Tiruvadi VR, Mayberg HS. Multimodal approaches to define network oscillations in depression. Biol Psychiatry 2015;77(12):1061—70.

TOOLBOXES

Network neuroscience is an emerging, interdisciplinary field. The goal of this book is to be helpful to anyone interested in learning about network neuroscience. To achieve this aim, this unit provides you with the vocabulary and fundamental concepts required to join the field—independent of whether you are training (or trained) as a neuroscientist, biologist, physicist, neurologist, psychiatrist, psychologist, mathematician, or another related academic discipline.

The unit is structured into *toolboxes*, short chapters that provide you with the sets of tools needed to make best use of the main part of the book. Each toolbox starts with the question *"Is this toolbox for me?"* to guide you to which toolboxes are likely to be most important for you. Although you may enjoy reading all of the toolboxes to get a refresher on things you already know, they are structured as standalone items for you to pick and choose from.

The toolboxes are grouped to further help you identify which ones are most relevant for you. If you have a quantitative background but are new to neuroscience, the toolboxes to focus on are: Neurons, Animal Model Systems, Neurology, and Psychiatry. If you do not have a quantitative background, the most important toolboxes are: Matlab, Electrical Circuits, Differential Equations, Dynamical Systems, Graph Theory, Modeling Neurons, Physics of Electric Fields, and Time and Frequency. Note that these toolboxes do not waste time and do not introduce general theoretical concepts that are not directly applicable to network neuroscience. Rather, they are optimized by focusing on the specific concepts that are recurring themes in network neuroscience and thus in this book.

Toolbox Neurons

IS THIS TOOLBOX FOR ME?

To understand how brain networks function, we need to understand their components, particularly individual neurons. If you are new to neuroscience (and biology), this toolbox will equip you with the basic language describing the fundamental biological properties of neurons.

BLUEPRINT OF A NEURON

As other cells, neurons consist of a *cell membrane* (a *phospholipid bilayer*) and intracellular content called the *cytosol*. The cell membrane delineates the cell and provides electrical insulation so that neurons exhibit a *membrane voltage*, defined as the difference between the intracellular and extracellular electric potential. Ions move between the intracellular and extracellular space by passing through pores embedded in the cell membrane called *ion channels*. The *ion currents* modulate the membrane voltage for neuronal signaling (see chapter: Dynamics of the Action Potential). The shape, that is, *morphology*, of individual neurons can vary greatly and presents one of the key criteria used to classify neurons. Most neurons consist of a set of structures that receive input from other neurons: there are *dendrites*, the actual cell body that contains the nucleus, the *soma*, and a branched structure, the *axon*, which sends out electric signals to other neurons (Fig. A1.1). Both dendrites and axons can exhibit complex branching patterns. As a result, the set of dendrites is often referred to as the *dendritic tree*. Dendrites are covered with *synapses*, electrochemical junctions at which presynaptic axons from other cells communicate with the postsynaptic cell. Dendrites are also endowed with *intrinsic*, that is, nonsynaptic, ion channels that modulate the integration of the incoming electric input. In the absence of input, a neuron exhibits a *resting membrane voltage* of around -70 mV, meaning that the electric potential inside the neuron is 70 mV less than the electric potential outside the cell. The degree of overall *depolarization* (change in membrane voltage toward less negative values) determines whether the neuron generates an output signal, that is, an *action potential*. Typically, a neuron "fires" an action potential when its membrane voltage reaches a value of about -50 mV. An action potential is a brief spike in the membrane voltage (1–2 ms in duration) that reaches a value of about 50 mV and that propagates both along the axon and back into the dendrites.

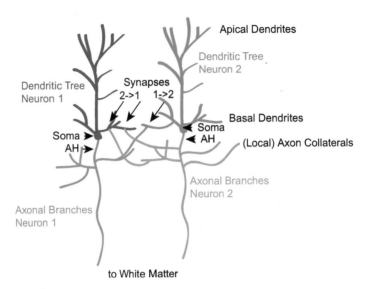

FIGURE A1.1   Schematic representation of two reciprocally connected pyramidal cells. The main morphological features are the apical and basal, local axon collaterals, and the axon projecting to other areas via white matter. Synapses are marked with an *arrow* (presynaptic → postsynaptic neuron). Dendrites are shown in *blue* and *green*. Axons are shown in *red* and *orange*. *AH*, Axon hillock.

DENDRITES

The morphology of dendrites differs by cell type (Fig. A1.2). Since the dendrites receive the majority of synaptic input, the location and size of the dendrites are often good indicators of the location and number of inputs that a neuron receives.

Transmission at one specific synapse causes a small electric transient in the post-synaptic membrane voltage near the synapse. It is important to remember that from the viewpoint of the postsynaptic soma, it is unknown from which of the potentially tens of

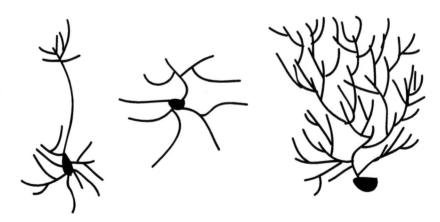

FIGURE A1.2   The dendritic arborization defines the morphology of the cell. *Left to right*: Pyramidal cell, stellate cell, and Purkinje cell. Only dendrites are drawn.

thousands of synapses the signal has originated. A neuron deals with all of these incoming signals by processing them locally where they arrive in the dendrites. Therefore extended dendritic trees allow neurons to receive different inputs and locally process them (since the postsynaptic signals rapidly decay with distance traveled in the dendrites). As discussed in the "chapter Synaptic Transmission," there are excitatory and inhibitory synapses that have opposite effects on the postsynaptic membrane voltage. Inhibitory synaptic input can cancel excitatory input if the synapses of both types are sufficiently close to each other. After this local processing, only the resulting, aggregated signal is forwarded to the soma and the axon hillock (Fig. A1.3). This propagation of the dendritic signal to the soma is associated with a suppression of the fast fluctuations in the signal (referred to as *low-pass filtering*). Also action potentials propagate back into the dendritic tree (backpropagating action potentials), where long-lasting dendritic action potentials can be initiated.

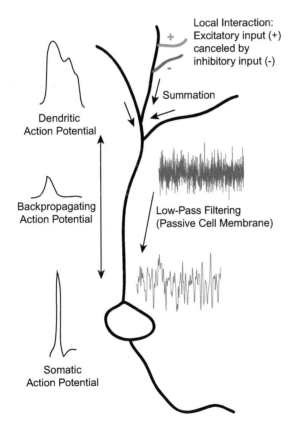

FIGURE A1.3   The dendritic tree performs a range of local and global computations (signal processing). Locally, excitatory, and inhibitory synapses can interact, allowing inhibitory input to cancel excitatory input. Signals are summed at dendritic branches. The dendritic membrane acts as a low-pass filter. In this process, the fast fluctuations in the dendritic membrane voltage (*green*) are removed and the signal arriving at the soma is smoothed (*blue*). Action potentials (from the soma) can propagate back into the dendrites (via voltage-gated ion channels). These back-propagating spikes can trigger long-lasting dendritic action potentials in the dendrites. The resulting depolarization of the dendrites modulates synaptic function (see chapter: Synaptic Transmission).

SOMA

The soma (plural: somata) contains the nucleus. The soma is the point of convergence of all incoming electric signals and the gateway to the axon. Typically, the soma is endowed with a rich set of intrinsic ion currents that determine the overall electric state of the neuron. Changes in the membrane voltage reflect processing of the input to the neuron and the resulting output. Neurophysiological recordings typically measure the somatic membrane voltage, since access with recording electrodes is comparatively easy because of the size of the soma (see chapter: Membrane Voltage). However, technical improvements in specialized techniques now allow for recording of the dendritic membrane voltage (at least from some of the larger main branches) and even axonal membrane voltage. Dendritic recordings make possible the direct measurement of the local dendritic computations. Axonal recordings can determine how the action potential is modulated by axonal ion channels.

AXONS

Action potentials are typically initiated at the proximal end of the axon, close to where it emanates from the soma (*axon hillock*). A characteristic of the axon hillock is the high density of ion channels that generate the action potential. Once an action potential is generated, it propagates down the axon and its branches. Axons can be very long (up to a meter) if they *project* (ie, are routed) to the spinal cord. The conduction velocity of uninsulated axons is quite low (up to 10 m/s). Electric isolation of the axon by *myelination* increases conduction velocity by up to a factor of 10. *Myelin* is a fatty white substance in the form of sheaths that surround the axon and is the source of the term *white matter* for axon bundles. Myelination is provided by glial (helper) cells, *oligodendrocytes*, in the case of the central nervous system. Both myelinated and unmyelinated axons exist in the central nervous system. The myelinated axon is endowed with regularly arranged zones with a high density of ion channels to actively regenerate the signal. These segments of the axon, *nodes of Ranvier*, lack myelination, which allows the flow of ion currents across the cell membrane (Fig. A1.4). Signal propagation in the myelinated axon is referred to as *saltatory conduction* (from the Latin *saltare*, to

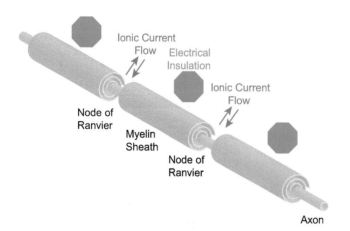

FIGURE A1.4 Myelinated axons are electrically isolated; the myelin wrapped around the axon acts as electrical insulation. At regular intervals (nodes of Ranvier, spaced by 1 mm), the axon is exposed such that ionic current can flow across the cell membrane (*green*) to amplify action potentials and enable fast propagation of the action potentials along the axon.

hop), since the action potential appears to "hop" from one node of Ranvier to the next node. Action potentials trigger transmission of signals between neurons at *synapses*, specialized sites where the axon of one cell contacts the dendrite of another cell.

Historically, the axon has been treated as a simple cable that propagates action potentials. But the axon includes more types of ion channels than the minimal set required for active signal transmission. These ion channels are also (as are other ion channels in neurons) susceptible to neuromodulation and plasticity. Axonal branches can project both locally (*axon collaterals*) and to long-distance targets (via white matter, ie, bundles of myelinated axons).

SYNAPSES

Although signaling in the nervous system is primarily based on electric impulses, the site of interaction between cells, the *synapse*, uses chemical signaling to transmit information from the presynaptic to the postsynaptic cell. Transmission of information at a *chemical synapse* follows a generic sequence of actions that converts the electric signal in the presynaptic cell into a chemical signal that is received by the postsynaptic cell and then transformed back into an electric signal. Action potentials travel down axons and reach the axonal terminals that form the presynaptic side of the synapse. The resulting depolarization of the axonal terminal opens ion channels that are selectively permeable to calcium ions, which flow into the cell. This transient and localized increase in intracellular calcium concentration triggers a sophisticated molecular process that eventually leads to the fusion of a *vesicle*—a small prepackaged sphere filled with a given type of neurotransmitter—to the presynaptic cell membrane. This fusion is followed by a subsequent release of the neurotransmitter into the synaptic cleft, which is typically only a few tens of nanometers wide. The neurotransmitter molecules diffuse across the synaptic cleft to postsynaptic receptors, which bind a single type of neurotransmitter. Excess neurotransmitter in the synaptic cleft is quickly removed by mechanisms on the presynaptic axon terminal, such as reuptake transport proteins. On the postsynaptic side, binding of the neurotransmitter to the receptor can trigger two distinct types of electric signaling mechanisms. For fast neurotransmission, the receptors are actually ion channels that change their state in response to the binding of the *ligand*, that is, the neurotransmitter. This allows for the flow of an ion current, which in turn generates an electric signal in the postsynaptic cell. Such channels are referred to as *ionotropic* receptors. In contrast, *metabotropic* receptors bind the neurotransmitter, but trigger an intracellular signaling cascade that can result in the modulation of other ion channels in the postsynaptic cell that generate the actual postsynaptic electric signal. In addition to chemical synapses, there are also electric synapses that form a direct electric connection between neighboring neurons. We discuss synaptic transmission in more detail in "chapter Synaptic Transmission."

SUMMARY AND OUTLOOK

In this toolbox, we reviewed the basic cellular biology of neurons. The main feature of neurons (in the context of network neuroscience) is their sophisticated electrical and chemical signaling. Synaptic inputs are small electrical signals caused by binding of neurotransmitter to receptors on the postsynaptic neuron. These inputs are processed and ultimately may or may not lead to an all-or-none electrical output signal, the action potential. Both the morphology of neurons and the ion channels in the cell membrane play an important role in shaping this process.

Toolbox Animal Models

IS THIS TOOLBOX FOR ME?

If you have a background in biology, you are already familiar with the concept of animal models and the questions related to choosing the right model system to answer a specific scientific question. However, if you have a different background and have not worked with animal models before, this toolbox will provide some helpful background on model systems frequently used in network neuroscience.

Animal models play a central role in biological and biomedical research. Unraveling biological complexity requires invasive procedures and measurements that ultimately benefit both humans and animals in the form of novel treatments for diseases. The choice of animal species to use to answer a given scientific question should not only be considered when designing experiments, but also when interpreting results from experimental studies. Here, we introduce some of the most common model species used in network neuroscience studies and discuss the advantages and limitations of their use in research. It is worth mentioning that the choice of animal model species encompasses more than just the underlying scientific rationale. Ethical, societal, and economic considerations play a role, as well.

In the United States, the number of animals used in research covered by the Animal Welfare Act (eg, cats, dogs, primates, but not rodents, such as mice and rats) has been steadily declining for the last two decades and has recently reached less than 1 million animals per year. To put this number in perspective, about 9 billion chickens are killed for food every year [1], and more than 350 million vertebrate animals are killed by cars every year in the United States alone [2]. Historically, a relatively broad set of animal species have been used for neuroscience and biomedical research. The advent of tools for genetic engineering has led to many researchers converging on common model species.

CAENORHABDITIS ELEGANS

The nematode *Caenorhabditis elegans* (*C. elegans*) is a harmless, 1-mm long worm that lives in the soil and feeds on bacteria (Fig. A2.1). Why does such a small worm play an important role in (neuro-)biology? Despite its small size, *C. elegans* exhibits many of the fundamental behaviors of interest to neuroscientists, and the small size of its nervous system makes it a highly tractable species. From a practical viewpoint, the short lifespan (ie, fast reproductive life cycle—development from egg to egg-laying adult in 3 days), the ease of growing it in a

FIGURE A2.1   Multiple *Caenorhabditis elegans* with several neurons labeled with green fluorescent protein. *Heiti Paves/shutterstock.com.*

Petri dish, and its transparent body that makes cells accessible to microscopy make *C. elegans* an attractive model species. Every adult *C. elegans* contains exactly 959 cells, of which 383 are neurons. The value of *C. elegans* as a model species was originally recognized by Sydney Brenner [3], who introduced this species to developmental biology and neuroscience and headed a research enterprise that (before the advent of modern computers) produced a complete circuit map of the *C. elegans* nervous system [4] based on reconstructions of electron microscopy data (together with John White).

The *C. elegans* genome is about 1/30th the size of the human genome in terms of number of base pairs and was the first to be fully sequenced. Limitations of *C. elegans* as a model system include the small size of its neurons and its hydrostatic skeleton (enclosing highly pressurized fluid), which make *C. elegans* nearly inaccessible for electrophysiological recordings. Indirect measurements of neuronal activity, such as optical measurements of intracellular calcium levels are typically used instead. Moreover, not all genetic pathways that guide development in the mammal are present in *C. elegans*.

In terms of neurobiological studies, *C. elegans* has been used to study the genes and mechanisms involved in development, sensory processing and behavior (eg, *thermotaxis*, motion along a heat gradient, *chemotaxis*, movement in response to chemical stimuli, mating rituals, feeding, and learning), and degeneration. The 383 neurons in the adult worm are connected by ~7000 chemical and electrical synapses.

Interestingly, none of the neurons express voltage-gated sodium channels (the genes for these channels are absent in the *C. elegans* genome). However, neurons express a wide variety of ion channels (including calcium channels that may generate action potentials) and employ a range of neurotransmitters, including glutamate, GABA, acetylcholine, dopamine, and serotonin.

FIGURE A2.2   Fruit fly (*Drosophila melanogaster*). *Studiotouch/shutterstock.com.*

FRUIT FLY

Similar to the nematode *C. elegans*, *Drosophila melanogaster* is an ideal model for the study of the genetic basis of brain structure and function (Fig. A2.2). The fruit fly develops from an egg into a larva, which undergoes three molts to emerge as an adult fly.

The use of *Drosophila* in neuroscience studies was initiated by Seymour Benzer [5], who leveraged the genetic tractability of the species for understanding the effect of mutations on behavior. Electrophysiological studies (both in the larva and in the adult) are easier to perform than in *C. elegans*. The fruit fly is a standard model for the study of neuronal development. In particular, the small number of neuronal stem cells (neuroblasts) in the larva, together with the wide array of genetic tools, fosters the study of differentiation into neurons in vivo. Uniquely identifiable synapses can be targeted and studied in the larval stages to understand synapse formation, stabilization, and removal. Beyond the study of development, fruit flies are also used for the study of visual processing, since they are highly visual animals (more than 50% of their brain is dedicated to visual processing). More complex behaviors and behavioral states, such as courtship, aggression, and circadian rhythms can also be studied in the fruit fly.

MOUSE

The house mouse (*Mus musculus*) is—together with humans—the most successful invasive species (Fig. A2.3). Mice have been used in biomedical research for centuries [6]. Currently, the mouse is perhaps the most popular model system in neuroscience research, for several reasons. The mouse genome is similar to that of humans (over 95% correspondence). Mice are a cost-effective model system and allow rapid scientific experimentation, since a year in the life of a mouse corresponds to about 30 human years. Today's laboratory mouse strains are a combination of different *M. musculus* subspecies, with the largest genetic contribution from

FIGURE A2.3   A mouse in its natural habitat. *Zoonar/shutterstock.com.*

the *M. musculus domesticus* from Western Europe. Typically, *inbred* mouse strains, for which all individuals are genetically identical, are used in research. The first inbred strains were created at the beginning of the 20th century as part of research that studied the inheritance of coat color. An inbred strain receives this designation after 20 or more consecutive brother–sister or parent–offspring matings. The availability of many hundred inbred strains enables exclusion of genetic differences as a driver of heterogeneity in experimental outcomes. Some of the most commonly used mouse strains in research are the BALB/c, the C3H, and the C57BL/6 strains. Furthermore, the variation across strains can be used to study how genetic differences account for differences in behavior (eg, social behavior, aggression), health outcomes (disease susceptibility), and many other variables of interest. These unique advantages of working with inbred animals come at a cost—mice are susceptible to *inbreeding depression*, defined as the loss of reproductive fitness as a result of inbreeding. Nevertheless, mice have become pervasive in neuroscience research because of the availability of a large number of genetically modified animals that enable the visualization and targeting of genetically defined cell populations. Typical values for the basic physiology of mice are shown in Table A2.1.

Mice consume up to 20% of their body weight in food every day. Feeding is spread over 200 small meals over 24 h. Grooming is another prominent behavior in mice. Self-grooming is important for hygiene, while grooming of others is used to communicate food preferences and to maintain social relationships. Mice avoid exposed areas where they are more likely to become the victim of predators; not surprisingly, the smell of carnivores elicits highly aversive behavior in mice. House mice are crepuscular (most active at dusk and dawn) or nocturnal. Mice sleep in short bouts and do not undergo prolonged periods of sleep as, for example, humans do. Mice are territorial animals, and male mice use urinary scent marks to convey information about social and reproductive status. Males are aggressive and defend their territory against unfamiliar intruders. In terms of the overall organization of brain areas, the mouse brain resembles the human brain. Aside from the obvious difference in size, however, there are many fundamental differences. For example, the mouse visual system

TABLE A2.1   Mouse Physiology

Body Weight	20–50 g
Body temperature	36.9°C
Respiratory rate	163 breaths per min
Heart rate	632 beats per min
Lifespan	Up to 3 years (typically 100 days in wild animals)
Sexual maturity	4 weeks
Estrus cycle	Every 4–5 days
Gestation	19–21 days
Litter size	Up to 10 pups

lacks the fine-grained organization of more visual mammals, such as humans. In addition, the mouse brain is lissencephalic (ie, it lacks cortical folding).

RAT

There are two wild rat species: the Norway rat (*Rattus norvegicus*), also referred to as the brown rat, and the black rat (*Rattus rattus*) (Fig. A2.4). The laboratory rat (derived from *R. norvegicus* [7]) has been prominently used in neuroscience research, since it is well suited for behavioral studies. In particular, the rat can be easily trained to perform tasks and respond to environmental stimuli. More recently, however, mice have replaced rats because of the availability of transgenic mice. Rats are poised to make a comeback with the advent of genetic manipulations in mammals other than mice. Similar to the mouse, there are hundreds of inbred rat strains. Among the most common strains are the F344 (SAS FISCH, albino), the Lewis rat (albino), and the SHR (spontaneously hypertensive rat). In addition, there are many strains used as disease models. For example, the genetic absence epilepsy rat from Strasbourg exhibits generalized, nonconvulsive epileptic seizures characterized by spike-and-wave discharges, behavioral arrest, and staring that resemble human absence seizures (see chapter: Epilepsy). Rats are semicontinuous feeders, and eat up to 5% of body weight every day. In contrast to mice, male rats are less prone to fighting. Rats are docile and very curious nocturnal animals. The most commonly used rat strains are *outbred* strains, namely, the Long Evans (black with white hood), Sprague–Dawley (albino), and the Wistar (albino) strains. Outbred animals are genetically heterogeneous since they are the offsprings of relatively unrelated animals that are bred with each other. Typical values for the basic rat physiology are shown in Table A2.2.

Traditionally, rodent behavior has been assessed with a number of basic tasks applied (with some variations) in both rats and mice (Fig. A2.5). Learning and memory are assessed with the *Morris water maze*, where the animal has to learn the position of a submerged, hidden (in cloudy water) platform relative to external environmental cues. The main outcome is the

FIGURE A2.4   Rat. *Vitalii Hulai/shutterstock.com.*

TABLE A2.2   Rat Physiology

Body Weight	500 g
Body temperature	37.0°C
Respiratory rate	75–115 breaths per min
Heart rate	260–400 beats per min
Lifespan	Up to 3.5 years
Sexual maturity	7 weeks
Estrus cycle frequency	Every 4–5 days
Gestation	20–22 days
Litter size	Up to 12 pups

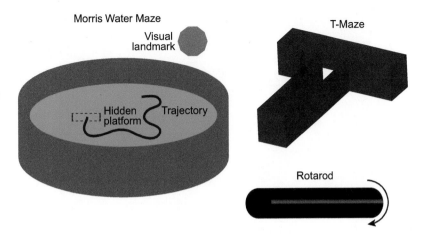

FIGURE A2.5 Apparatus for behavioral testing of rodents. The Morris water maze is used to assess memory since the animal needs to remember the position of the hidden platform relative to visual landmarks. The T-maze is a common experimental system in which the animal needs to make a decision at the junction between the stem and the bar of the T-shaped maze. The rotarod assesses functioning of the sensory—motor system, which is required to balance on the rotating rod.

time it takes the animal to swim to the platform. In the *T-maze*, the animal is required to navigate through a series of T junctions to find a reward. Sensory—motor function is assessed by the ability of the animal to balance on a rotating rod (the so-called rotarod performance test). The primary outcome is how long the animal manages to stay on the rod. Historically, mice and rats have not been widely used in vision research since their visual systems are poorly developed in comparison to other model species, such as the ferret, cat, and nonhuman primate. For example, acuity and contrast sensitivity of the laboratory rat are lower than in those species—rats do not have trichromatic color vision, lack a fovea (the area of highest density of photoreceptors in the retina), and detailed spatial organization in the visual system, such as orientation columns in the primary visual cortex is absent. Instead, the visual system, of the rat is optimized for low-light and low-resolution vision.

NONHUMAN PRIMATES

Rhesus macaques (*Macaca mulatta*) are the most frequently used nonhuman primate (NHP) species in neuroscience research (Fig. A2.6 [8]). Most macaque species live in southern and eastern parts of Asia, where they have successfully adapted to human civilization. This ability of macaques to adapt to changing environmental circumstances makes them an ideal model species since they are easy to breed and maintain. The typical values for the basic physiology of the rhesus macaque are shown in Table A2.3.

NHPs have brains that are very similar to the human brain. NHPs can be trained in complex tasks for the study of higher-order brain function in health and disease. For example, primates provide a far superior model system for studying Parkinson's disease than rodents. In addition, many advances in cognitive neuroscience would have been impossible without

FIGURE A2.6   Rhesus macaque (*Macaca mulatta*). *DPS/shutterstock.com.*

the use of NHPs in research. For example, the study of attention (the targeted allocation of processing resources to specific aspects of the sensory stream) in NHPs has delivered fundamental insights into the behavior of single neurons during otherwise inaccessible internal mental states. NHPs can be trained to allocate attention to a location in the visual field. The response properties (such as the firing rate) of the corresponding neurons are enhanced when the animal allocates attention to that location. Such a bridge from single-cell firing activity all the way to a higher-order state of "attention" is essential to understand the neural basis of behavior. Despite the rapid advances in using the mouse as a model species in neuroscience research, many important aspects of cognitive function (and overall behavior) cannot be studied in rodents, such as the complex neuronal mechanisms of gaze control, binocular vision, or motor control of complex limb and hand movements. As with any model species, the use of NHPs has its disadvantages. Their long life cycle makes the generation of genetically modified animals very difficult. It is unlikely that there ever will be such a broad

TABLE A2.3   Rhesus Macaque Physiology

Body Weight	7.7 kg (males), 5.4 kg (females)
Body temperature	37.3°C
Respiratory rate	37 breaths per min
Heart rate	215 beats per min
Lifespan	40 years
Sexual maturity	4.5–7 years (male), 2.5–4 years (female)
Estrus cycle	29 days
Gestation	165 days
Litter size	1

set of transgenic animals to allow precise targeting of genetically defined neuronal populations for recording and stimulation as in the mouse. In addition, the cost of maintaining primates for research is substantial. As a result, only a small number of laboratories use NHPs in their neuroscience research, to address questions that cannot be addressed in any other model systems.

Animal experiments, in particular in NHPs, pose complex ethical questions that deserve an open and honest conversation. Animal experiments are performed under very comprehensive and strict legal guidelines. Almost universally, animals used for scientific research are born and raised in captivity and are not captured from the wild. In the United States, all animal research needs to be carefully justified by investigators by demonstrating how they have applied the principles of reduction, refinement, and replacement (the *3Rs*) before permission is granted to perform an animal study under a heavily scrutinized and audited animal protocol.

SUMMARY AND OUTLOOK

In this toolbox, we introduced the most prominent model species for the study of brain networks. We considered the tradeoff between "simple" and "complex" model species. Simple species have only a small number of neurons and therefore provide the opportunity to characterize how behavior arises from the interactions within the entire nervous system. More complex animal species are used for the study of higher-order cognitive function. Feasibility for genetic manipulations has become one of the main driving forces for the choice model species. For example, neuroscience has shifted from the rat (ideal for studying behavior) to the mouse (ideal for genetic manipulations). With the advent of new genetic engineering tools, more major changes in the landscape of model species used in network neuroscience are likely to occur.

References

[1] Poultry — production and value 2014 summary, in National Agricultural Statistics Service. United States Department of Agriculture; 2015.
[2] Lalo J. The problem of road kill. Am For 1987;93(9):50−2.
[3] Brenner S. The genetics of *Caenorhabditis elegans*. Genetics 1974;77(1):71−94.
[4] Emmons SW. The beginning of connectomics: a commentary on White et al. (1986) 'The structure of the nervous system of the nematode *Caenorhabditis elegan*'. Philos Trans R Soc Lond B Biol Sci 2015;370(1666).
[5] Harris WA. Seymour Benzer 1921−2007 the man who took us from genes to behaviour. PLoS Biol 2008;6(2):e41.
[6] Hedrich HJ. The laboratory mouse. 2nd ed. Amsterdam, AP: Elsevier; 2012. xxi, 845 p.
[7] Suckow MA, Weisbroth SH, Franklin CL. The laboratory rat. In: American College of Laboratory Animal Medicine series. 2nd ed. Amsterdam, Boston: Elsevier; 2006. xvi, 912 p.
[8] Wolfensohn S, Honess P. Handbook of primate husbandry and welfare. 1st ed. Oxford, UK; Ames, Iowa: Horizontal Blackwell Pub; 2005. vi, 168 p.

Toolbox Neurology

IS THIS TOOLBOX FOR ME?

This toolbox introduces some of the most common neurological disorders and shows how neurologists assess the functioning of the nervous system. It pays to be aware of the differences in approach between neuroscience basic research and neurological clinical practice. Bridging the gap between these disciplines is the goal of translational research and requires a dialog between the fields. This toolbox aims to provide the required background to take the first step in this direction.

Important note: Nothing in this toolbox should be construed as medical advice. If you or someone you know has a health concern, please contact a medical provider. In the case of a medical emergency, call 911 (or the phone number for emergency services in your country) or visit an emergency room. Medications and their brand names are listed as examples and are protected by international patent and trademark laws.

Understanding brain networks and their pathologies is the starting point for developing novel, targeted therapeutic approaches. In this toolbox, we will learn the basics about common neurological illnesses and the field of neurology in general. Many (network) neuroscientists are familiar with epilepsy and Parkinson's disease. They are both network disorders characterized by a direct link between aberrant network dynamics and clinical symptoms. However, other common neurological disorders are becoming equally amenable to the tools of network neuroscience as we are learning more about these disorders. The overall aim of this toolbox is to paint the scene of how neurological disorders are diagnosed and treated in preparation for an accelerated, broader application of network neuroscience to neurology.

NEUROLOGICAL DISORDERS

To a neuroscientist, it seems surprising that disorders of the brain can fall into two seemingly distinct categories. *Neurological disorders* are defined as diseases of the central or peripheral nervous system. *Psychiatric disorders* are defined as diseases associated with abnormalities of thought, feeling, or behavior. This distinction is based on the historical assumption that the brain and the mind are separate entities. In reality, however, the brain and the mind are just two different ways to look at the same system. For example, changes in mood and personality have been historically assumed to reflect the emotional reaction to the life situation in patients with neurological disorders. Today, it has become increasingly clear that these psychiatric symptoms are (at least in part) caused by lesions or disturbances in specific brain circuits. But medical training and medical care are structured around the distinction between neurology and psychiatry. Nevertheless, there is recognition that some disorders defy such classification, because they include both "neurological" and "psychiatric" symptoms. These so-called neuropsychiatric disorders include epilepsy, dementia (including Alzheimer's disease), Parkinson's disease, multiple sclerosis, and migraines.

In addition, there are disorders and injuries that are primarily treated by specialists other than neurologists but can have neurological consequences (*sequelae*). This category includes cerebrovascular disease, infections of the brain (eg, meningitis), nutritional deficiencies, and physical traumas of the brain and spinal cord.

Next, we will discuss some of the most common disorders encountered by neurologists: dementia, headache disorders, multiple sclerosis, stroke, and traumatic brain injury. Reviewing these neurological disorders provides an understanding of the challenges ahead in finding new prevention and treatment approaches. Note that we omit discussion of Parkinson's disease and epilepsy in this toolbox, since they are covered by two separate chapters in the unit Network Disorders.

DEMENTIA

Alzheimer's disease (AD) is the most common form of dementia, but many patients exhibit mixed pathologies that also include signs of vascular dementia. AD is a slow-progressing disorder that includes memory difficulty as one of the prominent early symptoms. In later stages, more major behavioral changes occur, such as disorientation and confusion. In the final stages, basic motor functions, such as speaking, swallowing, and walking are also affected. The main features of neuropathology in AD are accumulation of beta-amyloid (plaques) in the extracellular space and twisted strands of the tau protein with intracellular localization. In only a very small subset ($\sim 1\%$) of patients, AD is caused by mutation in one of three genes (amyloid precursor gene, presenilin 1 and 2). Any such mutation will guarantee the development of AD, often much earlier than in the vast majority of patients who do not have any of the mutations. For the remaining 99% of patients, AD is likely to be the product of a multifactorial process, and several risk factors have been identified: age, family history of AD, apolipoprotein E-ε4 (ε3 is the most common form), cardiovascular disease, and low number of years of formal education. In 2014, an estimated 5.2 million people in the United States had AD, but only about half of them received a formal diagnosis from their physician [1]. To date, no treatment options are available that can change the disease course of AD.

HEADACHE DISORDERS

Headache disorders are subdivided into *primary* and *secondary*. Primary headache disorders include migraine and tension-type headache. Secondary headache disorders include headaches that result from other disorders (eg, head injury). Migraines are recurring headaches (termed *attacks*) that can include other symptoms, such as nausea and pathological sensitivity to light and sound. Migraines have a genetic component (the specific genetic loci for most patients are unknown), but the manifestation of migraine attacks is susceptible to environmental factors. Prevalence in women (18%) is much higher than in men (6%). This gender difference is not fully understood but it appears to be likely that hormonal fluctuations in women play a role. About a quarter of people with migraines experience an aura (often visual but can also include *aphasia*, that is, inability to speak, and other symptoms) before onset of the migraine. Trigger factors include certain foods, stress, change in sleep pattern, and sensory overstimulation. Migraines have been proposed to relate to a transient suppression of cortical activity caused by a breakdown in the potassium concentration gradient across the neuronal cell membrane (see chapter: Neuronal Communication Beyond Synapses). Tension-type headaches lack the specific features of migraine headaches and are described as pressure or tightness around the head. No mortality is associated with headaches, but the severe burden on affected patients and the high prevalence (approximately three-quarters of the population) make headache disorders an important frontier in neurological research [2].

MULTIPLE SCLEROSIS

Multiple sclerosis (MS) is a demyelination disorder of the CNS caused by a poorly un-derstood autoimmune process that attacks myelin (white matter) in the CNS. Diagnosis of MS is aided by magnetic resonance imaging (MRI) to probe for spatially distributed lesions of white matter that change over time. MS can take several different time courses: *relapsing/ remitting* MS (partial or complete remission after each attack) and *progressive* MS. Further-more, progressive MS is divided into *primary progressive* MS (lack of distinct attacks, gradual worsening) or *secondary progressive* MS (attacks associated with progressive worsening). People with MS exhibit a broad range of symptoms: fatigue, numbness or tingling, weakness, dizziness, gait difficulties, spasticity, vision problems, sexual problems, pain, bladder and bowel problems, cognitive changes, emotional changes, and depression. It is likely that MS is caused by a complex interaction between genes and the environment. The main puzzle and yet the best clue for identifying the cause of MS is that the prevalence scales with geographic latitude, with populations closer to the equator having lower prevalence. No treatment is available that alters the disease outcome. However, several drugs are available that signifi-cantly reduce disease burden in affected patients. These medications modulate the response of the immune system.

STROKE

Stroke is one of the most common causes of death in the western world [3]. Strokes can be classified into ischemic strokes (focal occlusion of a cerebral blood vessel, up to 80% of all cases) and hemorrhagic stroke (rupture of a blood vessel). In contrast to the other neuro-logical disorders discussed here, stroke is an acute medical emergency. The infarction process may take several hours, and therefore it offers a precious time window for early intervention that drastically improves overall outcomes. Stroke is largely preventable since the five clas-sical risk factors are: hypertension, smoking, physical inactivity, diabetes, and atrial fibril-lation (a form of irregular heartbeat). For example, aggressive efforts to reduce hypertension at the population level (community-based health screenings and education efforts) have led to a 70% reduction in strokes in Japan. Survival rate at 1 year after stroke is about 30%. Partial recovery occurs in multiple phases. The first phase is characterized by resolution of the ischemia and cerebral edema (swelling). The second phase is a rehabilitation phase (physical and occupational therapy), where neuronal plasticity enables relearning of lost function.

TRAUMATIC BRAIN INJURY

Application of external force to the head injures the brain. In mild cases there are no visible abnormalities in imaging; in moderate cases there is superficial bruising of the brain (*contusion*) while in severe cases there can be pronounced edema and collections of blood (*hematomas*). Globally, the leading causes for traumatic brain injury are traffic accidents, falls, and violence (eg, gunshot wounds). Similar to stroke, early and aggressive treatment is important to optimize long-term clinical outcomes.

DIAGNOSTIC TOOLS

In contrast to psychiatrists, neurologists have a larger set of tools available to arrive at the correct diagnosis, since most neurological disorders are associated with macroscopic, detectable damage to the CNS. With the advent of modern imaging technology (particularly MRI and computed tomography), diagnostic strategies have significantly shifted in neurology. A list of common diagnostic tools and strategies is provided in Table A3.1.

TABLE A3.1   Common Diagnostic Tools in Neurology

Exam	Medical history Neurological exam
Laboratory tests	Blood/cerebrospinal fluid by lumbar puncture. Detect signs of infection, antibodies, toxins, monitor blood level of drugs DNA testing
Imaging	Angiography. Insertion of catheter in artery in leg, threaded through the body into neck artery, radiopaque dye released. Detects blockages of arteries or veins Computed tomography. Composite 3D X-ray imaging. Detects bleeding, stroke, tumor Magnetic resonance imaging. Functional magnetic resonance imaging Ultrasound. Determines integrity of blood vessels
Electrophysiology	Electroencephalography. Noninvasive measurement of spontaneous cortical brain activity. Detects seizure activity and coma Electrocorticography. Invasive measurement of brain activity for localization of the origin of epileptic seizures and for identification of brain areas that would cause major neurological impairment if surgically removed Evoked potentials. Stereotyped cortical and subcortical responses to sensory input (same technology as in electroencephalography) Electromyography. Measurement of electric activity in skeletal muscles (typically maximal voluntary contraction). Surface or needle (intramuscular) recordings Nerve conduction velocity study. Stimulation and recording (at a distance) from a nerve to measure conduction velocity and thereby integrity of nerve

In spite of the growing role of technology, the neurological exam is still the centerpiece of neurological diagnosis. The neurological exam is a formalized, hands-on examination of the patient. It is divided into several aspects: general appearance, mini-mental status exam, cranial nerve testing, motor testing, sensory testing, reflexes, coordination, and gait. Assessment of general appearance includes observation of the patient and measurement of height, weight, and vital signs. The mini-mental status exam assesses cognitive impairment on a scale of 0–30 points and can be administered in less than 10 min. The testing of cranial nerves is summarized in Table A3.2. Sensory testing is performed by providing different sensory stimuli at different well-defined locations and asking the patient to describe the stimuli. Motor testing focuses on range of motion and strength of specific movements. Coordination is tested by specific motions, such as moving the index finger between two locations. Gait is examined by inspecting the patient standing (with eyes open and closed) and walking.

SUMMARY AND OUTLOOK

In this toolbox, we provided a brief overview of the most prevalent neurological disorders and discussed their etiology, symptoms, and diagnosis. For most disorders, effective treatment options are limited, and most available treatments address symptoms but do not provide a cure. We then looked at the diagnostic tools used by neurologists and found that—aside from the comprehensive neurological exam—several electrophysiology and imaging modalities used in research are also used clinically. This presents a valuable opportunity to bridge the gap between basic science and clinical activities. For example,

TABLE A3.2   Cranial Nerves and How Their Function Is Examined as Part of the Neurological Exam

Cranial Nerve (CN)	Tests
CN I (olfactory)	Identification of odor in test tube with eyes closed
CN II (optic)	Acuity (Snellen chart, letters of different size) Visual field (detect finger movement in periphery) Exam with ophthalmoscope (visual appearance of optic disk, retinal vessels, fovea)
CN III (oculomotor), IV (trochlear), VI (abducens)	Eye movements (follow pointer)
CN V (trigeminal)	Sensory branches: touch forehead, cheek, and jaw Motor branches: close and move jaw
CN VII (facial)	Look for facial asymmetries during smiling, frowning, etc.
CN VIII (vestibulocochlear)	Hearing test
CN IX (glossopharyngeal), X (vagus)	Swallowing and gag reflex
CN XI (accessory)	Shrugging shoulders, turning head laterally
CN XII (hypoglossal)	Tongue movement, deviation to one side

electrocorticography in humans is only used in a clinical context, but it has become a unique and powerful research tool (see chapter: LFP and EEG). Knowing the absolute basics of neurology as presented in this toolbox provides a starting point to learn more about neurology, to develop translational projects in the basic science laboratory, and to interact with neurologists.

References

[1] 2014 Alzheimer's disease facts and figures. Alzheimer's Association; 2014.
[2] Friedman D. Ten things that you and your patients with migraine should know. American Headache Society; 2015.
[3] Neurological disorders: public health challenges. World Health Organization; 2006.

Toolbox Psychiatry

IS THIS TOOLBOX FOR ME?

This toolbox will introduce you to the field of psychiatry by explaining how diagnoses are made and by briefly summarizing the standard treatments prescribed. It is beneficial to have basic knowledge of the clinical aspects of psychiatry, because many of the brain network disorders discussed in this book are psychiatric disorders.

Important note: Nothing in this toolbox should be construed as medical advice. If you or someone you know has a health concern, please contact a medical provider. In the case of a medical emergency, call 911 (or the phone number for emergency services in your country) or visit an emergency room. Medications and their brand names are listed as examples and are protected by international patent and trademark laws.

From a neuroscientific viewpoint, there are no differences between neurological and mental illnesses, because both result from pathological changes in brain structure and function. However, psychiatric diseases often do not cause physical symptoms. Historically, psychiatry has been much more influenced by nonmedical cultural norms and approaches. As a result, research of the biological mechanisms that cause psychiatric disorders has been significantly delayed in comparison to other medical disciplines. As we discuss the diagnostic and therapeutic tools in psychiatry, we will see that psychiatrists are forced to operate with a very limited toolset, and that more research on the neurobiological basis of these illnesses is urgently needed.

PSYCHIATRIC DIAGNOSIS VERSUS NEUROBIOLOGY

Psychiatric diagnoses are based on a comprehensive manual issued by the American Psychiatric Association. This comprehensive manual, the Diagnostic and Statistical Manual of Mental Disorders (Version 5, issued in 2013, abbreviated as DSM-5 [1]), is periodically updated and defines psychiatric illnesses by clustering symptoms and associating them with disease labels. The DSM-5 is divided into the following classes of disorders:

- Neurodevelopmental Disorders
- Schizophrenia Spectrum and Other Psychotic Disorders
- Bipolar and Related Disorders
- Depressive Disorders
- Anxiety Disorders

- Obsessive-Compulsive and Related Disorders
- Trauma- and Stressor-Related Disorders
- Dissociative Disorders
- Somatic Symptom and Related Disorders
- Feeding and Eating Disorders
- Elimination Disorders
- Sleep-Wake Disorders
- Sexual Dysfunctions
- Gender Dysphoria
- Disruptive, Impulse-Control, and Conduct Disorders
- Substance-Related and Addictive Disorders
- Neurocognitive Disorders
- Personality Disorders
- Paraphilic Disorders
- Other Mental Disorders
- Medication-Induced Movement Disorders and Other Adverse Effects of Medication

All diagnoses are made based on behavioral observations. To date, there have been no laboratory tests to establish a diagnosis of a specific psychiatric illness. This lack of objective metrics for diagnosis makes the DSM-5 vulnerable to being criticized as subjective and ambiguous. The nature of the DSM-5 also reminds us of the importance of the clinical skills of the psychiatrist who needs to establish correct diagnoses based on observations alone.

For example, the definition for *major depressive disorder* (often referred to as MDD, or just *depression*) lists the following criteria (cited verbatim from the DSM-5):

> Five (or more) of the following symptoms have been present during the same 2-week period and represent a change from previous functioning; at least one of the symptoms is either (1) depressed mood or (2) loss of interest or pleasure.
>
> **Note:** Do not include symptoms that are clearly attributable to another medical condition.
>
> 1. Depressed mood most of the day, nearly every day, as indicated by either subjective report (eg, feels sad, empty, hopeless) or observation made by others (eg, appears tearful). (**Note:** In children and adolescents, can be irritable mood).
> 2. Markedly diminished interest or pleasure in all, or almost all, activities most of the day, nearly every day (as indicated by either subjective account or observation).
> 3. Significant weight loss when not dieting or weight gain (eg, a change of more than 5% of body weight in a month), or decrease or increase in appetite nearly every day. (**Note:** In children, consider failure to make expected weight gain).
> 4. Insomnia or hypersomnia nearly every day.
> 5. Psychomotor agitation or retardation nearly every day (observable by others, not merely subjective feelings of restlessness or being slowed down).
> 6. Fatigue or loss of energy nearly every day.
> 7. Feelings of worthlessness or excessive or inappropriate guilt (which may be delusional) nearly every day (not merely self-reproach or guilt about being sick).
> 8. Diminished ability to think or concentrate, or indecisiveness, nearly every day (either by subjective account or as observed by others).
> 9. Recurrent thoughts of death (not just fear of dying), recurrent suicidal ideation without a specific plan, or a suicide attempt or a specific plan for committing suicide.

The symptoms cause clinically significant distress or impairment in social, occupational, or other important areas of functioning.
The episode is not attributable to the physiological effects of a substance or to another medical condition.

Reprinted with permission from the Diagnostic and Statistical Manual of Mental Disorders, 5th ed. (Copyright © 2013). American Psychiatric Association. All Rights Reserved.

Several things become apparent when reviewing the DSM-5's diagnostic criteria for depression. First, the diagnosis is an exclusion diagnosis, since the diagnosis of depression is only made once other (presumably better understood) physiological and pathological factors of depressed mood can be excluded as causes. Second, the vast majority of symptoms are behavioral symptoms that do not correspond in any obvious way to the underlying neurobiology, and as a result no biological measurements could be made to assist with the diagnosis. Third, the range of symptoms is quite diverse. Many different combinations of symptoms lead to a diagnosis of depression, and it is unclear whether or not all patients who receive this diagnosis exhibit similar underlying pathologies in the central nervous system. This observation highlights one of the main problems of psychiatry—the lack of objective diagnostic tools that are commonplace in other medical specialties. The National Institute of Mental Health has proposed a complementary framework that aims to organize psychiatric illness by the affected neurobiological substrate [2]. The reclassification of psychiatric illnesses according to this framework is still in its beginning stages. This framework is called the Research Domain Criteria (RDoC) project (Fig. A4.1). RDoC can be conceptualized as a matrix in which individual rows denote *constructs* that capture data on a specific aspect of behavior. Related constructs are organized into *domains*. RDoC consists of five domains: "Negative Valence Systems," "Positive Valence Systems," "Cognitive Systems," "Systems for Social Processes," and "Arousal/Regulatory Systems." For example, constructs for the domain "Negative Valence Systems" are: "acute threat, fear," "potential threat, anxiety," "sustained threat," "loss," and "frustrative non-reward." Each construct can be examined at different scales. These scales (units of analysis) correspond to the columns in the matrix: "genes," "molecules," "cells," "neural circuits," "physiology," "behaviors," and "self-reports." A column that lists established experimental paradigms to study each construct is also added to the matrix. The purpose of this approach is to provide a framework that guides and organizes neuroscientific research with the aim of understanding the underlying mechanisms that give rise to psychiatric symptoms.

TREATMENT FOR PSYCHIATRIC ILLNESSES

The main treatment modalities at the disposal of mental health care providers include medications (also referred to as *psychopharmacology*), psychotherapy, and brain stimulation. Often, combinations of these treatment approaches are used. All of these treatment approaches have their relative strengths and limitations, and it remains a formidable challenge for a health care provider to balance these treatment approaches to maximize the stabilization and recovery of people with mental illness.

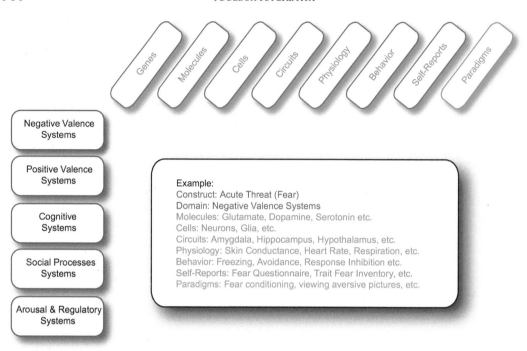

Negative Valence Systems

Positive Valence Systems

Cognitive Systems

Social Processes Systems

Arousal & Regulatory Systems

Example:
Construct: Acute Threat (Fear)
Domain: Negative Valence Systems
Molecules: Glutamate, Dopamine, Serotonin etc.
Cells: Neurons, Glia, etc.
Circuits: Amygdala, Hippocampus, Hypothalamus, etc.
Physiology: Skin Conductance, Heart Rate, Respiration, etc.
Behavior: Freezing, Avoidance, Response Inhibition etc.
Self-Reports: Fear Questionnaire, Trait Fear Inventory, etc.
Paradigms: Fear conditioning, viewing aversive pictures, etc.

FIGURE A4.1   Research Domain Criteria (RDoC) matrix with example construct "acute threat (fear)."

Psychiatric medications can be divided into the following classes: *antidepressants, antipsychotics, mood stabilizers, antianxiety drugs,* and *stimulants. Antidepressants* are used to treat depression, but they are also prescribed for a range of other psychiatric conditions, such as eating disorders and obsessive compulsive disorder. About 10% of the American population takes antidepressant medication. *Selective serotonin reuptake inhibitors* (SSRIs) are a relatively new group of antidepressants. Examples include fluoxetine (Prozac), sertraline (Zoloft), and paroxetine (Paxil). Possible side effects include flu-like symptoms, trouble sleeping, irritability, and weight changes. In adolescents, SSRIs have been associated with suicide risk. *Serotonin norepinephrine reuptake inhibitors* (SNRIs) represent an older, less specific group of antidepressant medications. SNRIs include venlafaxine (Effexor) and duloxetine (Cymbalta). Possible side effects and adverse reactions for SNRIs include altered mind states, such as thoughts of suicide, panic attacks, and hallucinations. Overall, the most frequent side effects of SSRIs and SNRIs are headaches, nausea, drowsiness, agitation, and reduced libido. The third type of antidepressant medication is the so-called *atypical antidepressants,* such as bupropion (Wellbutrin), which targets the dopaminergic system and therefore exhibits a different side-effect profile. Overall, antidepressant medications take several weeks to act. It is likely that their clinical effect is more complex than a simple change in a neuromodulator concentration, because those direct neuropharmacological effects are instantaneous and do not take weeks to develop. There are also older groups of antidepressant medications that have mostly fallen out of use, such as *tricyclics, tetracyclics,* and *monoamine oxidase inhibitors.*

Antipsychotic medications are used to treat psychotic disorders, such as schizophrenia and bipolar disorder. They are divided into two groups: *typical (first-generation) antipsychotics* and *atypical (second-generation) antipsychotics*. Examples of first-generation antipsychotics are D2 dopaminergic antagonists that also modulate other neuromodulatory systems, such as haloperidol (Haldol) and chlorpromazine (Thorazine). The second-generation antipsychotics are also D2 dopaminergic antagonists, but they have less severe movement disorder side effects. The reason for this difference is a point of ongoing debate. Examples of second-generation antipsychotics are aripiprazole (Abilify), quetiapine (Seroquel), and clozapine (Clozaril). These medications are still associated with other severe side effects. In particular, clozapine is a very effective medication for the treatment of psychotic symptoms, hallucinations, and breaks with reality. But it is associated with a life-threatening condition called *agranulocytosis*, in which the bone marrow fails to produce enough mature white blood cells (neutrophils). As a result, patients who take clozapine require frequent monitoring of their white blood cell counts. *Mood stabilizers* refer to a diverse set of agents that do not fall into any of the previously discussed categories and are predominantly used to prevent swings between mania and depression in bipolar disorder. Lithium is the classic mood stabilizer. Some of the medications developed for the treatment of epilepsy also stabilize mood. These so-called *anticonvulsant mood stabilizers* include lamotrigine (Lamictal) and valproic acid (Depakine). *Antianxiety medications* (tranquilizers) are used to treat severe anxiety. Most prominent are the benzodiazepines, which modulate synaptic inhibition (see chapter: Synaptic Transmission) and are potentially addictive. Psychotropic medications that fall into this category are clonazepam (Klonopin) and alprazolam (Xanax). Finally, *stimulants* are commonly used to treat attention deficit hyperactivity disorder. These medications stimulate the central nervous system and can be addictive. Examples are methylphenidate (Ritalin) and amphetamine (Adderall). These medications modulate the adrenergic and dopaminergic systems.

Aside from medication treatment, *psychotherapy* can also be highly effective for the treatment of mental illnesses. Psychotherapy refers to the interaction between a person with a mental illness and a mental health care provider in a therapeutic context. The interaction focuses on helping the patient to understand the factors that contribute to their illness and how to modify those factors, how to regain a sense of control over their lives, and how to learn both problem-solving and coping skills. Psychotherapy can assume many forms and can be administered in different contexts, such as individual, group, couple, or family therapy. Psychotherapy has been shown to be effective in clinical studies and is often combined with pharmacological treatment.

In addition to medication and psychotherapy, brain stimulation is used in psychiatry. With a few exceptions, most brain stimulation treatments are still under active investigation. Two main forms of brain stimulation are routinely used in psychiatric care, predominantly for the treatment of mood disorders. *Electroconvulsive therapy* (ECT) is administered to trigger electrographic seizures under anesthesia. ECT is very effective for the treatment of otherwise treatment-resistant depression. The underlying mechanism of ECT, which is often conceptualized as a "brain reset," remains unknown. ECT is only administered as a last resort because of the risk of cognitive side effects (in particular amnesia, ie, memory loss). Transcranial magnetic stimulation (TMS, discussed in detail in chapter: Noninvasive Brain Stimulation) is Food and Drug Administration approved for the treatment of depression. In a typical treatment course, the left dorsolateral prefrontal cortex is stimulated at 10 Hz in daily sessions of a few thousand pulses for 4—6 weeks. While TMS is an effective treatment, ECT is generally more effective, but it often has more adverse side effects.

SUMMARY AND OUTLOOK

In this toolbox, we discussed how psychiatric diagnoses are made based on symptoms using the DSM. We compared this approach with the RDoC initiative, which aims to organize research around neurobiological concepts that transcend the symptom labels of the DSM. We learned that this type of research is in its infancy and that psychiatry mostly lacks objective diagnostic tools. We reviewed some of the more common classes of neuropharmacological agents used in psychiatry: antidepressants, antipsychotics, mood stabilizers, antianxiety drugs, and stimulants. Most of these medications either interact with neuromodulatory systems, such as the serotonergic and dopaminergic system (see chapter: Neuromodulators), or they modulate excitatory or inhibitory synaptic transmission (see chapter: Synaptic Transmission). Psychotherapy and brain stimulation are also important treatment modalities in psychiatry. From a network neuroscience perspective, psychiatric disorders represent both a challenge and opportunity. Few new pharmacological agents are on the horizon, and the discovery of new treatment targets (such as specific network activity patterns) is urgently needed.

References

[1] American Psychiatric Association and American Psychiatric Association. *DSM-5 Task Force., Diagnostic and statistical manual of mental disorders: DSM-5*. 5th ed. Washington, DC: American Psychiatric Association; 2013. xliv, 947 p.
[2] Cuthbert BN, Insel TR. Toward the future of psychiatric diagnosis: the seven pillars of RDoC. *BMC Med* 2013;**11**:126.

Toolbox Matlab

IS THIS TOOLBOX FOR ME?

If you have seen Matlab or a similar (scientific) programming language, this toolbox probably contains no new information and can be safely skipped. However, if you have little or no prior training in scientific programming and you are joining the rapidly growing rank of neuroscientists who are facing more and more complex data sets to analyze, this toolbox is for you. The aim of this toolbox is to get you started. It will equip you to read more in-depth training materials offered online and to venture out on your own to practice with your own data.

Large data sets are routinely collected in studies of brain structure and function. For example, it is not unusual to record neuronal activity on hundreds of channels simultaneously, leading to very large data sets. As a result, spreadsheets for the analysis of biological data (eg, Microsoft Excel) have become obsolete and have been replaced with scientific programming languages that offer flexibility and computational power without requiring the user to learn a general-purpose programming language, such as C++ or Python. Key advantages of switching from a spreadsheet program to a scientific programming language for data analysis are (1) availability of more powerful analysis and visualization tools, (2) streamlining of analysis processes to make them easy to reapply and reproduce, and (3) easier sharing and collaborating on data analysis. There are several different scientific programming solutions. The commercial package Matlab (Mathworks, http://www.mathworks.com/) is one of the most prevalent. Many universities have a site license so that using Matlab comes at no cost for the end user, and there is also an affordable student version for readers without access to a site license. Furthermore, there is an open-source scientific programming environment called Octave; this program uses similar syntax (http://www.gnu.org/software/octave/). In this toolbox, we will introduce Matlab with a focus on the basic operations that are frequently used in data analysis, such as plotting and statistics. The code fragments shown in this chapter have been developed for Matlab R2015a. Note that Matlab commands do sometimes change with new versions of the program.

The Matlab main screen is divided into several windows (Fig. A5.1). The main window is the *command window* where the user interacts with Matlab. The window titled *current folder* displays the contents of the folder in which Matlab saves files and looks for files by default if not instructed otherwise (referred to as *working directory*). The *workspace* window displays all

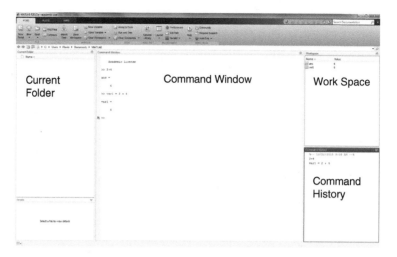

FIGURE A5.1 The main screen of Matlab is divided into windows. The most important window is the *command window*, in which the user interacts with the program. The windows *current folder* (content of the working directory), *workspace* (variables), and *command history* (previously executed commands) provide useful information when developing and debugging Matlab code.

variables (ie, data available to Matlab for processing; see later for a more formal definition). The *command history* window lists all previously executed commands.

Matlab can be used as a sophisticated calculator. In the command window, individual commands can be entered. Once the enter key is pressed, Matlab will perform the specified operation and provide the output on a new line in the command window. Most calculations or analyses will require more than just one line of code, and performing them line by line is inefficient; instead, Matlab *scripts* are used. Scripts are files (extension *.m) that contain all of the instructions to be carried out. "Running a script" will execute all of the commands in the script. There are two main advantages of using scripts instead of entering instructions in the command window. First, using scripts makes it very easy to rerun Matlab code, for example, on data from another experiment. Second, scripts make finding and fixing mistakes in the code (*debugging*) much easier.

In contrast to a calculator, Matlab has many more commands than just basic arithmetic functions. First, it allows the creation of data *variables*. A variable is a data storage location associated with a name. For example, if a script implements an algorithm that requires the sampling rate, we can create a variable that saves the value of the sampling rate. There are many advantages of using variables—sometimes a value is not known in advance, the same variable is used many times, or the value needs to be changed during calculations. Second, Matlab includes commands to direct the flow of execution of individual instructions in the script (*program flow*), like regular programming languages. For example, conditional statements (if...else...end) can be used to execute different pieces of code, and loops (for...end) can be used to efficiently automatize stereotyped analysis steps that need to be applied multiple times.

CREATING AND MANIPULATING VARIABLES

We start with a simple example of using the command window to execute a single line of code. The symbol >> denotes the location in the command window where the user enters code.

```
>> 2+4
ans = 6
>>
```

Matlab executes the command and assigns the response to the default variable, ans. You can also assign the output to a variable of your choosing. Note that the structure of the command is the variable name, followed by the equal sign, and finally the calculation we want Matlab to perform.

```
>> var1 = 2+4
var1 = 6
```

Now you can perform mathematic operations with this variable, for example, computing the square of var1.

```
>> var1^2
ans = 36
```

Often, the data of interest are more than just a single number (scalar). For example, a recording of the extracellular voltage of a single electrode is called a *time series*, in essence a 1 by N list of recorded values (samples) where N is the number of samples (see , toolbox: Time and Frequency). Such a list can be represented as a $1 \times N$ *matrix* (1 row, N columns) in Matlab. Generally speaking, a matrix is a rectangular array of numbers. An $M \times N$ matrix has M rows and N columns. If the matrix has only one row ($1 \times N$), it is also called a *row vector*. If the matrix has only one column ($M \times 1$), it is called a *column vector*. For example:

```
>> var2 = [1 2 3 3 0 -3]
var2 =          1        2        3        3        0       -3
```

Note that the square brackets [] are used to create a $1 \times N$ matrix (row vector). In case of several time series recorded simultaneously, we need an extra dimension so that each row corresponds to a given electrode. For M electrodes, the matrix will be an $M \times N$ matrix, such as (for $M = 3$, $N = 6$):

```
>> var3 = [1 2 3 3 0 -3; 2 4 5 5 4 3; 1 -4 -2 2 4 5]
var3 =

        1        2        3        3        0       -3
        2        4        5        5        4        3
        1       -4       -2        2        4        5
```

In this case, we use the semicolon to indicate where a new row of the matrix begins. Once such variables are set up, we can apply functions to the variables, such as the exponential function exp. These operations are executed on each element of the matrix individually.

```
>>exp(var3)
ans =

     2.7183       7.3891      20.0855      20.0855       1.0000       0.0498
     7.3891      54.5982     148.4132     148.4132      54.5982      20.0855
     2.7183       0.0183       0.1353       7.3891      54.5982     148.4132
```

Logical operations, that is, operations that have a false (0) or true (1) outcome, can also be applied to matrices. For example, if we want to determine which elements of the matrix are larger than zero in matrix var3, we can write:

```
>> var3>0
ans =
```

1	1	1	1	0	0
1	1	1	1	1	1
1	0	0	1	1	1

Note that we get back a matrix of the same size with 0s in elements that were less than or equal zero and 1s in the elements that are larger.

To read out specific elements of a matrix, we can use the following syntax: varName(x,y) where x is the row number and y is the column number. For example,

```
>> var3(2,3)
ans =
          5
```

Sometimes, we want to read out an entire column or an entire row. In this case, we replace the row or column number with a colon:

```
>> var3(:,1)
ans =
          1
          2
          1
```

Here, we specified that we want all rows and only the first column, thus getting back a column vector. Similarly:

```
>> var3(1,:)
ans =
```

| 1 | 2 | 3 | 3 | 0 | -3 |

gives us a row vector. We can also read out the values of a specific subset of elements. For example, if we want to read out only the first two rows, we can specify the range by the notation 1:2:

```
>> var3(1:2,:)
ans =
```

| 1 | 2 | 3 | 3 | 0 | -3 |
| 2 | 4 | 5 | 5 | 4 | 3 |

Sometimes, it is unknown ahead of time how large the matrix will end up being, but we know we want to access the last element. In this case, we can use the key word end:

```
>> var3(1,end)
ans =
         -3
```

to index the last element (in the example: the last element of row 1). There are many more tricks to address and reshape data matrices in Matlab; the documentation and help features that come with the program are very helpful to learn more about these strategies.

PLOTTING DATA

Matlab is great tool to quickly create a reasonable-looking plot. Here, we plot the data stored in matrix `var3`. For example, we can plot the data in `var3` as a set of lines. Note that Matlab assumes that each column corresponds to a line. If we want to plot each row of the matrix as a line, we can use the matrix transpose operation (`'`), which swaps rows and columns:

```
>>var3'
ans =

       1         2         1
       2         4        -4
       3         5        -2
       3         5         2
       0         4         4
      -3         3         5
```

Note that `var3` contains the values sampled at equal intervals and does not contain any information about time, that is, the time points at which the values were sampled. If we assume that each column in the original `var3` corresponds to a sample taken with a sampling rate of 10 Hz (100 ms apart), we can provide a second vector that will give us the correct time base (ie, labeling) on the x-axis, in milliseconds:

```
>> timeBase = [0 100 200 300 400 500]
timeBase =

       0       100       200       300       400       500
```

In fact, we can achieve the same thing by leveraging the colon operator:

```
>> timeBase = 0:100:500
timeBase =

       0       100       200       300       400       500
```

where we instruct Matlab to create a variable `timeBase` that has 0 as its first value and then increments with a step size of 100 until 500 is reached (same outcome, more convenient notation). When we use this with the plot function, we write:

```
>> plot(timeBase,var3')
```

The resulting plot is shown in Fig. A5.2. We can also quickly and easily make the plot more informative by changing some of its attributes:

```
>> plot(timeBase,var3','LineWidth',2)
```

Here, we set the attribute LineWidth to a value of 2 (many more attributes are listed in the documentation of the plot function). We wrap up this plot by adding a title and labels to the x- and y-axes:

```
>> title('Time Series Plot')
>> xlabel('Time [msec]')
>> ylabel('Amplitude [mV]')
```

Note that `title`, `xlabel`, and `ylabel`, are all built-in Matlab functions that we can call (execute). Functions typically require at least one argument, a piece of information that the

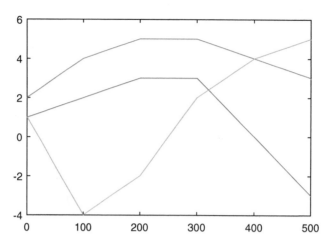

function uses. In the case of the title function, the argument is a string of letters, denoted with single apostrophes in Matlab. There are many more-sophisticated functions in Matlab useful for signal processing of neuronal data.

We next expand the range on the *y*-axis; again we call a function, ylim. This function takes a 1×2 matrix as an argument that contains the lowest and the highest value to be shown on the y-axis (note the square brackets that indicate a matrix as the argument):

```
>> ylim([-5 10])
```

As a last step, we add a legend (Fig. A5.3):

```
>> legend('Electrode 1','Electrode 2','Electrode 3').
```

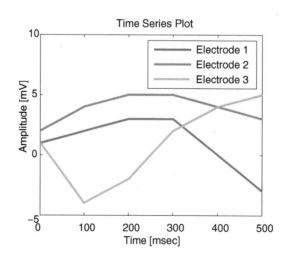

FIGURE A5.3   Matlab plot of time series for three electrodes (data saved in var3).

Once we are satisfied with the plot, save the plot as an *.eps file for further postprocessing in a drawing software tool:

```
>> print(gcf,'-depsc2','TimeSeriesPlot')
```

Print is a function that we call with three arguments. The first argument points to the figure (gcf gets us this information; it stands for "get current figure"). The second and third arguments are strings that specify the file type (color postscript) and the filename. If we do not provide a directory as part of the filename argument, the plot will be saved in the working directory.

We can now save all of these instructions; the next time we want to make a similar plot, we can simply execute the corresponding script. We create a new empty script with Ctrl + N or by choosing New → Script in the toolbar:

```
% Script to plot multiple time-series
% Version 1.0
% Data to plot
var3 = [1 2 3 3 0 -3; 2 4 5 5 4 3; 1 -4 -2 2 4 5];
% Time-base vector
timeBase = 0:100:500;
% Plot data
plot(timeBase,var3','LineWidth',2)
% Enhance plot
xlabel('Time [msec]')
ylabel('Amplitude [mV]')
title('Time Series Plot')
ylim([-5 5])
ylim([-5 10])
legend('Electrode 1','Electrode 2','Electrode 3')
print(gcf,'-depsc2','TimeSeriesPlot')
```

We added extra lines of text, *comments,* by starting the line with the percent sign (%). These lines are ignored by Matlab but make it easier to read and thus share code. For visualization, Matlab displays comments in green by default. We can save and run the script by pressing F5 or clicking the "Run" button in the toolbar.

Of course, there are also other ways to display the same data. One type of plot often used in network neuroscience is a *heatmap.* In this type of plot, the data depends on two variables (such as, in our case, electrode number and time point) and is represented with color in a matrix-like display:

```
>> imagesc(timeBase, [1 2 3], var3)
>> colorbar
```

Note that the function imagesc here is provided with the values for the x-axis and the y-axis (we had to create a vector with the numbers 1 to 3 to number the electrodes). The command colorbar provides a colorscale to interpret the colors in the plot (Fig. A5.4).

PROGRAM FLOW

Now that we have encountered matrices, plotting, functions, and basic scripting, we will turn to what makes Matlab more a programming language than a calculator. We will first

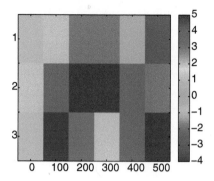

FIGURE A5.4   Heatmap representation (imagesc) of the data shown in Fig. A5.3. The colorbar to the right shows how numeric values are mapped onto colors.

consider *loops*, and then we will look at *branch points*. Loops are a convenient way to apply the same instructions over and over again. Although you could just copy-paste the code multiple times in a script, creating a loop not only increases readability, but it is also very helpful if we do not know beforehand how many iterations will be needed. For example, if we want to create an individual plot for every row vector in our variable var3, we would use the for...end instruction:

```
for iRow = 1:3
        figure(iRow)
        plot(timeBase,var3(iRow,:),'LineWidth',2)
end
```

Following the for keyword, a counter variable iRow is created that is incremented each time the code within the loop is executed (the commands between the for and the end statement). In this case, we count from 1 to 3 and therefore execute the code inside the loop three times. Within the loop, we have access to the counter variable. On the first row of code inside the loop we use the command figure to make sure all plots go on a new figure. We number them by the value of iRow (1 on the first iteration, 2 on the second, and so on). With the second line inside the loop, we plot the data as before. The one difference is that we only plot one of the three rows using the indexing strategy discussed earlier. Instead of writing a number for the row, we use the variable iRow, which stands for 1, and thus the first row on the first iteration of the loop and so on. We use the colon operator to address the entire row (ie, all column indices).

We can make the code more general for the case where we do not know ahead of time how many rows of data there are to plot (and therefore the number of needed plots). We use the command size, which we can use to ask for either the number of rows or columns by using the number "1" or "2" as the second argument. Here we use (var3,1) to get the number of rows:

```
>> size(var3,1)
ans =
      3
```

The new, more general code then reads:

```
for iRow = 1:size(var3,1)
        figure(iRow)
        plot(timeBase,var3(iRow,:),'LineWidth',2)
end
```

The `for` loop is ubiquitous in Matlab (and most types of) code. In addition to loops, branch points are another important tool for the control of program flow. Branch points are used to provide a means to execute different code(s) as a function of information that is not available when the code is written, but will be when the code is executed (called run-time).

For example, if the row vectors correspond to measurements of extracellular voltage, each row may stem from either an electrode in the prefrontal cortex (PFC) or an electrode in the hippocampus (HC). This information can change from data set to data set; we want to be able to process data from different recordings and experiments with the same code. Let us assume that the location information is provided as an array of labels (character strings defined by bracketing single apostrophes '), a so-called cell array:

```
>> location = {'PFC','HC','HC'}
location =     'PFC'       'HC'        'HC'
```

Cell arrays are very similar to normal data matrices with the exception that there is more flexibility as to what can be stored in an individual element (here, eg, entire strings). They are created by using curly brackets and also addressed by using curly brackets {}. Cell arrays are typically used when the dimensions of the data to be stored in individual elements are different (eg, the two characters of 'HC' versus the three of 'PFC'). The indexing scheme is the same as for regular matrices:

```
>> location{1,2}
ans = HC
```

With this information, we can execute different code for data from the two brain areas. We use the function `strcmp` to compare two strings. The function returns a value of 1 if they are identical and 0 if they are different:

```
>> strcmp(location{1,2},'HC')
ans = 1
>> strcmp(location{1,1},'HC')
ans = 0
```

A statement that equals logical 1 after the key word `if` will execute the subsequent code until the occurrence of (1) the key word `else` that defines the code that gets executed if the statement instead equals logical zero or (2) the key word `end` that concludes the `if` statement. We use this functionality to create the correct location titles for the plots:

```
for iRow = 1:3
        figure(iRow)
        plot(timeBase,var3(iRow,:),'LineWidth',2)

        if strcmp(location{1,iRow},'HC')
                title('Hippocampus')
        else
                title('Prefrontal Cortex')
        end
end
```

BASIC STATISTICS

Next, we will look at how to perform basic statistics in Matlab. *Descriptive statistics* summarize a data set—the most common metrics are mean, median, and standard deviation. *Statistical tests* are used to make quantitative decisions about data by determining if a given hypothesis can be rejected.

We will again create a mock data set to work with for further explanation. We assume that we have measured the reaction time (RT) in a working memory task in two conditions (with "sham" and "verum" transcranial direct current stimulation of the left frontal cortex). We further assume that 10 participants completed the study such that we have an average reaction time for both conditions for all participants. This corresponds to a "repeat measure, within subject" design, which generates two numbers per participant:

```
>> RT_verum = [346      370      314      346      376      354      344      365
384      322];
>> RT_sham = [435      392      436      375      389      388      397      370
360      411]
```

Starting with the descriptive statistics, we compute the mean and the median for both conditions:

```
>> mean(RT_sham)
ans = 395.3000
>> median(RT_sham)
ans = 390.5000
>> mean(RT_verum)
ans = 352.1000
>> median(RT_verum)
ans = 350
```

Similarly, we determine the standard deviation for both conditions:

```
>> std(RT_sham)
ans = 25.5693
>> std(RT_verum)
ans = 22.5509
```

We visualize the outcomes as histograms created with the command hist:

```
subplot(2,1,1)
hist(RT_sham, [300:20:460])
xlabel('[msec]')
yalebl('Count')
subplot(2,1,2)
hist(RT_verum, [300:20:460])
xlabel('[msec]')
yalebl('Count')
```

We introduced the command subplot that allows for plotting several graphs in the same figure. These subplots are numbered from left to right, top to bottom. The first two arguments of this command denote the arrangement of the individual plots (here: 2 rows and 1 column),

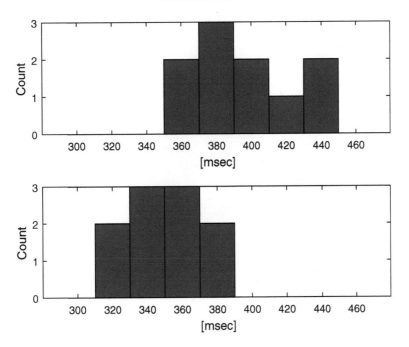

FIGURE A5.5   Plot of distributions using the hist command.

and the third argument is the number of the subplot used for the subsequent plotting command. The hist function is used with two arguments, where the first is the data vector and the second defines the binning of the histogram (Fig. A5.5).

We can now perform a paired t-test that determines with what probability the two matched distributions (sham and verum) are of equal mean. Typically, if this probability is lower than 5% ($p < 0.05$) the difference is denoted as statistically significant:

```
>> [h p] = ttest(RT_sham, RT_verum)
h = 1
p = 0.0139
```

The function ttest provides a variable h, which denotes whether the difference is significant (value 1, ie, true) and the associated *p*-value. We find that the *p*-value is smaller than 0.05, and therefore we can state that the reaction times for the stimulation (verum) condition were significantly lower than for the sham condition.

If the data were not normally distributed (Fig. A5.6), a so-called *nonparametric test* would be used. Nonparametric tests do not assume the data to be normally distributed. For matched samples, the Wilcoxon signed-rank test is used (signrank):

```
>> [p h] = signrank(RT_sham, RT_verum)
p = 0.0137
h = 1
```

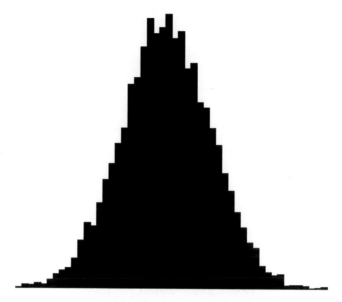

FIGURE A5.6   Data with a *normal distribution*. The normal distribution is also referred to as *Gaussian distribution*.

If the samples came from two different groups of participants, the unpaired versions of these tests would need to be applied. The parametric version for unpaired data is the Matlab function ttest2:

```
>> [h p] = ttest2(RT_sham, RT_verum)
h = 1
p = 8.2685e-04
```

For parametric data, the unpaired version is the Wilcoxon rank sum test (ranksum):

```
>> [p h] = ranksum(RT_sham, RT_verum)
p = 0.0019
h = 1
```

These and many other functions are described in the Matlab documentation. This documentation can be accessed by typing doc followed by the function name in the command window. For example, doc ranksum will show us that the order of p and h are switched in this function in comparison to the order used for ttest2.

SUMMARY AND OUTLOOK

Knowing a scientific programming language is essential in network neuroscience. In this toolbox, we provided a brief introduction to Matlab, a general-purpose platform for scientific computing. We discussed basic programming concepts, such as variables and functions. We created plots and calculated basic statistics of a sample data set. Typically, once these first hurdles are overcome, Matlab is quite intuitive and ideal for rapid development of code for data analysis. There is a large user community online that provides support. Note that multiple excellent alternatives to Matlab are emerging. We have not discussed any Matlab alternatives (eg, Python) in this toolbox because the overall concepts are very similar to those used in Matlab.

NOTES

- An excellent follow-up to this toolbox is the book *Matlab for Neuroscientists* [1]. This book introduces a broad set of Matlab capabilities in the framework of neuroscience in terms of both data analysis and computational modeling.
- Microsoft, Excel, Matlab, and Mathworks are registered trademarks. Matlab and Excel are protected by a number of US and international patents and copyrights.
- GNU Octave is provided under the GNU General Public License and is free software supported by the community.

References

[1] Wallisch P. MATLAB for neuroscientists: an introduction to scientific computing in MATLAB. 2nd ed. Amsterdam: Academic Press; 2014. p. 550.

Toolbox Electrical Circuits

IS THIS TOOLBOX FOR ME?

If resistors, capacitors, and current sources are not part of your everyday vocabulary, you will benefit from reading this toolbox to learn a powerful method to understand electrical signaling in neurons: modeling them as electrical circuits.

The brain is an electrical organ. Absence of electric signaling in neurons is a sign of deep anesthesia, coma, or death. It therefore makes sense as a neuroscientist to borrow strategies originally developed for designing electric devices. In this toolbox, we will discuss the basics of electric circuits. Our efforts focus on learning about electric circuit diagrams, which are symbolic abstractions of electrical circuits. Understanding how to read a basic electric circuit diagram, relate current to voltage, and ultimately model ion channels and cell membranes as electric components is incredibly helpful for understanding the neurophysiology of individual neurons and networks of neurons (see chapters: Membrane Voltage and Dynamics of the Action Potential). The main strength of this approach is that it allows for abstraction and quantification of neuronal and synaptic properties—meaning that we will be able to focus on functionally relevant mechanisms. Such electrical circuit modeling has been very successful for understanding what makes a neuron fire an action potential or respond in a specific way to incoming synaptic input. Electrical circuit "equivalents" (ie, models and representations) of neurons enable us to directly compute, and therefore predict, the behavior of neurons.

The main strength of electric circuit diagrams is that they abstract from the underlying concepts from physics and provide a mathematical representation of how voltage V and current I relate to each other. Voltage is measured in volts (V, or more commonly in millivolts, mV, in electrophysiology) and is defined as the difference in electric potential between two points in space. Current is measured in amperes (A, nanoamperes, nA, in electrophysiology) and is defined as the amount of charge per time that flows through a point. In neurophysiology, voltage represents our signals of interest. For example, in the case of neurons, we are most interested in membrane voltage V_m, measured as the difference in *electric potential* between inside and outside the cell. V_m tells us the state of the neuron. Most importantly, the action potential is a brief spike in the V_m and represents the output signal of neurons. If we want to understand how the action potential is generated, we study the ion channels that permit ionic currents to flow across the cell membrane, which affects the membrane voltage.

This toolbox is structured in three parts. First, we will introduce the main electric components used to model neurons. Second, we will discuss the rules that guide current and voltage when we connect these electric components. Third, we will apply our knowledge of what we have learned by building a model of a passive cell membrane (ie, not containing voltage-gated ion channels). Electrical models of ion channels are discussed in the toolbox: Modeling Neurons.

ELECTRIC COMPONENTS

Inspecting the circuit diagram of any electronic device, we find hundreds of symbols that stand for different electric and electronic components. These symbols tell us how current and voltage relate to each other, while they ignore other information about the components, such as manufacturer, size, and color. These symbols enable a trained engineer to understand the function of any circuit by inspection of the circuit diagram. The good news is that there are only four components we need to understand to model neurons as electric circuits. First, we will discuss two *passive* components (defined as not producing power), the resistor and the capacitor. Then, we will learn about two *active* components (defined as producing power), the voltage source and the current source.

RESISTOR

A *resistor* is a simple electric device that impedes the flow of current. The official symbol of a resistor used in electrical circuit diagrams is a zigzag line. For a given voltage drop "across" the resistor (ie, electric potential difference between the two ends or terminals of the resistor), only a well-defined amount of current flows through the resistor (Fig. A6.1).

The relationship between voltage and current is given by Ohm's law:

$$I = \frac{V}{R} \quad V = IR \tag{A6.1}$$

The constant R is called the *resistance* (electric property of a resistor). The physical unit of resistance is the ohm, abbreviated with the Greek letter Ω. The larger the resistance, the less

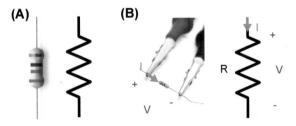

(A)   **(B)**

FIGURE A6.1   (A) Resistor (*left*) and symbol for resistor (*right*). The colored rings on the resistor encode the resistance *(Leo Blanchette/shutterstock.com)*. (B) Voltage V is measured across the resistor (positive voltages are usually applied using red wires while zero or negative voltages are black); the current I flows through the resistor R (*left*: physical reality, *right*: circuit diagram abstraction) *(Ziga Cetrtic/shutterstock.com)*.

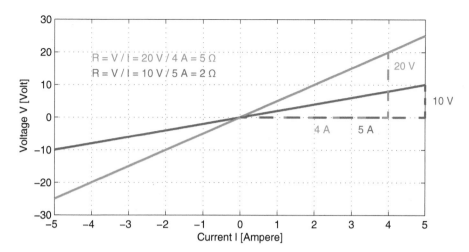

FIGURE A6.2 Ohm's law. Voltage and current are proportional to each other with constant R (resistance). Voltage as a function of current is shown for $R = 2\,\Omega$ (*blue*) and $R = 5\,\Omega$ (*red*).

current can flow. If the resistance is infinite, the current is zero (perfect insulation). If the resistance is zero, the current is infinite, which cannot occur in real situations. This case is called an "electric short" and what happens practically is that as much current as possible flows until something explodes, melts, or otherwise interrupts current flow. Mathematically, the relationship between current and voltage in the resistor is linear; voltage plotted as a function of current is a straight line with slope R (Fig. A6.2).

Importantly, the relationship between voltage and current is *instantaneous* at a resistor (Fig. A6.3). Any change in current will be immediately reflected in the voltage (and vice versa).

In neurobiology, we often use the term *conductance*, G, which is defined as the inverse of resistance:

$$G = \frac{1}{R} \tag{A6.2}$$

The physical unit of conductance is the siemens, abbreviated with the letter S. The definition of conductance is intuitive—if conductance is high, charge carriers can move more freely and the resistance is low. Alternatively, if conductance is low, resistance is high and prevents movement of charge carriers. When using conductance instead of resistance, Ohm's law remains the same (Fig. A6.4), with the only difference being that voltage V is treated as the independent variable (plotted on the x-axis) and the current I as the dependent variable (plotted on the y-axis). The slope of the line in this so-called $I-V$ plot is G:

$$I = GV \tag{A6.3}$$

Although there is no electric current generated by freely moving electrons in the brain, the framework of resistance and conductance is extremely useful for understanding cellular

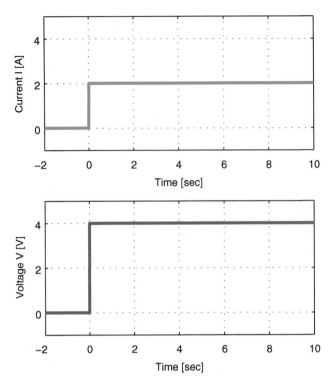

FIGURE A6.3   Time courses of current (*red*) and voltage (*blue*) at a resistor.

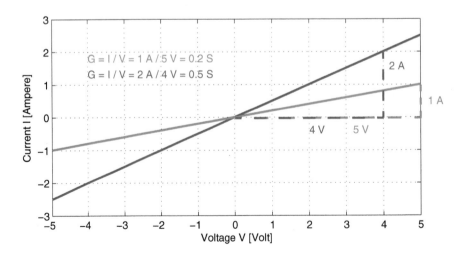

FIGURE A6.4   *I–V* plot. Voltage and current are proportional to each other with constant *G* (conductance). Current as a function of voltage is shown for $G = 0.2\,S$ (*red*) and $G = 0.5\,S$ (*blue*).

neurophysiology. Instead of electrons, the charge carriers in the brain are ions—charged molecules, such as potassium and sodium. We use conductance to model ion channels, which are pores in the cell membrane that allow for the selective flow of ions. Trumping human-built systems, neurons are more complex in the sense that there are different ions that require their own selective channels. For example, sodium ions flow through specific sodium ion channels that are highly selective for sodium ions. We deal with this by modeling each type of ion channel with its own conductance. In fact, we often do not discuss the conductance of an individual ion channel, but rather conductance of the overall assembly of all ion channels of a given type, and model the total current provided by that type of ion channel.

CAPACITOR

The capacitor is another small, simple electric component (Fig. A6.5), which behaves very differently from a resistor. A capacitor consists of two conducting plates separated by a very small but insulated gap through which no electrons can pass. As we will see, this gap is of fundamental importance.

The two plates are so close to each other that movement of charge on one plate causes a matched movement of charge on the opposite plate. This happens because charge moves until there is no net excess positive or negative charge, which would otherwise create an electric potential that causes a current to flow. Thus, the current on one side is equivalent to the current on the other side (terminal) of the capacitor. This explains why there is a capacitive current, despite the fact that electrons cannot physically pass between the plates. If no charge is moved to or from the plates (ie, no change in voltage), the resulting capacitive current is zero. However, the more change in voltage there is, the more current occurs, because of movement of electrons to and from the two plates. Therefore, in contrast to the resistor, the current I is proportional to the rate of change (the temporal derivative) of the voltage across the capacitor with capacitance C:

$$I = C\frac{dV}{dt}$$

Capacitance describes the electric properties of a capacitor and indicates how much charge a capacitor can store for a given voltage. The larger the capacitance C, the larger the current, since it scales with C (measured in farad, abbreviated with the letter F, typically nanofarad, nF, in neuroscience). C is a constant—the larger the surface area of the plates, the larger the C of the capacitor. If the voltage is constant across a capacitor, that is, the temporal derivative is

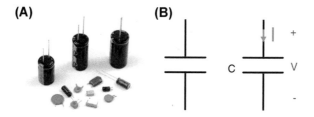

FIGURE A6.5   (A) Capacitors as used in electrical circuits *(Yura Zaga/ shutterstock.com)*. (B) Electric circuit equivalent of capacitor.

zero (no change), the capacitive current is zero. Thus, in "steady state," a capacitor is equivalent to an "open circuit" (two points in the circuit that are not connected). Capacitors make circuits more interesting since, in contrast to the resistor, the voltage–current relationship is not instantaneous. If the voltage changes at the capacitor, charge is moving on and off the capacitor. In addition, if we find that the voltage changes quickly (large dV/dt), the current will be accordingly large. If we know the current that flows but want to know how the voltage changes, we need to solve the equation that relates voltage and current. However, this equation is not a standard algebraic equation, since it contains a temporal derivative and the solution is a function of time $V(t)$. We can undo the differentiation by applying an integral, which is the opposite operation of differentiation. The integral corresponds to the summation of all values of a function, that is, it determines the area under the curve.

$$\frac{dV}{dt} = \frac{I}{C}$$

$$\int \frac{dV}{dt} = V = \int \frac{I}{C} dt = \frac{1}{C} \int I \, dt$$

(A6.4)

Voltage is determined by how much charge was inserted into the capacitor over time. So, if you provide a capacitor with constant current, the voltage at the capacitor will linearly increase (it will accumulate, as per the integral, Fig. A6.6).

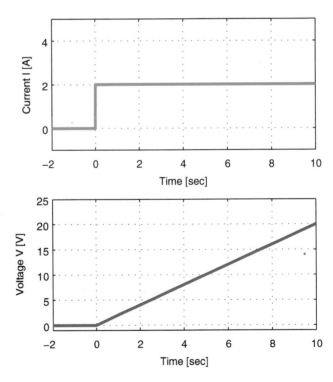

FIGURE A6.6 Time course of current and voltage in a capacitor. *Top*: Time course of current I (*red*). *Bottom*: Time course of voltage V (*blue*).

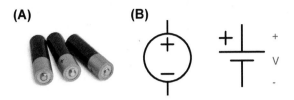

FIGURE A6.7   (A) Batteries are voltage sources *(Maxim Ibragimov/ Shutterstock.com)*. (B) Electric circuit symbol of a voltage source *(left)* and a battery, that is, a constant voltage source *(right)*.

The cell membrane provides very good electric insulation (ignoring the ion channels that form pores modeled by conductances), but the charge inside and outside the cell interacts because of their close proximity. Therefore, the cell membrane acts as a capacitor, and the capacitance is proportional to the surface area of the cell.

VOLTAGE SOURCE

Both resistors and capacitors are passive components, meaning that they do *not* provide energy to the system. Without energy, no current flows through resistors and capacitors. The most common approach is to power these components by using a battery, which ideally provides a constant voltage independent of the device or circuit (referred to as *load*) connected to its terminals (Fig. A6.7).

For example, a standard AA battery provides 1.5 V. Unless it is empty, you will always measure 1.5 V between the two terminals. If nothing is connected to the battery, the air provides a nearly infinite resistance such that the current is zero, according to Ohm's law. However, if a resistor is attached between the two terminals, the resistor determines how much current flows. Voltage sources do not necessarily need to provide a constant voltage—a battery is a special case of a voltage source that provides a constant voltage. The criteria for voltages sources are that the time course of the voltage source is specified and independent of the electric load. For example, power outlets in the United States are voltage sources that provide a sine-wave voltage time course with an amplitude of 110 V and a frequency of 60 Hz. The general symbol of a voltage source is a circle with a "+" and "−" sign; a battery is drawn by two parallel lines of unequal length (not to be confused with the capacitor, which is represented by two equally long parallel lines).

In neurons, the concentration of the various ion species is different inside and outside the cell membrane. As a result, if you open a pore (or channel), the ions will want to spontaneously flow across the membrane as the results of diffusion (to normalize concentrations) and drift (caused by the electric fields from the ions). This "push" for ions to move across the cell membrane corresponds to what a battery does in an electrical circuit. Therefore we can model (and thus abstract) the electrochemistry of ion concentration gradients with a battery, that is, a voltage source.

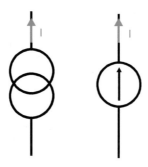

FIGURE A6.8　Commonly used symbols for a current source.

CURRENT SOURCE

An ideal current source is conceptually very similar to a voltage source, except that it provides a predefined current independent of the load connected to it (instead of a predefined voltage, in case of the voltage source). Current sources are denoted either as two overlapping circles or as a circle with an arrow within the circle (Fig. A6.8).

Current sources are ubiquitous in electronic circuits, but are not household items, such as voltage sources (in the form of batteries)—they are a vital tool in the context of neurophysiology. We will encounter current sources throughout this book, for example, when we discuss manipulating the membrane voltage of individual neurons (chapter: Membrane Voltage) and noninvasive brain stimulation with transcranial current stimulation (chapter: Noninvasive Brain Stimulation). In many ways, a current source is the conceptual opposite of a voltage source. First, if we turn on a current source and nothing is connected to it (infinite impedance), it will still try to push out the required current, but it will fail (creating a condition called *overload*). In contrast, a voltage source with no load connected provides zero current and is not an issue at all. Second, shorting a current source by connecting the two ports with a wire is perfectly fine, since only a predefined amount of current will flow (and cause zero voltage across the wire that connects the two ports of the current source). A shorted voltage source, however, will try to provide an infinite amount of current since it encounters no resistance, and that will cause an overload condition.

CONNECTING COMPONENTS TO FORM CIRCUITS

Electric components can be connected to each other, forming meaningful circuits through which electric current flows. Knowing how voltage and current relate to each other at individual components is fundamental for understanding a circuit. We also need to know how to string components together. We use two simple rules called the *Kirchhoff circuit laws*.

Kirchhoff's current law states that charge cannot magically appear or disappear in a circuit. We restate this concept by using the notion of a *node*, which is a branch point of electric wiring. The total current flowing into a node must equal the current flowing out of that node. For example, at a junction in a wire, current will split up but the sum of the two currents

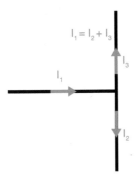

$I_1 = I_2 + I_3$

I_3

I_1

I_2

FIGURE A6.9   Kirchhoff's current law applied to a node.

flowing out of the node into the two branches is equal to the current flowing into the junction (Fig. A6.9). Mathematically, this means that the sum of all currents at a node has to be zero:

$$\sum_{k=1}^{N} I_k = I_1 + I_2 + \cdots + I_N = 0 \tag{A6.5}$$

Practically, at a node of a circuit, all currents pointing inward (toward the node) are added, while all currents pointing outward (away from the node) are subtracted.

Kirchhoff's voltage law states that voltage (and therefore electric potential) is conserved in a closed loop of electric components. This law is derived from the fact that voltage is defined as energy per charge, and both charge and energy are conserved. As a result, the sum of the voltages in a closed loop equals zero:

$$\sum_{k=1}^{N} V_k = 0 \tag{A6.6}$$

Therefore, for any closed loop in a circuit, an arbitrary starting point is chosen and a loop is traced. Every component has a voltage drop drawn as a plus and a minus sign. All of these voltages are added together for a given loop in the circuit. We add the voltages for which we encounter a plus sign first and subtract the voltages when we encounter a minus sign first (Fig. A6.10).

EXAMPLE ELECTRIC CIRCUITS

Knowing the voltage–current relationship of the individual electric components and Kirchhoff's laws enables us to determine current flow and voltages in any electric circuit. At least, we can write down the equations, though solving them may be tricky since they involve derivatives (dV/dt) when there are capacitors in the circuit (see toolbox: Differential Equations). However, we can get a sense for the response if we keep in mind that capacitors are high resistance for low-frequency input and vice versa. Let us analyze a few basic circuits in preparation for those we will encounter in neurophysiology.

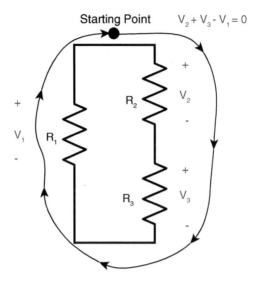

FIGURE A6.10   Application of Kirchhoff's voltage law to a circuit of three resistors. The *line with arrow heads* indicates the loop used to sum the voltages.

VOLTAGE SOURCE AND RESISTOR

Connecting a resistor to a battery gives us a simple circuit (Fig. A6.11) where the voltage across the resistor V_R equals the voltage V_S of the battery by application of Kirchhoff's voltage law (ie, the two voltages need to add to zero). Now that we know V_R, we can use Ohm's law to determine the current that flows through the resistor $I_R = V_R/R$. We do not encounter any derivatives dV_S/dt in the equations for this circuit (because there is no capacitor), and any change in V_S is instantaneously reflected in V_R and I_R.

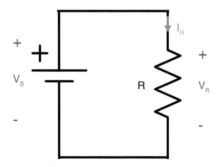

FIGURE A6.11   Voltage source and resistor form a simple circuit in which the voltage across the resistor V_R equals the voltage V_S of the voltage source. The current is determined by Ohm's law: $I_R = V_R/R$.

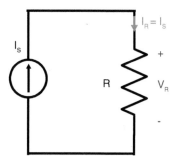

FIGURE A6.12 Current source and resistor form a simple circuit in which the current I_R through the resistor equals the current I_S of the current source. The voltage V_R across the resistor R is determined by Ohm's law: $V_R = I_S R$.

CURRENT SOURCE AND RESISTOR

Connecting a current source instead of a voltage source to the resistor (Fig. A6.12) changes little about the behavior of the system. The current from the source I_S can only flow through the resistor, therefore the voltage across the resistor is $V_R = I_S R$.

VOLTAGE SOURCE AND CAPACITOR

Do not try this at home! This circuit makes little sense and in fact would potentially be dangerous to build. In Fig. A6.13, we can determine how much current flows at the moment the battery is connected to the capacitor. As previously, the voltage at the capacitor is $V_C = V_S$. The current is then determined by the component equation for the capacitor: $I = C \, dV/dt$. However, as we connect the battery, there is an instantaneous increase (jump) in voltage that has by definition infinite slope, so in theory the current would be infinite.

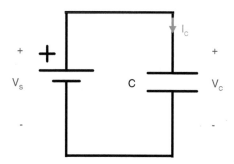

FIGURE A6.13 Do not build this circuit. The source voltage V_S must equal the voltage V_C across the capacitor C, which would require infinite current I_C at the time of connection of the two components.

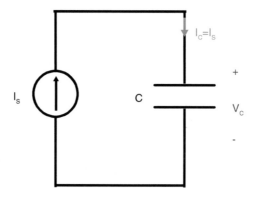

FIGURE A6.14 Do not build this circuit, because the capacitance C is charged with I_S with no bounds to the voltage V_C.

CURRENT SOURCE AND CAPACITOR

This is another circuit not to be implemented at home, because it would also be dangerous. As demonstrated earlier, the voltage at a capacitor is proportional to the integral of the current. Therefore, if a constant current source is connected to a capacitor, the voltage across the capacitor will linearly increase until the system physically breaks (eg, the capacitor explodes) (Fig. A6.14).

CURRENT SOURCE, RESISTOR, AND CAPACITOR

For this circuit, we connect both a resistor and a capacitor to the current source (Fig. A6.15), such that the three components are *in parallel* since the current is split between these components. Specifically, there is a branch point, where supplied current I is split up between the resistor and the capacitor (together they must equal the current that flows into this branch point). Therefore, we find:

$$\frac{V}{R} + C\frac{dV}{dt} = I_S$$

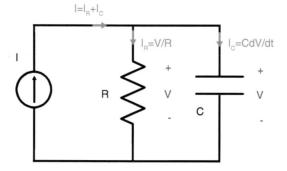

FIGURE A6.15 Electric circuit equivalent of the passive cell membrane.

Importantly, this circuit is the standard model of the passive cell membrane in the absence of voltage-gated ion channels. The toolbox "Differential Equations" shows how to solve this equation for $V(t)$.

SUMMARY AND OUTLOOK

In this toolbox, we have learned how to draw and analyze electric circuit diagrams. With these tools, we derived the differential equation that governs the membrane voltage of a passive cell membrane. We will use the same circuit modeling strategy when we discuss how action potentials are generated (see chapter: Dynamics of the Action Potential) and how synaptic potentials are generated by neurotransmitter release (see chapter: Synaptic Transmission).

Toolbox Differential Equations

This toolbox introduces one of the most useful tools to understand dynamical systems, that is, systems that change their state over time. The brain is a highly dynamical system that undergoes changes at all spatiotemporal scales, from the level of subcellular processes all the way to large-scale organization of electric activity across networks of neurons. Understanding the brain, particularly at the level of networks, means understanding these dynamics. In this toolbox, we will familiarize ourselves with the mathematics that enables us to model dynamical systems, *differential equations*.

Algebraic equations have numbers as solutions, for example:

$$2x + 1 = 0 \tag{A7.1}$$

for which the solution is $x = -0.5$. In contrast, differential equations have solutions that are functions of time, denoted as $x(t)$. Differential equations have received their name because they contain derivatives of $x(t)$, here denoted as $dx(t)/dt$. The derivative is the slope of $x(t)$ and quantifies how fast a signal changes over time. Positive values of $dx(t)/dt$ indicate an increase in the value of $x(t)$, and negative values correspond to a decrease of $x(t)$. The derivative itself is again a function of time t, and the slope for any moment in time can be determined by plugging in the desired value of t. For example, the function:

$$x(t) = t^2 \tag{A7.2}$$

has the derivative (slope):

$$\frac{dx(t)}{dt} = 2t \tag{A7.3}$$

Thus, the slope of function $x(t)$ increases linearly with time. As we can see from the plot of $x(t)$ in Fig. A7.1 (left), the slope is negative for values of t smaller than zero and positive for values of t larger than zero (Fig. A7.1, right). The larger the value t, the larger the slope (derivative).

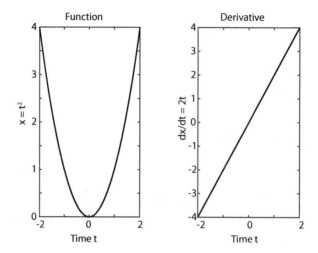

FIGURE A7.1   (*Left*) Function $x(t) = t^2$. (*Right*) Derivative $dx(t)/dt = 2t$.

We will focus on what are called *ordinary differential equations*, which include only derivatives with respect to time. Here is an example:

$$C\frac{dV(t)}{dt} + \frac{V(t)}{R} = I(t) \qquad (A7.4)$$

This is the differential equation that corresponds to the previously introduced circuit that models the passive cell membrane (toolbox: Electrical Circuits); the equation models a capacitor with capacitance C and a resistor with resistance R. The current source injects a defined current $I(t)$ that can be a function of time. $V(t)$ is the unknown function in this differential equation for which we want to find a solution.

How do we solve differential equations? Unfortunately, as opposed to polynomial algebraic equations, which we know how to solve, no general approach or recipe is applicable to all differential equations. However, many differential equations we encounter in neuroscience are of the same form as the one for the passive cell membrane [Eq. (A7.4)]. We multiply Eq. (A7.4) with R:

$$RC\frac{dV(t)}{dt} + V(t) = RI(t) \qquad (A7.5)$$

and rewrite the equation as

$$\tau\frac{dx(t)}{dt} + x(t) = u(t) \qquad (A7.6)$$

where $\tau = RC$ denotes the *time constant* of the system (for reasons discussed later), $x(t)$ represents the function for which we want to find a solution, and $u(t) = RI(t)$ represents the external input. The function $x(t)$ is called the *state variable* since it describes the state of the system. In essence, the left-hand side of the equation describes the internal dynamics of the system and the right-hand side represents the external contribution to the dynamics. First, we will develop an intuition for the dynamics described by Eq. (A7.6), and then we will proceed to solve it. Second, we introduce numerical solvers as a tool to solve more complicated differential equations.

SOLVING THE DIFFERENTIAL EQUATION FOR THE PASSIVE CELL MEMBRANE

The time constant τ received its name from the fact that it determines how fast $x(t)$ changes over time. This becomes more obvious when we rearrange Eq. (A7.6) as follows:

$$\frac{dx(t)}{dt} = \frac{-x(t) + u(t)}{\tau} \tag{A7.7}$$

This equation tells us what the derivate of $x(t)$ is, in other words the rate of change of $x(t)$. The larger the value of τ, the smaller the term on the right-hand side, and therefore the smaller the rate of change of $x(t)$. This is why τ is called the time constant that denotes the *time scale* of the state variable $x(t)$.

We next consider a simplified version of Eq. (A7.6) that does not contain any input term $u(t)$:

$$\frac{dx(t)}{dt} = -\frac{1}{\tau}x(t) \tag{A7.8}$$

At this point, we use calculus to guess the general solution (referred to as *Ansatz* in mathematics). An exponential function has the unique property of having a derivative that is also an exponential function. Eq. (A7.8) states that the derivative of the solution $x(t)$ needs to equal $x(t)$ multiplied with the constant $-1/\tau$, and therefore the exponential function is a good guess. To find the most general solution, our Ansatz is

$$x(t) = Ae^{Bt} \tag{A7.9}$$

where A and B are free parameters that need to be determined and do not change with time t. We now verify whether this is the correct solution by plugging the Ansatz in Eq. (A7.9) into the differential equation [Eq. (A7.8)] and using the chain rule of differentiation:

$$BAe^{Bt} = -\frac{1}{\tau}Ae^{Bt} \tag{A7.10}$$

We now need to determine if there are values of constants A and B for which Eq. (A7.10) is true. For this to be the case, we find that A cancels out (meaning it can assume any value for our Ansatz to be a solution) and that $B = -1/\tau$. Therefore the solution is:

$$x(t) = Ae^{-\frac{t}{\tau}} \tag{A7.11}$$

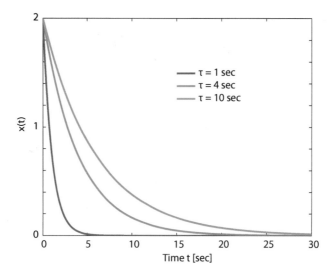

FIGURE A7.2   Solutions to the first-order differential equation for different values of time constant τ. Despite the differences in time constants, all lines converge on zero when t approaches infinity. The parameter A was arbitrarily set to 2 for this example.

Several solutions are plotted in Fig. A7.2. As determined earlier without solving the equation, τ determines how fast $x(t)$ changes, that is, in case of no input, how fast $x(t)$ converges to zero.

Next, we will take into account the input $u(t)$, which we have ignored thus far. We will consider the simple case where input $u(t)$ assumes a constant value and does not vary as a function of time. More precisely, we assume that $u(t)$ is zero for all times $t \leq 0$ and that at time $t = 0$, $u(t)$ switches to u_0 such that $u(t) = u_0$ for all values of $t > 0$. The response to this type of input is called *step response*. Since the input is constant, we assume that if we wait long enough, $x(t)$ will eventually converge to a final, constant value x_∞. At that point, $x(t)$ does not change anymore, so its derivative $dx(t)/dt$ equals 0 (this represents the steady state) and the differential equation is reduced to:

$$0 = \frac{-x_\infty + u_0}{\tau} \tag{A7.12}$$

to show that $x_\infty = u_0$. We now add the original solution of the equation with no input [Eq. (A7.11)] to this steady-state solution:

$$x(t) = Ae^{-\frac{t}{\tau}} + u_0 \tag{A7.13}$$

By plugging this solution into Eq. (A7.6), we can confirm that this is the correct solution. However, we still need to address the free parameter A. We can determine A by requesting that our solution assumes a value of 0 at time point zero before the input is turned on:

$$x(0) = Ae^{-\frac{0}{\tau}} + u_0 = A + u_0 \tag{A7.14}$$

Therefore $A = -u_0$ and we have arrived at the final solution:

$$x(t) = -u_0 e^{-\frac{t}{\tau}} + u_0 = u_0\left(1 - e^{-\frac{t}{\tau}}\right) \tag{A7.15}$$

Inspecting this equation, we see that $x(t)$ converges with rate τ to its final value of u_0, as shown in Fig. A7.3.

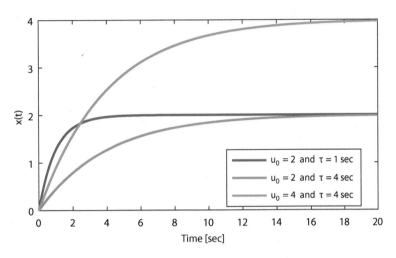

FIGURE A7.3   Solutions to the first-order differential equation with a constant input u_0. The solutions for three different parameter combinations are shown. All solutions converge to the value of the constant input term u_0.

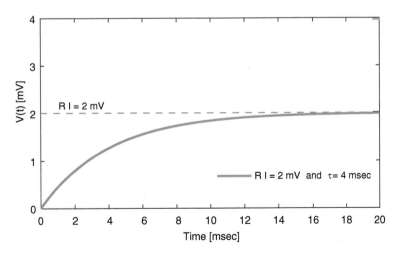

FIGURE A7.4   Solution to the first-order differential equation that describes the dynamics of the passive cell membrane in response to a constant step current injection.

Lastly, we can simply replace variable names to write down the solution to our original equation of the passive cell membrane:

$$V(t) = RI\left(1 - e^{-\frac{t}{RC}}\right)$$ (A7.16)

Recall that R, C, and I are the cell membrane resistance, capacitance, and the injected current, respectively. The dynamics of the state variable $V(t)$ are shown in Fig. A7.4.

NUMERICAL SOLVERS

The strategies discussed here are not always usable for solving more complex differential equations. Instead of finding the *analytical solution* to a differential equation as done earlier, we often use *numerical solvers*, algorithms that compute the solution $x(t)$. Many different numerical solvers can be used, and they have their unique advantages and disadvantages (eg, precision, speed). However, we will only focus on a straightforward but powerful approach, called *Euler's method*. This method provides numerical solutions to differential equations of the form:

$$\frac{dx}{dt} = f(x, t)$$ (A7.17)

This form describes any differential equation for which the derivative of $x(t)$ equals a function of both $x(t)$ and t, such as in the previous example [Eq. (A7.6)] where:

$$f(x, t) = -x(t) + u(t)$$ (A7.18)

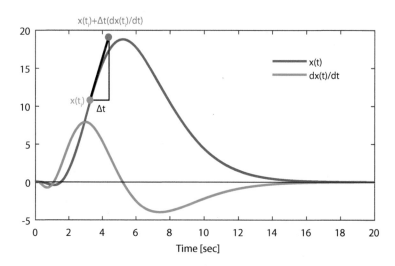

FIGURE A7.5   Euler method for solving the differential equation for $x(t)$. The differential equation defines the slope at any given point (*red*) and can be used to stepwise linearly extrapolate (in *black*) to determine the solution $x(t)$ (*blue*). Note that the time step Δt in this example is too large, and as a result the value $x(t_{i+1})$ is not accurate when compared to the true solution $x(t)$.

The main principle of Euler's method is that if we know the value of a function $x(t)$ at time point t_i, we can approximate (extrapolate) the value at time point $t_i + \Delta t$. We can do this by assuming that the function $x(t)$ is linear with a slope of the derivative of $x(t)$ at t_i:

$$x(t_i + \Delta t) \approx x(t_i) + \Delta t \frac{dx(t = t_i)}{dt} \tag{A7.19}$$

Thus the solution is determined by starting at $t = 0$ where $x(0) = x_0$, the *initial condition*. Then, the next value at $t = \Delta t$ is determined. The derivative of $x(t)$ is given by the differential equation (Fig. A7.5). This process is repeated at each time point $n\Delta t$ where n is the step number that iterates from 0 to the maximal step number required. Choosing a sufficiently small time step Δt is crucial, since the linear approximation of $x(t)$ will deviate more and more from the actual time course of $x(t)$ for larger values of Δt.

SUMMARY AND OUTLOOK

In this toolbox, we introduced differential equations, which describe the dynamics of state variables as a function of time. Differential equations are ubiquitous in quantitative descriptions of biological process. For example, the electric properties of the cell membrane are described by differential equations. We focused on this specific differential equation since it builds the foundation for understanding both how action potentials are generated (chapter: Dynamics of the Action Potential) and how synaptic input is processed (chapter: Synaptic Transmission).

Toolbox Dynamical Systems

IS THIS TOOLBOX FOR ME?

This toolbox provides an intuitive introduction to the basic concepts of system theory. Quite likely, if your background is different from physics, mathematics, or engineering you will learn important new material in this toolbox. The toolbox "Differential Equations" is closely related to the content of this toolbox.

Almost all measurable quantities are *dynamic*, meaning that they change with time. In the toolbox "Differential Equations," we learned how to solve equations that describe the dynamics of state variables. One example was the differential equation that describes the membrane voltage of the passive cell membrane. Here, we will discuss general properties of dynamical systems. We will develop an intuition for their behavior without solving the governing differential equations. The methods introduced here will be helpful for understanding the dynamics of individual and networks of neurons. Specifically, we learn how to predict the effect of perturbations on the behavior of dynamical systems. Typical examples of perturbations to neurons and networks of neurons are electrical or optical stimulation and sensory input.

STABLE EQUILIBRIUM

We first consider a linear system described by Eq. (A8.1)

$$\frac{dx(t)}{dt} = -x(t) + A \tag{A8.1}$$

where $x(t)$ is the state variable and A is a constant. Note that this is a linear system, since the derivative $dx(t)/dt$ is a linear function. We have solved this equation in the toolbox "Differential Equations." In this toolbox, we will instead develop an intuitive understanding of the behavior of $x(t)$ without solving the differential equation. The strategy we develop can also be applied to more complicated dynamical systems for which finding an analytical solution is not straightforward or is even impossible.

First, we ask if there is a certain value of $x(t)$ for which—once reached—$x(t)$ will no further change. To determine what this *steady-state value* x_∞ is (assuming that it exists), we require that the derivative of $x(t)$, that is, $dx(t)/dt$, is zero, such that $x(t)$ does not further change.

In other words, we set the derivative to zero in Eq. (A8.1) and solve the remaining algebraic equation (where we replaced $x(t)$ with x_∞):

$$0 = -x_\infty + A \qquad (A8.2)$$

We find that $x_\infty = A$. This value is also referred to as the *equilibrium* of the system. We next determine what happens if we apply a small perturbation that moves $x(t)$ away from its equilibrium A. We plug in $x = A + \Delta x$, where Δx is our small perturbation. We find that the derivative dx/dt is

$$\frac{dx(t)}{dt} = -(A + \Delta x) + A = -\Delta x \qquad (A8.3)$$

Therefore, the rate of change, dx/dt, is negative and x will decrease; in other words, a small perturbation that increases x beyond its value A will cause a transient response of the system that takes $x(t)$ back to A. Similarly, a negative perturbation will make the derivative dx/dt positive, and will also move the state variable $x(t)$ back to A. This can be visualized in a representation where $x(t)$ is on the x-axis and dx/dt is on the y-axis. This is shown in Fig. A8.1 as an example of a linear differential equation similar to Eq. (A8.1):

$$\frac{dx(t)}{dt} = -2x(t) + 1 \qquad (A8.4)$$

In Eq. (A8.4), the equilibrium is determined the same way, by setting the derivative to zero. We find that $x_\infty = 0.5$.

Note that Fig. A8.1 is not a plot of the solution $x(t)$ as a function of time. Instead, it shows how the derivative of the state variable depends on the state variable itself. This visualization will be critical to our characterization of the equilibrium. As long as dx/dt is smaller than zero for perturbations that increase x to a value larger than its equilibrium $x_\infty = 0.5$, and larger than zero for perturbations that decrease x to values below 0.5, the equilibrium is referred to

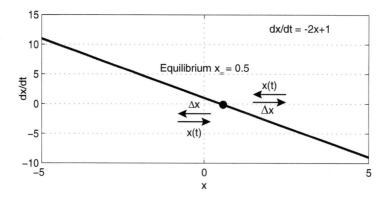

FIGURE A8.1   The plot of the derivative of $x(t)$ crosses the x-axis at the equilibrium $x_\infty = 0.5$. The value of dx/dt in the vicinity of the equilibrium defines the response of the system to a small perturbation to state variable $x(t)$. The equilibrium is stable, since $x(t)$ moves back to the equilibrium point in response to small perturbations.

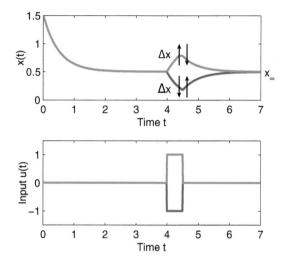

FIGURE A8.2 (*Top*) Time course of $x(t)$ with two perturbations (*red*: positive deflection; *blue*: negative deflection). After initial convergence to the equilibrium at $x = 0.5$, both perturbations transiently deflect $x(t)$ before the trajectories return to stable equilibrium at $x = 0.5$. (*Bottom*) Time course of transient perturbations used.

as *stable*. A stable equilibrium is surrounded by a *basin of attraction*, which is defined as the set of values around the equilibrium from which the system will return to the equilibrium. Fig. A8.2 shows the response dynamics for perturbations to the stable equilibrium $x_\infty = 0.5$ in Eq. (A8.4).

UNSTABLE EQUILIBRIUM

In contrast, if the state variable is at an unstable equilibrium and is exposed to a perturbation, the system will not return to the equilibrium, but will move away from the equilibrium (ie, it will *diverge*). We explore this by considering a differential equation similar to Eq. (A8.4):

$$\frac{dx(t)}{dt} = x(t) - A \tag{A8.5}$$

This equation has also an equilibrium at $x_\infty = A$. However, if the same perturbation is applied, we find that $dx/dt = \Delta x$ for positive perturbations Δx away from A, so $x(t)$ will continue to grow and move away from A (ie, diverge). Similarly, a negative perturbation, $-\Delta x$, will move x away from A since the resulting derivative is negative. This can be visualized as in Fig. A8.3, which shows dx/dt as a function of $x(t)$. In this graphical representation, the main difference between Eqs. (A8.1) and (A8.5) is the slope at the equilibrium. When comparing Figs. A8.1–A8.3, we see that positive slope corresponds to an unstable equilibrium and negative slope corresponds to a stable equilibrium.

Fig. A8.4 shows the response dynamics for perturbations to an unstable equilibrium.

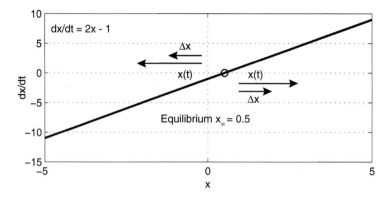

FIGURE A8.3   Same representation as in Fig. A8.1, but for unstable equilibrium.

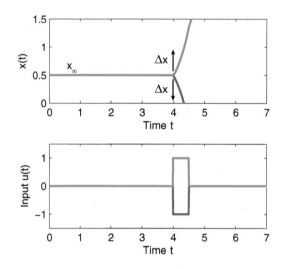

FIGURE A8.4   Same representation as in Fig. A8.2, but for unstable equilibrium. (*Top*) Time course of $x(t)$ with two perturbations (in *red* and *blue*). After initial convergence to the equilibrium at $x = 0.5$, both perturbations cause the trajectories to diverge to plus or minus infinity since the equilibrium is unstable. (*Bottom*) Time course of transient perturbations used.

MULTISTABLE SYSTEMS

Linear systems, which are modeled by a differential equation of the form:

$$\frac{dx(t)}{dt} = -Bx(t) + A \tag{A8.6}$$

exhibit a single equilibrium. However, in the case of nonlinear differential equations, for which the derivative is a more complicated function of $x(t)$, multiple (or in some cases no) equilibria can occur. As an example, consider the following differential equation:

$$\frac{dx}{dt} = -x^2 + 1 \tag{A8.7}$$

By using the same approach as for the linear systems, we see that the system has two equilibria at $x_0 = -1$ and $x_1 = 1$. The plot of dx/dt as a function of $x(t)$ shows that the slope at $x_0 = -1$ is positive and therefore is indicative of an unstable equilibrium (Fig. A8.5).

In contrast, $x_1 = 1$ is a stable equilibrium since the slope is negative. Since these are the only two equilibria, we predict that for all initial conditions (starting value of x) $x(t = 0) > -1$,

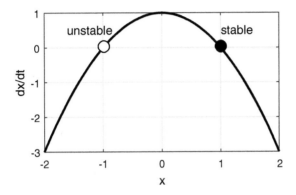

FIGURE A8.5 $dx/dt = -x^2 + 1$ has two equilibria at $x_0 = -1$ (unstable) and $x_1 = 1$ (stable).

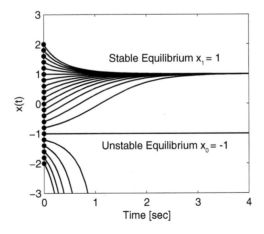

FIGURE A8.6   Time courses of $x(t)$ for different initial conditions indicated by *black circles*. For all initial values of $x(t)$ larger than -1, $x(t)$ converges to the stable equilibrium $x_1 = 1$. Only if the initial value is -1 will $x(t)$ remain at the unstable equilibrium $x_0 = -1$.

the system will converge to $x = x_1$. For $x(t = 0) = -1$, the system will stay at $x = -1$ as long as no perturbation is applied to the system. For $x(t = 0) < -1$, the system will exhibit runaway dynamics and diverge (toward minus infinity). We can test this by numerical simulations (Fig. A8.6).

SUMMARY AND OUTLOOK

In this toolbox, we developed strategies to qualitatively understand the behavior of dynamical systems without solving the governing differential equations. We discussed how to find the equilibrium points and determine their stability. This approach allows us to understand the overall behavior of the system. In particular, we introduced how to distinguish between stable and unstable equilibrium points. The intuition developed here for multistable systems is a powerful tool to understand the behavior of both individual neurons and networks of neurons, which both often exhibit two equilibria (bistability). For example, a neuron can be conceptualized as a bistable system (two states: "resting membrane voltage" and "firing action potentials").

NOTES

- The book *Dynamical Systems in Neuroscience* by Eugene Izhikevich provides an easy-to-read and intuitive expansion of the topic introduced in this toolbox [1].
- The book *Nonlinear Dynamics and Chaos* by Steven Strogatz offers a highly accessible, in-depth introduction to dynamical systems [2].

References

[1] Izhikevich EM. Dynamical systems in neuroscience: the geometry of excitability and bursting. In: Computational neuroscience. Cambridge, Mass: MIT Press; 2007. xvi, p. 441.
[2] Strogatz SH. Nonlinear dynamics and chaos: with applications to physics, biology, chemistry, and engineering. 2nd ed. Boulder, CO: Westview Press, A Member of the Perseus Books Group; 2015. xiii, p. 513.

Toolbox Graph Theory

IS THIS TOOLBOX FOR ME?

In this toolbox, we review the basic terminology used to describe the structure of networks. The concepts covered in this toolbox are instrumental in quantifying both structural and functional neuronal networks. If you have studied graph theory, you can safely skip this toolbox.

Understanding the structure of how neurons are connected to each other is fundamental for explaining the signaling of neuronal networks. Networks are not unique to neuroscience. There is an entire mathematical theory dedicated to describing and analyzing networks, often referred to as *graph theory*. In this toolbox, we will introduce the basic terminology in preparation for the application of graph theory to microscopic (individual neurons) and macroscopic (entire brain areas) structural networks, as well as functional networks, which describe "connectivity" based on the relative coupling of activity in different network locations (chapter: Network Interactions).

DEFINING GRAPHS

Graphs describe pairwise relationships between discrete entities. Each entity is called a *node* (in our case, nodes are individual neurons or brain areas), and nodes are connected by *vertices* or *edges*. Edges are *directed* if we are able to define the direction of the interaction, or *undirected* in cases where the interaction between the two nodes has no direction (Fig. A9.1). For example, a network of two neurons is drawn as a directed graph, because synaptic

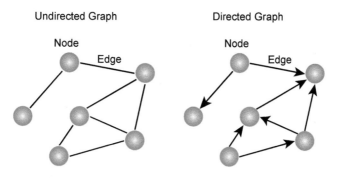

FIGURE A9.1   Undirected (*left*) and directed (*right*) graph.

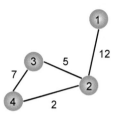

FIGURE A9.2   Graph with nodes numbered and weights assigned to edges.

TABLE A9.1   Adjacency Matrix for the Undirected Graph in Fig. A9.2

	Node 1	Node 2	Node 3	Node 4
Node 1	0	12	0	0
Node 2	12	0	5	2
Node 3	0	5	0	7
Node 4	0	2	7	0

connectivity has a direction caused by the unidirectional nature of chemical synapses (ignoring retrograde transmission). In addition, the edges can be of different strengths, which are referred to as *weights* (Fig. A9.2).

A graph can be either drawn with nodes (eg, circles) and edges (eg, arrows or lines), or it can be represented as a table that lists all connections. In such a table, called an *adjacency matrix*, there is a row and a corresponding column for each node. In this n-by-n matrix (where n is the number of nodes), row i lists the nodes that node i connects to ("targets"). The values in Table A9.1 denote the strength of each connection for the example graph in Fig. A9.2. For example, if node 1 is connected with node 2 with strength 12, the value where the first row and second column meet is 12. Since the example graph is undirected, the adjacency matrix is symmetrical, since no distinction could be made between the connection from node A to B and the connection from node B to A.

ANALYZING GRAPHS

Neighbors are nodes connected by an edge (and therefore have a nonzero entry in the corresponding location of the adjacency matrix). The *degree of a node* quantifies to what extent a node is connected to the remainder of the network. In an undirected graph, the degree of a node corresponds to the number of connections formed by that particular node. In a directed graph, the distinction between the number of incoming edges (*indegree*) and outgoing edges (*outdegree*) is made (Fig. A9.3).

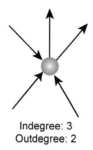

Indegree: 3
Outdegree: 2

FIGURE A9.3 Node with three incoming (*indegree* = 3) and two outgoing (*outdegree* = 2) connections.

For any given graph, the degree of all nodes can be determined and subsequently represented by a distribution called *degree distribution*. The correlation between degrees of connected nodes is referred to as *assortativity*. Positive assortativity indicates that nodes with high degrees are likely to connect to each other. If nodes are not neighbors, they can be indirectly connected by paths, defined as an ordered sequence of nodes and edges. If the edges do not have different strengths, meaning if the only information provided is the absence (indicated by 0) or presence (indicated by 1) of an edge for all entries in the connectivity matrix (*binary graph*), the length of a path is defined as the number of edges it contains. For every pair of nodes, the length of the shortest path can be determined and represented in a distance matrix comprised of rows and columns, similar to an adjacency matrix. The *path length* denotes the average of all the values represented in the distance matrix.

With these definitions in hand, we can now describe both microscopic (local) and macroscopic (global) properties of graphs. First, we look for local organization of nodes that are highly interconnected. Groups of nodes with high connectivity among their members are referred to as *neighborhoods* or *clusters*. For every node, we can compute a *clustering coefficient* that describes the connectivity of all neighbors of the node. First, the total number of possible connections between the neighbors is a function of the number of neighbors k_i of node i, since in principle each node can connect with all of other nodes, but not itself.

$$k_i(k_i - 1) \tag{A9.1}$$

In the case of an undirected graph, Eq. (A9.1) is divided by 2, since every pair of directed edges between two nodes is replaced with a single, undirected edge.

$$\frac{k_i(k_i - 1)}{2} \tag{A9.2}$$

Second, we count the number of actual connections in the cluster and divide it by the total number of possible connections to determine the cluster coefficient. At the scale of an entire graph, there can be clusters (groups of nodes with high clustering coefficients) that are connected by few edges that connect nodes from different clusters. This property of the

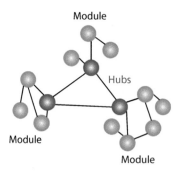

FIGURE A9.4   Network with three modules. Hubs (*blue*) have a higher node degree and are part of the shortest path between any pairs of nodes, as long as the two nodes are part of different modules.

network being divided into interconnected clusters (so-called *modules*) is referred to as *modularity*. The relative importance of a node can be defined by considering its role within a network. Nodes that are "centers" of the network are called *hubs*. For example, nodes with high degrees are often considered hubs (Fig. A9.4). Alternatively, if a graph is partitioned into modules, hubs can also be defined by the number of connections that connect different modules (*participation index*). Furthermore, *centrality* can be determined by the number of shortest paths (between any two nodes in the network) a node participates in (*betweenness centrality*).

SMALL-WORLD NETWORKS

With this technical vocabulary in place, we can now discuss different types of graphs and their structure (*topology*). Random graphs are constructed by randomly assigning connections between nodes. The key characteristics of a random graph are a normal degree distribution, short path lengths, and low levels of clustering. In a certain sense, the opposite of a random graph is a lattice graph, in which nodes connect to their neighbors. Such graphs have longer paths but higher clustering than random graphs. One of the most important concepts in network science is so-called small-world connectivity [1]. Mathematically, a small-world network is generated by starting with a lattice network (only local connections) and replacing a certain fraction of those local connections with random connections that originate from the same node but target a randomly chosen node. The fraction of connections subject to this process is called the *rewiring probability*. If the rewiring probability is zero, we are left with a lattice graph. If the rewiring probability is one, we have transformed the graph into a random graph. For very small rewiring probability values, the graphs combine short path lengths and high clustering, the two features of small-world networks (Fig. A9.5). Notably, this is not the only algorithm that can be used to generate small-world networks. Small-world networks are prevalent in the brain. There are multiple advantages for brain networks to exhibit small-world structure, such as the comparably small number of long-range connections needed to synchronize networks and the ability for local processing by means of the local connections.

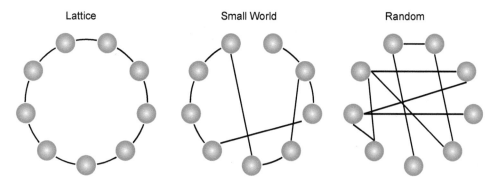

Lattice   Small World   Random

FIGURE A9.5 Transformation of a lattice network (*left*) into a small-world network (*middle*) and ultimately into a random network (*right*). The number of paths remains constant between graphs. But the average path length and clustering coefficients of the graphs are different. *Adapted from Watts DJ, Strogatz SH. Collective dynamics of 'small-world' networks. Nature 1998;393(6684):440–442.*

POWER LAWS

In reality, few networks exhibit the structure of a random network, which has a normal degree distribution. Rather, networks often have a *power law* as their degree distribution, such that the probability of a node to have degree x is

$$p(x) = \frac{c}{x^k} = cx^{-k} \tag{A9.3}$$

where k is an integer number, c is a constant, and x needs to be at least as large as x_{min} (to avoid a division by zero):

$$x \geq x_{min} \tag{A9.4}$$

Unlike most other distributions typically considered, random variables that follow a power distribution do not cluster around a "typical value" (Fig. A9.6). Instead, power distributions are referred to as *heavy-tailed distributions*, because they are much more likely to produce values that are orders of magnitude larger than others. As a result, the mean of the distribution is infinite for distributions where $1 < k < 2$. For $2 < k < 3$, the mean exists, but the standard deviation is infinite. The power law is referred to as *scale-free* since the probability of a small event occurrence and a large event occurrence always has a constant ratio. Mathematically, we can show this by comparing the likelihood of a random process to generate values x and mx, with m being a constant:

$$\Pr(mx) = (cmx)^{-k} = cm^{-k}x^{-k} \sim cx^{-k} = \Pr(x) \tag{A9.5}$$

One way to generate graphs that have a power-law degree distribution is to use the rule of *preferential attachment*. This growth rule (referred to as a *generative model* by mathematicians) starts with a single node. Then, nodes with outdegree 1 are added. With probability $\alpha < 1$, this one outbound connection of the newly added node is formed with a randomly chosen

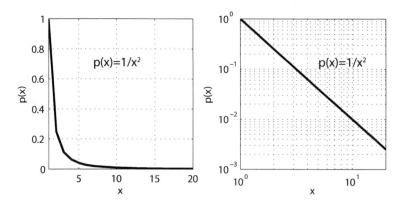

FIGURE A9.6   Power law distribution (*left*) corresponds to a straight line when plotted in a log–log plot (*right*). In such a plot, the logarithm of the values is plotted on both axes.

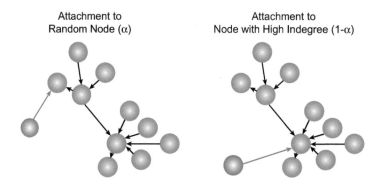

FIGURE A9.7   (*Left*) Attachment of a new node (*green*) to a random existing node (*red*), here a node with an indegree of 1. (*Right*) Attachment to an existing node with high indegree, here a node with an indegree of 6.

node in the already existing network. With probability $1 - \alpha$, the new node connects to a node that is not randomly chosen but is selected by a rule that favors nodes with high indegree (Fig. A9.7). This way, nodes that are already highly connected are bound to receive even more connections. Mathematically, in approximation, the indegree distribution then follows a power law with exponent $k = 1/(1 - \alpha)$.

SUMMARY AND OUTLOOK

Graph theory provides a conceptual framework to describe the structure of networks. In this toolbox, we introduced the basic concepts we will encounter throughout the book. For example, the "chapter Imaging Structural Networks With MRI" describes how noninvasive imaging of white matter can be used to determine how different brain areas are connected to each other. This structural connectivity can be represented and quantified by using the graph theoretical approaches discussed here. Similar analysis strategies can be applied to functional networks, as described in the "chapter Imaging Functional Networks With MRI." Similarly, the methods discussed in the "chapter Network Interactions" are used to determine network graphs based on functional interactions between different measurement locations.

NOTES

- The book *Networks of the Brain* by Olaf Sporns provides a highly readable and comprehensive introduction to the use of graph theory in network neuroscience [2].

References

[1] Watts DJ, Strogatz SH. Collective dynamics of 'small-world' networks. *Nature* 1998;**393**(6684):440−2.
[2] Sporns O. *Networks of the brain*. Cambridge, Mass: MIT Press; 2011. xi, 412 p., 8 p. of plates.

Toolbox Modeling Neurons

IS THIS TOOLBOX FOR ME?

If you have not yet encountered computational models of neurons, this toolbox will get you started. You will learn about the computational modeling of neurons, and will become acquainted with several neuronal models used in neuroscience studies. The information covered in the toolboxes on "Differential Equations," "Electrical Circuits," and "Matlab" prepares you for this toolbox. If you have picked up this book with a solid computational neuroscience background, you may want to skip ahead.

Understanding complex processes requires the ability to integrate experimental measures, replicate the phenomenon under study, and make novel predictions that can be verified by new experiments. Mathematical models uniquely offer that ability, and have for many decades played a key role in understanding electric signaling in single neurons and network of neurons. The most famous example of using mathematical models to explain a neurobiological phenomenon is the work by Alan Hodgkin and Andrew Huxley [1]. Together, those researchers performed a series of elegant experiments on the squid giant axon that enabled them to build a mathematical model of how sodium and potassium currents generate action potentials. To date, their model has not only provided many predictions that have been verified in subsequent experiments, but it has also remained the canonical model of electric signaling in neurons, often simply referred to as the Hodgkin—Huxley model (discussed in more detail in the chapter: Dynamics of the Action Potential). Here, we will first introduce a conceptual framework for computational modeling ("Modeling Process"), and then we will review several models commonly used in computational neuroscience. In the last section, we will derive the cable equation to determine the membrane voltage not only as a function of time but also as a function of space. This section is marked with an asterisk (*) due its mathematical nature that requires more background than the remainder of the book.

MODELING PROCESS

A single neuron is so complex that even the most sophisticated mathematical model captures only a very small fraction of the processes that contribute to neuronal activity. Furthermore, even if feasible, making a model as complicated as possible without considering the type of scientific question to be answered may not be very helpful. Instead, the first step in the modeling process is to define the scientific question. The second step is to choose a *level of abstraction* for the model that facilitates answering the question of interest. For example, if the study concerns calcium signaling in individual postsynaptic spines, clearly a detailed, spatial model of (part of) the neuron is required. In contrast, if the study concerns the synchronization properties of a large population of neurons, such subcellular details of individual cells are probably less important. After the choice of the class of model, up to millions of parameter values (depending on the model class) need to be assigned. Ideally, in this third step, biological data are used to constrain and choose the parameter values. Then, computer simulations are performed to solve the model equations and to chart the behavior of the model over time. In the fourth step, the behavior of the model is compared to the available biological data on the phenomenon under study and parameters are adjusted (*parameter tuning*) to optimize the model for biological accuracy. At this point, the fifth step, the model is ready to serve as a tool to answer biological questions. More precisely, models do not fully answer questions but at a minimum provide targeted hypotheses that can then be tested in experiments.

FROM SIMPLE TO COMPLEX MODELS OF NEURONS

Threshold Model

Models of neurons span a wide range of types. At one end of the spectrum of model complexity, we find extremely simple model neurons that sum all of the inputs they receive and compare that value to a threshold. If the value is above the threshold, the neuron is called active and assumes a state of one; if the value is below the threshold, the state is zero. We define the state of the neuron as x and the inputs it receives u_1 and u_2:

$$
\begin{aligned}
x &= 1 \quad \text{if } u_1 + u_2 > \theta \\
x &= 0 \quad \text{if } u_1 + u_2 \leq \theta
\end{aligned}
\tag{A10.1}
$$

where θ is the threshold. We can make Eq. (A10.1) look more elegant by introducing the *Heaviside step function H*, which is zero for arguments smaller than zero and one for arguments larger than zero:

$$
x = H((u_1 + u_2) - \theta) = H\left(\sum_{i=1}^{2} u_i - \theta\right)
\tag{A10.2}
$$

Knowing the complexity of real neurons, this approach may seem too simple to deepen our understanding of how neurons work. But such models have helped neuroscientists to understand fundamental properties of neuronal signaling.

Before we explore this model in more detail, we need to review an important concept from engineering: a linear system is defined as a system with the following property: the sum of the two output values x_1 and x_2 separately obtained for two individual inputs u_1 and u_2 is the same as the output obtained for a single input $u_1 + u_2$, which is the sum of the two input values:

$$
\begin{aligned}
f(u_1) &= x_1 \\
f(u_2) &= x_2 \\
f(u_1 + u_2) &= x_1 + x_2
\end{aligned}
\tag{A10.3}
$$

Most of us are familiar with linear systems since they are mathematically easy to deal with and often provide reasonable approximations of real-world processes. Here is an example of a linear system with output x and input u:

$$x = 2u \tag{A10.4}$$

In this example model, the output is two times the input, where the parameter with value 2 is called the *gain* of the system. We can plot how x and u relate to each other and get a simple straight line with a slope of 2 and a y-axis intersect at 0. If we now go back to our definition and use inputs u_1 and u_2, we see that

$$2u_1 + 2u_2 = 2(u_1 + u_2) \tag{A10.5}$$

However, neurons fundamentally do not follow this rule since they exhibit a threshold (or, more precisely, a dynamic mechanism that under most circumstances closely resembles a threshold). If the two inputs are below threshold, neither of them will generate an action potential on its own (ie, an output of one in our simple model neuron). Therefore the sum of the output values for these two inputs is zero. However, together the two input values may be strong enough to exceed the threshold and therefore trigger a nonzero output. Concretely, if we assume a threshold $\theta = 3$ and $u_1 = 2$ and $u_2 = 2$, we find the response to the individual inputs to be zero and the response to the sum of the two inputs to be one. Clearly, this violates our definition of a linear system. Such systems are called nonlinear, and their behavior can be quite tricky to intuitively grasp. Therefore, even such seemingly simple models as the one in Eq. (A10.2) can be crucial to fundamentally understand (networks of) neurons. Indeed, such models have enabled several decades of fundamental work, and as a by-product exciting engineering applications have emerged, often referred to as *neural networks*.

Of course, such a model will be absolutely inadequate if the question to answer is, for example, how synaptic inputs that target different parts of a complex dendritic tree integrate and propagate to the soma. Those and other questions require more detailed, biologically plausible models. Next, we will look at several common approaches to model individual neurons that are nonlinear to capture the threshold-like behavior of action potential initiation.

(Leaky) Integrate-and-Fire Neuron

We first expand the concept of a neuron that sums its inputs by explicitly introducing time into the model. In the simplest form of the *integrate-and-fire* model, the neuron integrates (ie, sums over time) the incoming input $u(t)$ [2]:

$$x(t) = \int_0^t u(\tau) \, d\tau \tag{A10.6}$$

If the threshold for action potential generation (explicitly specified as parameter θ) is crossed, the neuron is said to have "fired" an action potential, and the membrane voltage is reset to its resting value x_0 immediately after the action potential (Fig. A10.1):

$$x(t) = x_0 \quad \text{if } x > \theta \tag{A10.7}$$

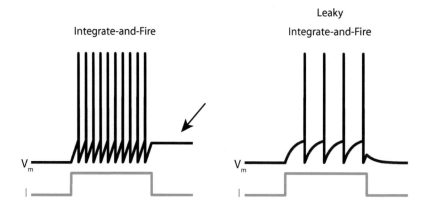

FIGURE A10.1 Integrate-and-fire neuron (*left*) and leaky integrate-and-fire neuron (*right*). *Black*: Membrane voltage V_m. *Red*: Injected current I. The *arrow* points to the issue that arises in the absence of leak term (lack of "forgetting" previous input). Note that both models do not generate actual spike waveforms. Instead, a vertical line is plotted each time the model neuron fired an action potential.

Note that the integrate-and-fire neuron does not model the time course of the action potential itself.

We can write the integration as a differential equation by taking the derivative on both sides:

$$\frac{dx}{dt} = u(t) \tag{A10.8}$$

Examining the output for a time-varying input $u(t)$ reveals that instantaneous changes in the input take some time to take effect in the state variable x. Indeed, the equation is the same as for the behavior of the voltage at a capacitor with capacitance C as a function of the current I (see toolbox: Electrical Circuits):

$$C\frac{dV}{dt} = I \tag{A10.9}$$

Therefore, the integrate-and-fire model combines a purely capacitive membrane that describes the subthreshold behavior and a simple reset for modeling the behavior of a single neuron. Accordingly, we interpret the voltage in Eq. (A10.9) as the membrane voltage V_m. To make this model more realistic, we include a resistive component to the passive cell membrane (see also toolboxes: Differential Equations and Electrical Circuits). The added resistor (input resistance of the cell) adds a decay or *leak term* ["$-V_m$" in Eq. (A10.10)] such that previous input is "forgotten" over time. This is referred to as a *leaky integrate-and-fire* model:

$$RC\frac{dV}{dt} = -V + RI \tag{A10.10}$$

The integrate-and-fire model does not have a memory beyond the time point of an action potential, since the membrane voltage and therefore the neuron is reset. However, biological neurons do have a memory of previous spiking activity such that for a constant step-current injection, the firing rate of many types of neurons decreases over time. This effect is mediated by multiple factors, for example, by accumulation of intracellular calcium that activates a calcium-activated potassium current, which hyperpolarizes the cell. One way to include such spike frequency adaption is to dynamically alter the spike threshold so that each action potential transiently elevates the threshold for a subsequent action potential (in essence, adding a second differential equation for the dynamics of the threshold). Another fundamental limitation of integrate-and-fire models is that real neurons (or at least some types of neurons) can fire action potentials when stimulated with a hyperpolarizing current injection. The biophysics of this phenomenon are explored in the chapter: Dynamics of the Action Potential (in essence mediated by deinactivation of voltage-gated sodium channels).

Quadratic Integrate-and-Fire and Izhikevich Model Neuron

A modification of the integrate-and-fire model that has become a broadly used neuronal model is the *Izhikevich model*, named after Eugene Izhikevich [3]. This model is a generalization of the so-called *quadratic integrate-and-fire* model. We will first examine the quadratic integrate-and-fire model, and then we will review the Izhikevich model. Note that we ignore all constants to make the equations more readable (at the price of omitted, mismatched physical units):

$$\frac{dV_m}{dt} = V_m^2 + I \tag{A10.11}$$

$$V_m = V_{reset} \quad \text{if } V_m > V_{spike}$$

When comparing this model with the leaky integrate-and-fire neuron, it becomes clear that the leak term $-V_m$ is replaced with the term V_m^2. Moreover, the reset does not occur anymore at the threshold for spike generation, but rather for a voltage that corresponds to the peak of the action potential (ie, peak cut-off). This subtle difference allows us to model the dynamics around the threshold with the differential equation instead of the hard reset (Fig. A10.2). Stability analysis (toolbox: Dynamical Systems) shows that for $I > 0$ there is no equilibrium, and the cell is therefore firing an infinite train of action potentials (tonic firing). For $I < 0$, there are two equilibrium points at

$$V_1 = -\sqrt{-I}$$
$$V_2 = +\sqrt{-I} \tag{A10.12}$$

with V_1 and V_2 being stable and unstable, respectively. In this case, the behavior depends on the reset value V_{reset}. The input will need to push the membrane voltage V_m beyond V_2 to generate an action potential (otherwise it will return without a spike to V_1, the resting value). If V_{reset} is smaller than V_2, then after a spike V_m converges back to V_1 (quiescence). If V_{reset} is larger than V_2, the neuron will continue to fire, since it is repelled from the unstable equilibrium point V_2 toward more positive values. In addition, if the perturbation is large enough to move V_m below V_2, the neuron becomes silent. Thus the neuron exhibits two stable states (quiescent and tonic firing, Fig. A10.3).

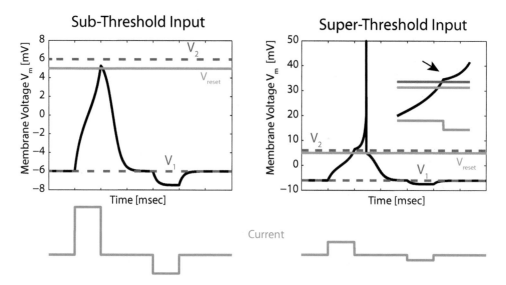

FIGURE A10.2   Subthreshold (*left*) and superthreshold (*right*) behavior of quadratic integrate-and-fire model neuron. The equilibrium points are at $V_1 = -6$ (stable) and $V_2 = 6$ (unstable, corresponds to threshold of neuron). If the current injection (*red*) moves the membrane voltage (*black*) above the threshold V_2 (*right, inset*), an action potential is generated. The reset voltage V_{reset} was set to 5 to enable return to the resting membrane voltage V_1 (stable equilibrium).

The Izhikevich model is a related, expanded version that uses two state (differential) equations. The first is similar to the quadratic integrate-and-fire neuron; the second adds a slower dynamic variable, u, which models slower ionic currents:

$$\frac{dV_m}{dt} = V_m^2 + I - u$$

$$\frac{du}{dt} = a(bV_m - u)$$

(A10.13)

where variables V_m and u are set to values c and $u + d$, respectively, if $V_m \geq 1$. The values of parameters a and b determine if the model cell behaves like an integrator (similar to the quadratic integrate-and-fire neuron) or not (Fig. A10.4). In other words, these extra parameters allow us to model neurons that integrate their synaptic input (eg, layer V pyramidal cells) and neurons that resonate (eg, fast-spiking inhibitory interneurons).

Hodgkin–Huxley Model Neuron (Point Model)

Here, we briefly introduce the overall structure of Hodgkin–Huxley-type mathematical models. The details are discussed in the chapter: Dynamics of the Action Potential. The main

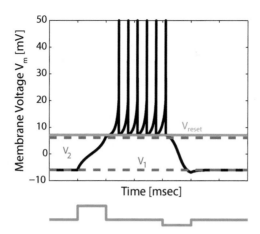

FIGURE A10.3   Bistable behavior of quadratic integrate-and-fire neuron. A depolarizing current pulse switches the neuron to tonic spiking (state switch). Similarly, a hyperpolarizing current pulse switches the neuron back to rest. For this bistability to occur, the reset voltage V_{reset} needs to be more depolarized than the unstable equilibrium point (ie, the *green line* above the *blue line*).

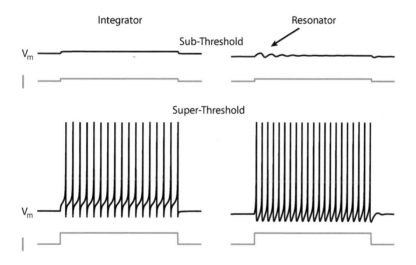

FIGURE A10.4   Subthreshold (*top*) and superthreshold (*bottom*) behavior of integrator (*left*) and resonator (*right*) Izhikevich model. The *arrow* points to subthreshold oscillations characteristic of resonators [4].

strategy in this approach is to model ionic currents. Therefore, in contrast to the previously discussed models, parameters in such models directly reflect measurable biological quanti-ties. In general, the subthreshold behavior remains the same (modeled as a resistor and a capacitor). However, the nonlinear amplification of sufficiently large input is now driven by voltage-gated currents modeled as:

$$I = G(V_m)(V_m - V_{eq}) \qquad (A10.14)$$

where $G(V_m)$ is the conductance that is a function of V_m, and V_{eq} is the equilibrium potential defined by the gradients of the ions that can pass through the modeled type of ion channel (Fig. A10.5). Therefore there are fundamentally two ways to see a zero net current through a given type of ion channel. First, if the conductance $G(V_m)$ is zero, there is no current. An example is the case of voltage-gated sodium channels at hyperpolarized membrane voltage values. When the cell is hyperpolarized, the conductance of these channels is zero, caused by deactivation of the channel. Second, if the membrane voltage V_m equals the equilibrium potential V_{eq}, then by definition the corresponding current is zero. An example is the case of synaptic inhibition for which the resting membrane voltage and the equilibrium potential can assume the same value such that there is no net hyperpolarizing current flowing through GABA$_A$-type receptors (see chapter: Synaptic Transmission). Typically, the dependence of conductance $G(V_m)$ can be experimentally determined by pharmacologically blocking all other ion channels. Alternatively, a specific ion channel is expressed in an otherwise electrically quiet cell, such as the Human Embryonic Kidney 293 cell that serves as an expression system. The term Hodgkin—Huxley model can refer to both the original model and to a larger class of neuronal models that employ the same modeling strategy (often also called *conductance-based models*).

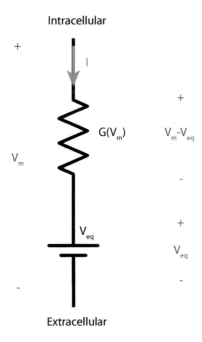

FIGURE A10.5   Ion channels are modeled as a conductance G (which can be voltage dependent) in series with a battery for the equilibrium potential V_{eq}. Current and voltage are drawn per standard convention. Note that the voltages are determined by the Kirchhoff law of voltages (see toolbox: Electrical Circuits).

Multicompartment Model: Cable Equation*

Given the elaborate morphology of some types of neurons (such as layer V pyramidal cells in the neocortex), some questions about electric signaling cannot be reduced to conductance-based *point models* that ignore space and only include time. Instead, the morphology of the cell needs to be explicitly taken into account. The main strategy is to divide the cell into individual cylindrical compartments, each equipped with a set of ionic conductances and passive cell membrane properties (Fig. A10.6). Every cylindrical compartment has an axial current I_a that follows Ohm's law:

$$\frac{V_m(x + \Delta x, t) - V_m(x, t)}{\Delta x} = r_a \Delta x I_a \qquad (A10.15)$$

where Δx denotes the length of the cylinder that has an axial resistivity r_a (units of $\Omega{}^*m$). The current that flows into a compartment is split into axial current, I_a, which leaves the compartment on the other side, and a transversal current, I_t, that crosses the cell membrane. The latter is described by the same equations as for a point model:

$$I_t = C_m \frac{\partial V_m}{\partial t} + \frac{V_m}{R_m} \qquad (A10.16)$$

$$I_a(x + \Delta x, t) = I_a(x, t) + I_t$$

We can then reshape this equation and divide by Δx:

$$\frac{I_a(x + \Delta x, t) - I_a(x, t)}{\Delta x} = \frac{\left(C_m \frac{\partial V_m}{\partial t} + \frac{V_m}{R_m}\right)}{\Delta x} \qquad (A10.17)$$

Remembering that

$$\frac{\partial f}{\partial x} = \frac{f(x + \Delta x) - f(x)}{\Delta x} \qquad (A10.18)$$

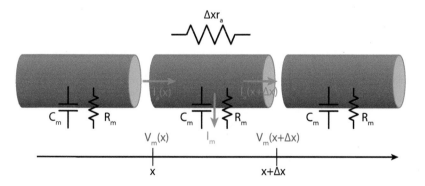

FIGURE A10.6   Electric circuit diagram of multicompartment neuron used to derive the cable equation, which describes the spatiotemporal dynamics of the membrane voltage V_m.

for very small Δx, we can rewrite Eq. (A10.15) as

$$\frac{dV_m}{dx} = r_a \Delta x I_a \tag{A10.19}$$

and Eq. (A10.17) as

$$\frac{dI_a}{dx} = \frac{\left(C_m \frac{\partial V_m}{\partial t} + \frac{V_m}{R_m} \right)}{\Delta x} \tag{A10.20}$$

By replacing the capacitance and the resistance by their normalized values:

$$c_m = \frac{C_m}{\Delta x} \tag{A10.21}$$

$$r_m = R_m \Delta x \tag{A10.22}$$

we find

$$\frac{dV_m}{dx} = r_a I_a \tag{A10.23}$$

$$\frac{dI}{dx} = c_m \frac{\partial V_m}{\partial t} + \frac{V_m}{r_m} \tag{A10.24}$$

Taking the spatial derivative of the first equation and then combining them gives us

$$\frac{d^2 V_m}{dx^2} = r_a \left(c_m \frac{\partial V_m}{\partial t} + \frac{V_m}{r_m} \right) \tag{A10.25}$$

which we can rewrite after some algebra:

$$\tau \frac{\partial V_m}{\partial t} = \lambda^2 \frac{\partial^2 V_m}{\partial x^2} - V_m \tag{A10.26}$$

$$\tau_m = c_m r_m \tag{A10.27}$$

$$\lambda = \sqrt{\frac{r_m}{r_a}} \tag{A10.28}$$

Eq. (A10.26) is the so-called *cable equation* that describes the behavior of the membrane voltage as a function of time and space. To develop an intuition for the behavior, we can set the temporal derivative to zero:

$$\lambda^2 \frac{\partial^2 V_m}{\partial x^2} = V_m \tag{A10.29}$$

which has the following solution:

$$V_m(x) = b_1 e^{\frac{x}{\lambda}} + b_2 e^{-\frac{x}{\lambda}} \tag{A10.30}$$

where b_1 and b_2 are constants. If we assume that for very large distances ($x \rightarrow \pm\infty$), the voltage should be zero (boundary condition), we find (ignoring some mathematical details):

$$V_m(x) = \frac{1}{2}e^{-\frac{|x|}{\lambda}}$$

(A10.31)

Therefore, in steady state, the potential in a cable decays with space constant λ. The larger the membrane resistivity r_m, the less charge is lost across the cell membrane, and therefore the larger the space constant. Conversely, the larger the axial resistivity, the larger the current loss across the cell membrane.

SUMMARY AND OUTLOOK

In this toolbox, we discussed how computational models of neurons are developed, and we then reviewed several classes of neuronal models. These models differ in terms of how biologically plausible they are and in terms of how complex they are. Even very simple models that do not model the dynamics of neurons can capture essential phenomena of neuronal signaling, such as the nonlinear nature of the transformation of synaptic input into action potential output. More abstract models also have the advantage of having a lower number of parameters to fit. For example, the Izhikevich neurons exhibit dynamics very similar to real neurons, despite the fact that they use very few parameters. Lastly, we examined conductance-based models, which have remained the gold standard for biophysically plausible neuronal models. Knowing the main neuronal models and understanding the main advantages and disadvantages are prerequisites for the successful use of computational neuroscience strategies, which represent an important aspect of network neuroscience.

References

[1] Hodgkin AL, Huxley AF. A quantitative description of membrane current and its application to conduction and excitation in nerve. *J Physiol (London)* 1952;**117**(4):500–44.

[2] Gerstner W, Kistler WM. *Spiking neuron models: single neurons, populations, plasticity.* Cambridge, U.K.; New York: Cambridge University Press; 2002. xiv, p. 480.

[3] Izhikevich EM. Simple model of spiking neurons. *IEEE Trans Neural Netw/Publ IEEE Neural Netw Counc* 2003;**14**(6):1569–72.

[4] Izhikevich EM. Dynamical systems in neuroscience: the geometry of excitability and bursting. In: *Computational neuroscience.* Cambridge, Mass: MIT Press; 2007. xvi, p. 441.

Toolbox Physics of Electric Fields

IS THIS TOOLBOX FOR ME?

If your background is not physics, this toolbox will help you understand the main concepts of electric fields. The toolbox is not a comprehensive introduction to electromagnetism but rather focuses on the simplified and specific concepts that apply to electric fields in the brain and the head in general. The material introduced here is fundamental for understanding electrophysiology.

Electrophysiology enables the study of brain dynamics with high temporal resolution by measuring electric signals caused by the flow of ions in the brain. Fundamentally, these ionic currents generate electric fields, the topic of this toolbox. Understanding the physics of electric fields is helpful for mastering neurophysiological recording techniques that rely on and are constrained by the basic principles of electromagnetic theory. The larger the spatial scale of a recording technique, the more relevant the physics of electric fields. For example, understanding the basics of whole-cell patch-clamp recordings only requires a good grasp of lumped-element electrical circuits that do not model "space" but only "time." In contrast, an electroencephalogram (EEG) records brain signals on a spatial scale of up to billions of neurons and therefore many square centimeters of cortical tissue. Thus this technique fundamentally rests on the principles of how the measured macroscopic fields are generated in the brain and propagated through tissue. In this toolbox, we will introduce the fundamentals of electric fields, but unlike an physics textbook, we embed this discussion in the relevant application of understanding endogenous electric fields caused by neuronal activity. Specifically, we focus on current flow in three spatial dimensions. Therefore the approach here differs from the electric circuit theory introduced for the study of the dynamics of individual neurons (Fig. A11.1).

Given the physical properties of the head and the electric signals generated by the brain, several assumptions are typically made. First, we can simplify the study of electric fields in the brain. We can safely assume electric and magnetic fields to be uncoupled because of the low frequencies of brain activity. Therefore we can limit ourselves to electrostatics and need not worry about electromagnetism. Note that neurons still generate a weak magnetic field (as measured by magnetoencephalography), which is unrelated to the fact that electric and magnetic fields are not coupled for low frequencies (see chapter: LFP and EEG). Second, we can focus on the flow of current in conductive media and can, for a first approximation, abstract from fields in insulators (nonconducting medium). Nevertheless, we will start by discussing

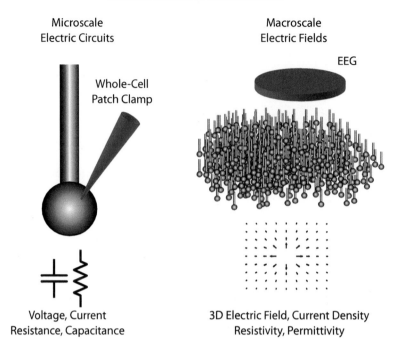

Microscale
Electric Circuits

Whole-Cell
Patch Clamp

Voltage, Current
Resistance, Capacitance

Macroscale
Electric Fields

EEG

3D Electric Field, Current Density
Resistivity, Permittivity

FIGURE A11.1   (*Left*) Whole-cell patch-clamp electrophysiology can be modeled using electric circuit theory (see toolbox: Electrical Circuits) that ignores space. (*Right*) An electroencephalogram (EEG) can be modeled using a simplified version of the Maxwell equations, which describe how electric fields and electric current relate to each other in three dimensions.

electric fields in a nonconducting medium (insulator). As we will see, there is a strong analogy between the electric fields in the two types of media. The two "Sections Electric Potential and Current Density" are labeled with an asterisk (*) due to their more mathematical nature than the remainder of the book.

ELECTRIC FIELDS

Positive and *negative charge* are defined by their effect on each other. Charges of the same sign repel, and opposite charges attract. The force between two charges q_1 and q_2 in a vacuum is determined by *Coulomb's law*:

$$F = \frac{q_1 q_2}{4\pi\varepsilon_0 r^2}e_{12} = \frac{q_1 q_2}{4\pi\varepsilon_0 |r_2 - r_1|^2}e_{12} \tag{A11.1}$$

Several notations are introduced in this equation. First, bold printed letters correspond to vectors (here in two- or three-dimensional space). Conceptually, a vector has a direction and a length. For example, force vector F (Fig. A11.2) points in a given direction and has a given length (ie, strength), denoted as $|F|$. Mathematically, in three dimensions, vector F consists

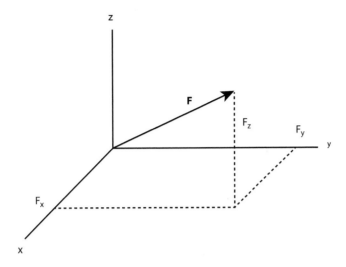

FIGURE A11.2   Force vector F consists of three scalars that define the strength along the x-, y-, and z-axes.

of three components (scalar numbers) that indicate the relative strength along the three axes, x, y, and z:

$$F = \begin{pmatrix} F_x \\ F_y \\ F_z \end{pmatrix} \tag{A11.2}$$

The length of the vector is computed as

$$|F| = \sqrt{F_x^2 + F_y^2 + F_z^2} \tag{A11.3}$$

The direction of the force is given by unit vector e_{12}, a vector that has length 1 (called unit length) and points along the direction of the direct line between q_1 and q_2 (Fig. A11.3). This vector is determined by first subtracting the vector r_1 (which points to q_1) from vector r_2 (which points to q_2), and then dividing it by the length of the resulting vector:

$$e_{12} = \frac{r_2 - r_1}{|r_2 - r_1|} \tag{A11.4}$$

The strength (ie, the length) of the force vector is given by the product of the charges q_1 and q_2 divided by the square of the distance r between the charges scaled with the factor $1/4\pi\varepsilon_0$ (Fig. A11.4).

The constant ε_0 denotes the *permittivity* of empty space ($\varepsilon_0 = 8.85 \times 10^{-12}$ F/m). The electric field E is defined as the force on a unitary charge and is measured in V/m:

$$E(r) = \frac{q_1}{4\pi\varepsilon_0 |r - r_1|^2} e \tag{A11.5}$$

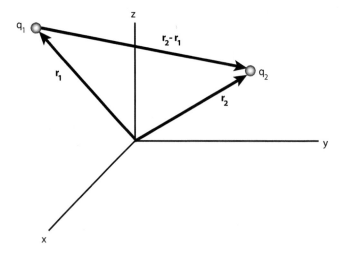

FIGURE A11.3   Vectors r_1 and r_2 point to charge q_1 and q_2, respectively. The force between the charges points in the direction of the vector from q_1 to q_2.

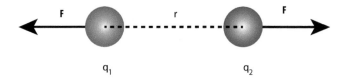

FIGURE A11.4   Coulomb's law allows us to calculate the force (vector) between two charges q_1 and q_2 (shown here: opposite charges with distance r).

where e is the unit vector on a line between charge q_1 and the unitary test charge (ie, $q_2 = 1$ C). The term $|r - r_1|^2$ denotes the radial distance between the unitary test charge and the charge q_1. The electric field again is a three-dimensional vector (in Cartesian coordinates) and is a function of location r of the test charge. Mathematically, this corresponds to a vector field and can be illustrated by a plot where an arrow is drawn on regularly sampled intervals corresponding to the field vector (for a test charge at that location). For a positive charge q_1, the resulting electric field points radially away from the charge and decreases by a factor of one over the distance square (Fig. A11.5).

If more than one charge is present, the resulting electric field can simply be determined by linear *superposition* of the field vectors that scale with one over the square of the distance between the charge and test charge, and point along the direct line between the charge and the test charge:

$$E(r) = \frac{1}{4\pi\varepsilon_0} \sum_{n=1}^{N} \frac{q_n}{|r - r_n|^2} e_n \qquad\qquad (A11.6)$$

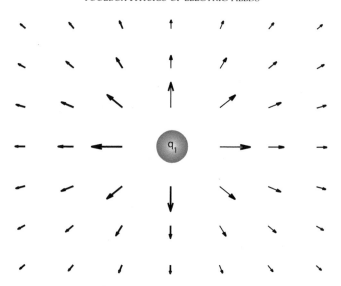

FIGURE A11.5   Electric field caused by a single charge q_1. With increased distance from the charge, the electric field is decreased. At any given point, the electric field points radially away from the charge q_1.

We sum (denoted by the capital sigma) the vectors that point in the direction of the direct line between the individual charges q_n and the test charge at location r (unit vectors e_n) scaled with the strength, which is computed as the charge divided by the square of the distance between the charge and the location r. The vector sum is again scaled (ie, multiplied) by the same $1/4\pi\varepsilon_0$ as in the case of a single charge. For two point charges (also referred to as *dipole*), we graphically demonstrate the superposition (Fig. A11.6).

ELECTRIC POTENTIAL*

For mathematical convenience, we then define the *electric potential* φ as a scalar function that depends on location r. Using electric potential greatly simplifies things, since we can abstract from vector fields. In the case of electrophysiology, we use the following definition:

$$E(r) = -\left(\frac{\partial\varphi}{\partial x}e_1 + \frac{\partial\varphi}{\partial y}e_2 + \frac{\partial\varphi}{\partial z}e_3\right) \tag{A11.7}$$

In essence, this equation states that the electric field is the spatial derivative of the electric potential; for every dimension (as indicated by unit vectors e_1, e_2, and e_3), the derivative with respect to that direction gives the corresponding magnitude of the electric field in that dimension. Mathematically, there is a symbol called "grad" for gradient and is drawn as an inverted triangle:

$$E(r) = -\nabla\varphi(r) \tag{A11.8}$$

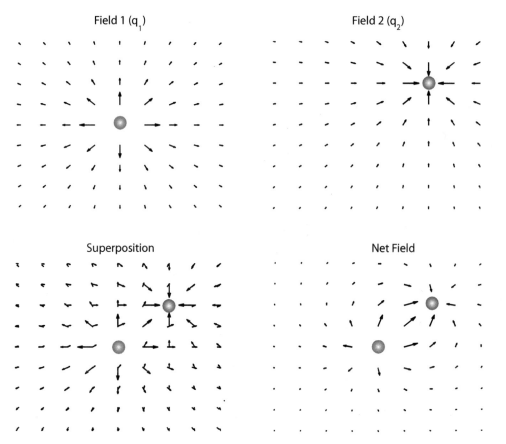

FIGURE A11.6   Electric field for two charges q_1 and q_2 of opposite sign (dipole). (*Top*) Individual fields caused by two point charges of opposite charge. (*Bottom*) Superposition of electric fields (*left*) and net electric field determined by summation (*right*).

The electric potential for multiple point charges is (derivation not shown):

$$\varphi(\boldsymbol{r}) = \frac{1}{4\pi\varepsilon_0} \sum_{n=1}^{N} \frac{q_n}{|\boldsymbol{r} - \boldsymbol{r}_n|} \tag{A11.9}$$

We can use this equation to calculate the electric potential of the dipole, an arrangement of two nearby charges of opposite charge, a frequently encountered arrangement for neurons and sheaths of neurons:

$$\varphi(\boldsymbol{r}) = \frac{1}{4\pi\varepsilon_0} \left(\frac{1}{r_1} - \frac{1}{r_2} \right) \tag{A11.10}$$

where r_1 and r_2 are the two distances from the two poles to location $\boldsymbol{r}$.

CURRENT DENSITY*

So far, all of our considerations applied to empty space (vacuum) and can be easily adjusted for any insulator with no free charge. To understand electric fields in the brain, however, we need to consider the fact that the brain is a reasonable conductor that permits the flow of electric current. As a result, any test charge will be surrounded by charge of the opposite sign and, as a result, the potential because of a test charge in biological tissue is extremely small at the distances of interest for the study of electrophysiology. Instead, in the case of biological tissue we need to consider electric current (sources and sinks) instead of static charge—we define the current density J, which is measured as current per area (typically A/cm^2). Two different contributions to J need to be considered. First, in the case of an applied electric field, there is a current flowing that is proportional to the electric field (in essence, Ohm's law):

$$J_E = \sigma E \tag{A11.11}$$

where σ is the conductivity (in siemens/meter) of the medium. Table A11.1 lists the conductivity of different materials. In addition, if there is current flowing across the cell membrane, there is a second current source, J_s. The total current is the sum of the two (in essence, Kirchhoff's current law):

$$J = \sigma E + J_S \tag{A11.12}$$

The last piece we need is a basic statement that charge is conserved in a closed volume such that no "magic" new charges are generated or lost at any given location (called *charge conservation*):

$$\frac{\partial J_x}{\partial x} + \frac{\partial J_y}{\partial y} + \frac{\partial J_z}{\partial z} = 0 \tag{A11.13}$$

TABLE A11.1   Conductivity of Different Materials

Material	Conductivity (S/m)
Silver	6.3×10^7
Iron	1.0×10^7
Sea water	4.8
Cerebrospinal fluid	1.6
Cortex	0.43
Bone	0.0063–0.013
Pure water	5×10^{-6}

where J_x, J_y, and J_z are the three components of the vector J. This equation states that if there is a nonzero gradient (change) of current density in one direction, it needs to be compensated by a current density gradient in another direction such that the net change in current is zero. Combining these two equations, remembering the relationship between electric field and electric potential, and assuming that conductivity is constant and linear, we find that

$$\sigma\left(\frac{\partial^2\varphi}{\partial x^2} + \frac{\partial^2\varphi}{\partial y^2} + \frac{\partial^2\varphi}{\partial z^2}\right) = -\frac{\partial J_S}{\partial x} - \frac{\partial J_S}{\partial y} - \frac{\partial J_S}{\partial z} \tag{A11.14}$$

We skip ahead to the solution to this equation for the special case of several point current sources I_n ($n = 1, ..., N$) in an infinite, homogeneous (ie, conductivity the same everywhere), isotropic (ie, conductivity the same in all three dimensions), and purely resistive volume conductor:

$$\varphi(r) = \frac{1}{4\pi\sigma}\sum_{n=1}^{N}\frac{I_n}{|r - r_n|} \tag{A11.15}$$

Of note, this equation for the electric potential in a volume conductor is analogous to the previously introduced equation for the electric potential of charge in a vacuum (a perfect insulator).

ELECTRIC MODELING OF THE HEAD

The head does not represent a homogeneous environment, but is a complex, multilayered structure (eg, skin, skull, cerebrospinal fluid) with the individual layers exhibiting very different conductivities (Table A11.1). In this case of an inhomogeneous medium, the equation for electric potential can be separately solved for each compartment. At the interface between the two compartments, a continuity argument defines the so-called *boundary conditions*. The first boundary condition is that the current normal to the interface needs to be the same on both sides (otherwise some charge would be "magically" created). Mathematically, the current density is expressed as the conductivity multiplied with the electric field component normal to the surface (expressed as dimension n):

$$\sigma_1\frac{\partial\varphi_1}{\partial n} = \sigma_2\frac{\partial\varphi_2}{\partial n} \tag{A11.16}$$

The second boundary condition derives from Maxwell's equations (not discussed here) and states that the tangential field at the interface needs to be identical (with dimension w representing the tangential direction):

$$\frac{\partial\varphi_1}{\partial w} = \frac{\partial\varphi_2}{\partial w} \tag{A11.17}$$

Using these boundary conditions, any tissue to be modeled can be segmented into volumes with different conductivity, and there is a unique mathematical solution for the electric potentials in each region. This is referred to as the *forward problem*. Practically, this corresponds to knowing where neural activity is generated and calculating the resulting electric field measured by EEG. The *inverse problem* refers to the inverse operation, where the EEG signal is known (measured) and the underlying neural activity (source currents) is computed. Unlike the forward problem, there is no unique solution to the inverse problem and therefore many assumptions need to be made.

SUMMARY AND OUTLOOK

In this toolbox, we discussed how electric charge gives rise to electric fields. Fundamentally, neuronal signaling is mediated by the flow of ions in the brain. These currents give rise to electric fields. We reviewed how electric fields can be represented as electric potentials, which are easier to deal with since they are not vectors but scalar values and correspond to the quantities measured by electrophysiology. Such a conceptual understanding of the physics of electric fields in the brain is instrumental for the "Chapter LFP and EEG," which discusses measurements of those signals. In addition, electric fields may be more than just side products of neuronal activity, but instead play an active role in shaping network dynamics, as presented in the "Chapter Neuronal Communication Beyond Synapses."

NOTES

• *A more in-depth treatment of the material in this toolbox is presented in* Electric Fields of the Brain *by Paul Nunez and Ramesh Sinivasan [1].*

Reference

[1] Nunez PL, Srinivasan R. Electric fields of the brain: the neurophysics of EEG. 2nd ed. Oxford, New York: Oxford University Press; 2006. xvi, 611 p.

Toolbox Time and Frequency

IS THIS TOOLBOX FOR ME?

This toolbox introduces how physical signals are measured and analyzed. Given the prevalence of rhythmic patterns in neuronal network dynamics, we introduce these main strategies to study the presence of oscillatory structure in signals. This toolbox is of fundamental importance for the material covered in the book. However, if you have an engineering or physics background, you are likely to already be familiar with the concepts introduced here.

MEASURING TIME SERIES

Knowing *when* an observation occurred greatly increases its value. Fundamentally, all measurements assume the format of a number (or a set of numbers), a physical unit, and a time stamp of when the measurement was performed. To understand dynamical processes (ie, time course of quantities of interest), it makes sense to take measurements in regular intervals over a prolonged amount of time to capture enough data. The intervals between equally spaced measurements are called the *sampling period* T_s (in neuroscience typically with units of milliseconds). The frequency at which these samples are recorded is called *sampling frequency* (f_s), and is calculated as the inverse of T_s:

$$f_s = \frac{1}{T_s} \tag{A12.1}$$

with unit hertz (Hz), which stands for 1/second abbreviated as 1/sec. Choosing the right sampling frequency is crucial to extract the desired information about the system observed. For example, if we want to understand the time course of surface air temperature, we could take a measurement at 8 am every day for 5 days (Fig. A12.1, left). In such an experiment, all samples (measured values) would be very similar, supporting the clearly incorrect conclusion that surface temperature is constant on Earth (and does not exhibit a 24-h cycle). Another researcher could perform a very similar experiment but take all measurements at 2 pm, again with a sampling frequency of once per day. The conclusion about the surface temperature value would be different (since it is warmer in the early afternoon), but the conclusion about the lack of temporal fluctuations (dynamics) would be identical. In reality, both experiments completely overlooked the key phenomenon, the 24-h cycle of temperature, because of their inappropriately low sampling (measurement) rate of once a day. A much better choice for the sampling rate would have been, for example, once per hour (instead of once per day), so that 24 samples would have been collected per day (Fig. A12.1, right).

From this example, we learn that choosing an appropriate sampling frequency is vital in the study of dynamical systems, and that the chosen sampling rate needs to exceed the

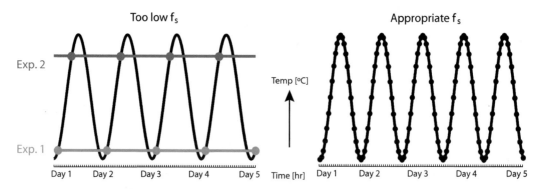

FIGURE A12.1   Time course of air temperature. (*Left*) Two experiments for which the sampling rate f_s was too low. (*Right*) More appropriate, higher sampling rate captures daily fluctuations of the temperature.

rate of change of the system under observation. The *Nyquist theorem* theoretically determines the required sampling rate, which should be at least twice as fast as the fastest process under observation. In the example, the Nyquist theorem would require sampling of the surface temperature twice a day. In practice, it is advisable to choose a sampling rate at least five times the highest frequency of interest in the signal; there is little downside to collecting data at higher sampling frequencies.

FREQUENCY ANALYSIS

Periodic sampling of a dynamic variable provides a *time series*, a list of consecutive measurements at equal time intervals. In neuroscience, examples of time series are the recorded traces of the membrane voltage of a neuron and scalp electric signals measured with electroencephalography. One way to communicate these results is to plot these time series, with time on the *x*-axis and the measurements of voltage on the *y*-axis. This approach to data presentation allows us to draw our own conclusions about the quality and meaning of the measurement. A disadvantage of this approach, however, is that space constraints will be prohibitive for studies that include multiple, possibly longer time series. A good solution to this problem is to extract fundamental features of the recorded time series that are (ideally) uniformly present throughout the recording, that is, they are stable over time (also referred to as *stationary*). To capture temporal fluctuations, one important question we can ask is about the presence of periodic structure in the signal of interest. Most signals occurring in nature are inherently periodic, meaning that measurements show the same or a similar value again after some amount of time has elapsed. Mathematically, this is denoted as

$$x(t_0) = x(t_0 + T) \tag{A12.2}$$

indicating that a signal repeats itself after time T has elapsed. The parameter T is called the *oscillation period*, and the inverse of T is referred to as the frequency f. The sine function is a prominent example of a periodic function. Sine waves are described by their amplitude (strength, sometimes also measured as peak-to-peak amplitude), frequency, and phase

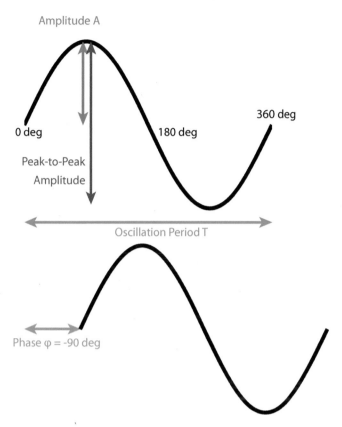

FIGURE A12.2 Any sine wave is described by its amplitude A, oscillation period T, and phase φ. Note that the value of the phase is negative in the example shown since the sine wave is delayed, that is, the top sine wave leads the bottom sine wave.

(Fig. A12.2). The phase denotes the temporal offset of the sine wave and is measured in degrees or radians. One oscillation cycle spans 360 degrees or 2π radians:

$$x(t) = A \sin(2\pi f t + \varphi) \tag{A12.3}$$

Returning to our initial example of surface temperature measurement, if we acquire measurements at an appropriate sampling frequency and for an appropriate duration of time, we will find that the key characteristic of the signal is its periodicity at a frequency of 1/24 h. At this point, we can reduce the entire data set to one value that captures the essence of the signal in Fig. A12.1: oscillation period $T = 24$ h.

This approach is called *frequency analysis*, where time series are transformed into the *frequency domain* such that they are plotted as a function of frequency, not as a function of time. This plot is termed the *power spectral density* plot. When the signal is a sine wave with oscillation period T and amplitude A (Fig. A12.3, left), the corresponding power spectral density is zero, except for the frequency $f = 1/T$, for which the value of the power spectral

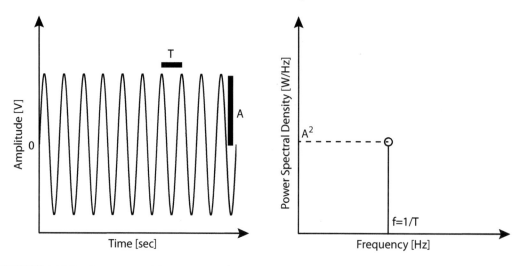

FIGURE A12.3   Power spectral density (also referred to as power spectrum) for a sine wave is a single peak at the frequency of the sine wave. For a signal with amplitude A, the power spectral density is A^2. Strictly speaking, if the amplitude is measured in volts, the signal is squared and normalized by a 1-Ω resistor to determine the power.

density is A^2 since the plot expresses power and not amplitude (Fig. A12.3, right). Thus the representation of the temperature time series in the frequency domain would essentially have one peak at the frequency of 1/24 h. Note that technically, for the units to be correct, an additional normalization would be required for the power spectral density to have the correct units. For practical purposes, this is not a concern and thus is omitted from further discussion in this toolbox.

Other periodic processes, such as the yearly cycle (stronger irradiation in summer and weaker irradiation in winter in the northern hemisphere) also contribute to the temperature. Therefore the *frequency spectrum*, the representation of the time series in the frequency domain, will have a second, peak at 1/365 days (Fig. A12.4). The relative amplitude of this periodic modulation will depend on where the measurements were recorded. The corresponding peak in the frequency spectrum will be low for measurements taken at the equator and become more prominent the more the measurement location is moved toward either of the poles of the globe.

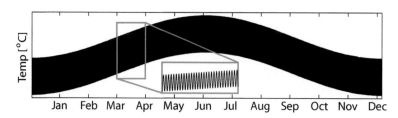

FIGURE A12.4   Second, slow time scale (year) of temperature time series (for the northern hemisphere). *Inset (red box)*: Daily fluctuations. Note that the real data for this type of measurement are more complex; they are simplified here to focus on the concept of frequency analysis.

Importantly, if provided with the frequency spectrum, one can easily reconstruct the original time series. Since the peaks in the spectra denote the relative strengths (amplitudes) of periodic processes, we can rebuild the signal by summing sine functions that have amplitudes and frequencies corresponding to the values in the spectrum.

$$x(t) = A_1 \sin(2\pi f_1 t) + A_2 \sin(2\pi f_2 t) \tag{A12.4}$$

In fact, any reasonable signal with periodic structure can be approximated by a sum of sine waves. In particular, we expect that the oscillation structure does not change over time, that is, that the signal is *stationary*. We will return to this later, making some modifications to include temporal offsets between the individual sine waves by introducing *phase*. For now, we want to develop an intuition for how the frequency spectrum is calculated. In our temperature example, the periodic processes and their relative strength were obvious from visual inspection of the time series, but this may not be true for real-world signals, such as the ones recorded from neuronal networks. However, we still want to know to what extent our time series resembles sine waves at particular frequencies. If our time series highly resembles a sine wave with a particular frequency, then the corresponding sine wave amplitude in our reconstructed signal will be high and the spectrum will exhibit a peak with that amplitude at the same frequency. Mathematically, we determine the degree of resemblance by multiplying our time series by the sine wave point by point and summing all resulting values. This number is maximal when a perfect match occurs (if our time series itself were a sine wave of the exact same frequency) and lowers as a function of the amount of mismatch between the time series and the sine wave. Repeating this process for all frequencies will provide the entire frequency spectrum. Mathematically, this corresponds to the sum of the product of the time series $x(t)$ and sine waves at different frequencies f and phase offsets φ_f:

$$X(f) = \sum_t x(t)\sin(2\pi f t + \varphi_f) \tag{A12.5}$$

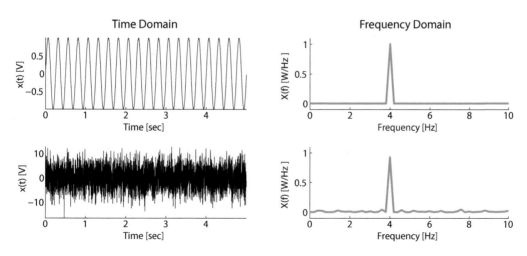

FIGURE A12.5 Clean (*top*) and noisy (*bottom*) 4-Hz signal. (*Left*) Time domain. (*Right*) Frequency domain.

where we denote the Fourier transform with uppercase $X(f)$. Squaring the frequency spectrum provides the power spectrum.

We conclude this section with another example (Fig. A12.5) where we compute the spectrum for a 4-Hz sine wave and for a 4-Hz sine wave to which we added *white noise* (ie, noise where all frequencies are equally represented such that there is no peak frequency). Notably, the 4-Hz structure is masked by the noise and is not visible in the time series plot. However, inspection of the power spectrum still reveals a clear 4-Hz peak. Thus the peak of the sine wave remains mostly intact, since the power of the noise is distributed over all frequencies.

TIME-RESOLVED FREQUENCY ANALYSIS

The limitation of this approach is that it only works well for signals that are of (at least in theory) infinite duration and constant frequency content (stationary signals). This assumption holds well for our examples here, but it fails for many other real-world processes. Attempting to study the presence of different frequencies without explicitly incorporating how this structure changes over time is not a good representation of the time series of interest. Instead, we should be interested in not only the spectrum but also how the spectrum changes over time. To visualize this, we need a representation that shows frequency content as a function of time. This requires three axes—the first for time (x-axis), the second for frequency (y-axis, which used to be the x-axis for the spectral analysis just described), and the third for the amplitude values (z-axis, typically represented as a color map). Such time-dependent representation of the spectrum is called a *spectrogram*. Fig. A12.6 shows an example of a signal that abruptly changes its frequency (top) and its corresponding spectrogram (bottom).

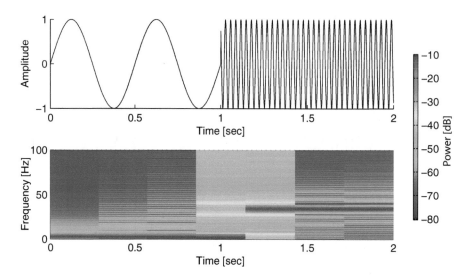

FIGURE A12.6 *(Top)* Time series. *(Bottom)* Spectrogram. Units are in dB (decibels), which are computed as 10 times the logarithm of the spectral power.

The simplest and most obvious way to create the spectrogram is to divide the time series into shorter intervals and to compute the spectrum for each of these intervals (called *windows*). All spectra are then collated together to form the spectrogram. There are several concerns with this method, however. First, the shorter the duration of the window is, the poorer our ability to determine the spectrum is. If the duration of the window is too short, we will be unable to determine the spectrum. Ideally, there should be several oscillation cycles within a single window. For example, if we are interested in a 1-Hz oscillation (1 cycle per second), our window should be several cycles, and thus seconds, long. The problem with longer windows is that temporal resolution is lost, since we are only able to determine the presence of the oscillation with a resolution of multiple seconds. This tradeoff between time and frequency resolution is inherent to all such methods, but some methods minimize this issue much better than others. For example, wavelet analysis provides an approach that elegantly deals with the *time—frequency uncertainty*. Wavelet analysis is based on the idea that in reality we are interested in time-limited occurrences of periodic signatures. Instead of artificially dividing the signal, we use the same fundamental approach of matching our signal to periodic "reference signals." Instead of sine waves, we use amplitude-modulated sine waves that wax and wane over a defined number of oscillation cycles. One of the most straightforward ways to create such wavelets is to multiply a sine wave by a Gaussian function. These *Morlet wavelets* exhibit a Gaussian amplitude envelope and the oscillatory structure of the sine wave. The decay rate of the Gaussian function is a function of the wavelet frequency such that wavelets of different frequencies contain the same number of oscillation cycles (Fig. A12.7).

In wavelet analysis, the signal of interest is "compared" to a family of wavelets with different frequencies. Every wavelet is compared to the signal of interest by determining its match with the signal of interest at a given time point. This process is repeated for all time points by shifting the wavelet in time to determine how the presence of the wavelet frequency changes over time (Fig. A12.8). Mathematically, this process is described by pointwise multiplication of the signal $x(t)$ and the wavelet $w(t)$, followed by summation to determine how well the wavelet matches the signal. This calculation is then repeated for all different

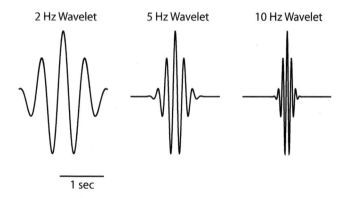

FIGURE A12.7 Three examples of Morlet wavelets with different frequencies. The higher the frequency of the wavelet (*from left to right*), the shorter the duration of the wavelet. As a result, temporal resolution increases with wavelet frequency.

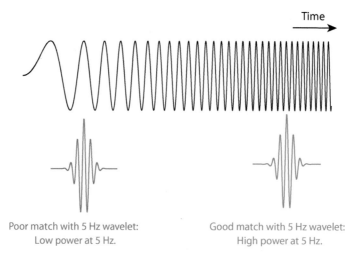

Poor match with 5 Hz wavelet:
Low power at 5 Hz.

Good match with 5 Hz wavelet:
High power at 5 Hz.

FIGURE A12.8   The wavelet of a given frequency (here 5 Hz) is shifted in time to determine its match to the signal of interest as a function of this time shift. Two time points are illustrated—one with poor frequency match (*red*) and one with good frequency match (*green*).

time shifts of the wavelet relative to the signal $x(t)$. The complete process is called *convolution* of the signal $x(t)$ and the wavelet $w_f(t)$ with frequency f.

$$X(f, \tau) = \sum_t x(t) w_f(t - \tau) \tag{A12.6}$$

SUMMARY AND OUTLOOK

In this toolbox, we discussed how periodic time series can be represented in the frequency domain, that is, they can be decomposed into a sum of sine waves with different frequencies and phases. This process of transforming time-domain signals into frequency-domain signals is perhaps the most important signal processing method for network neuroscience. As we extensively discuss in the unit Cortical Oscillations, network activity in the brain is organized into different rhythmic patterns. Thus the analysis of electrophysiological measurements is based on the strategies introduced in this toolbox. In particular, the time-resolved frequency analysis (which provides spectrograms) is key, because of the nonstationary nature of electric signals in the brain that rapidly change as a function of behavioral demands.

Index

'*Note:* Page numbers followed by "f" indicate figures, "t" indicate tables.'